C
Chemistry

28 Projects, from the Creation of Fire to the Production of Plastics

Kevin M. Dunn

Caveman Chemistry: 28 Projects, from the Creation of Fire to the Production of Plastics

by Kevin M. Dunn

www.CavemanChemistry.com

Universal Publishers /uPublish.com

www.upublish.com/books/dunn.htm

ISBN 1-58112-566-6

Revision History

Revision 1.9 2003/07/11 First Printing

Table of Contents

List of Tables

List of Figures

List of Equations

Prologue

Bottom: Peter Quince!

Quince: What sayest thou, bully Bottom?

Bottom: There are things in this comedy of Pyramus and Thisby that will never please. First, Pyramus must draw his sword to kill himself; which the ladies cannot abide. How answer you that?

Snout: By'r lakin, a parlous fear.

Starvling: I believe we must leave the killing out, when all is done.

Bottom: Not a whit! I have a device to make all well. Write me a prologue, and let the prologue seem to say we will do no harm with our swords, and that Pyramus is not kill'd indeed; and for the more better assurance, tell them that I Pyramus am not Pyramus, but Bottom the weaver. This will put them out of fear.

Quince: Well; we will have such a prologue, and let it be written in eight and six.

— William Shakespeare, *A Midsummer Night's Dream*, *ca.* 1596 AD [1]

1 ☿

Bottom: No; make it half; let it be written in four and three.

Snout: Will not the readers be confused by the fictions?

Starvling: I fear it, I promise you.

Bottom: Masters, you ought to consider with yourselves: to bring in—God shield us!—a fiction among facts, is a most dreadful thing and there is no room for equivocation; we ought to look to it.

Snout: Therefore another prologue should probably tell each fact from each fiction.

Bottom: Nay, you must clearly distinguish the one from the other, giving each fiction a typographical symbol of some kind. And the author himself must speak the facts without embellishment or subterfuge of any kind and tell the reader plainly which ones are the facts.

1. Reference [24], Act III, Scene 1.

Quince: With such an inspiration, all is well. Come, sit down, every mother's son, and rehearse your parts; and so every one according to his cue.

2 🜍

There are three kinds of people in the world. The first kind believe what they see; they prefer to experience life as it unfolds, holding pre-conceived expectations in check until observations make all things plain. If you belong to this tribe, a Prologue or preface might spoil some of the fun of figuring things out for yourself. I advise you to skip it altogether and proceed without delay to Chapter 1. The second kind see what they believe; they prefer to experience life with structure, pre-conceived expectations providing a plane from which to make observations. If this is your family, I wrote this Prologue to provide such a vantage point. The third kind don't believe that seeing is worth the effort; they have skipped the Prologue already, presuming that it is merely the conventional place for the author to thank his cronies for their invaluable support. If you are of this ilk, I need say nothing to you at all.

I teach chemistry at Hampden-Sydney College, a small liberal-arts college in central Virginia. The students here, by and large, do not come equipped with insatiable curiosity about my discipline and experience has convinced me that the profession of professing has more to do with motivation than with explanation; a student who is not curious will resist even the most valiant attempts at compulsory education; conversely, inquiring minds want to know. A great deal of my time, then, has been spent devising tricks, gimmicks, schemes and plots for leading stubborn horses to water, knowing full well that I can't *make* them think.

One day as I scuttled across campus, I overheard a tour guide gushing over the Federalist architecture; "If these walls could speak, what stories they would tell." They teach them to say things like that, you know. The phrase brought to mind a lyric from an old song; "If you could read my mind Love, what a tale my thoughts would tell." You know the one. I noticed myself humming it on the way to the post office. "Just like a paperback novel," it continued as I attempted to concentrate on my grading. "I never thought I could act this way, and I've got to say that I just don't get it." The tune wouldn't let go. It's probably playing in your head too, by now. "And I will never be set free. . . " as long as there's this song inside of me. It had begun to mutate. It ended up, "If I could read my own mind, what a tale this song would tell."

That song began life as a simple phrase in the head of Gordon Lightfoot. The phrase combined with others to form lyrics. The lyrics enchanted family members, friends and record executives and eventually came to possess millions of radio listeners. And our heads are full of such things: songs, stories, plays, instructions. From the Gettysburg Address to the recipe for the perfect martini, from one mind to another traveling through time, if only they could speak...

Look into your own mind and grab the first recognizable bit of such thought-stuff you come across. Where did you get it? Where did the person you got it from, get it? Who was the first person to get it, or, to be more precise, to be got *by* it? Wouldn't it be interesting if the thought itself could tell you its story? Wouldn't that be an interesting premise for a book? "And you would read that book again..." unless the idea's just too hard to take.

3 Θ

> The actors are at hand; and by their show,
> You shall know all, that you are like to know.
>
> — *A Midsummer Night's Dream*[2]

∇ You are probably wondering what the little symbol at the beginning of this sentence means. I will tell you. The book, as you will no doubt recall from the first section of this Prologue, is written in four and three. Four spirits, Fire, Earth, Air and Water narrate the chapters, which could get confusing were it not for the presence of these symbols intended to identify which spirit is speaking at any particular point. When the water spirit speaks, for example, the text will begin with the alchemical symbol for water: ∇. The astute reader will instantly surmise that since this paragraph began with the water symbol it is, in fact, narrated by the spirit, Water. In other words, I am that part of Doctor Dunn's mental inventory having to do with *watery* things, those things having come to him from parents and teachers, and their parents and teachers in a long and steady stream back through history and into pre-history where

2. Reference [24], Act V, Scene 1.

we find the very first watery thought. Moving forward, this first watery thought passed from the first person who had it to the second person, where it accumulated new watery bits like a watery snowball until at last it dribbled, bit by bit, into the Doctor's mind. And yours, I might add. It is my job to provide a first-hand account of some of the major events in the watery quadrant of the history of chemical technology.

🜃 Well, *that* was clear as mud. I'm afraid that Water tends to run at the faucet sometimes, so if you want a firm foundation for understanding this book, you're better off listening to Earth. This book is about digging stuff out of the ground and making it into other, more valuable stuff. There are twenty-eight chapters and at least seven of them will teach you something useful. Each chapter starts with a section explaining how that particular chapter's stuff got invented. Then there's a section telling you what you need to know about why the stuff is the way it is. Each chapter wraps up with a section showing you how to make the stuff that the chapter's about. If you're not interested in making stuff, it'd be a waste of time for you to read this book because you're not going to get anything out of it if you're not willing to get your hands *earthy.*

🜁 If Water and Earth haven't convinced you that the Author is off his nut, I'm afraid there's not much I can do to help. I'm supposed to represent the element, air, in case you haven't figured it out. I know it's confusing, but there we are. The Author wanted to write an unusual book, an interesting book, a book that would entertain as well as instruct, but I'm sorry to say that *unusual* is as far as he got. It takes most readers until Chapter 5 even to figure out that the book has characters, like actors in a play. To keep potential readers from chucking his masterpiece in with their empty pop bottles and pizza boxes, the Author has written this Prologue as a "device to make all well," but for that to work they would have to actually get through the Prologue without drifting off into a midsummer's daydream.

Δ Let them slumber; this book is not for the lazy or the timid. It is a book of secrets to be carefully tended like an eternal flame, not casually browsed like a four-year-old fishing magazine in a dentist's office. Everyone who lays hands on it and often tries it out will think that a kind of key is contained in it. For just as access to the contents of locked houses is impossible without a key, so also, without this commentary all that appears in the Emerald Tablet of Hermes Trismegistos will give the reader a feeling of exclusion and darkness. The text is composed in four and three; Fire, Earth, Air, and Water; Mercury, Sulfur, and Salt. In this way was the book created. From this there will be amazing applications, for this is the pattern.

> Truly, without deceit. Certainly and absolutely. That which is Above corresponds to that which is Below and that which is Below corresponds to that which is Above in the accomplishment of the miracle of One Thing.
>
> And just as all things come from One, so all things follow from this One Thing, in the same way.
>
> Its father is the Sun; its mother is the Moon. The wind has carried it in its belly. Its nourishment is the Earth.
>
> It is the father of every completed thing in the Whole World. Its strength is intact if it is turned towards the Earth. Separate the Earth by Fire, the fine from the gross, gently and with great skill.
>
> It rises from the Earth to Heaven and descends again to the Earth, and receives power from Above and from Below. Thus thou wilt have the glory of the Whole World. All obscurity shall be clear to thee.
>
> This is the strong power of all powers for it overcomes everything fine and penetrates everything solid. In this way was the World created. From this there will be amazing applications, for this is the pattern.
>
> Therefore am I called Hermes Trismegistos, having the three parts of wisdom of the Whole World.
>
> Herein have I completely explained the Operation of the Sun.
>
> — *The Emerald Tablet of Hermes Trismegistos*

🜁 Right. I'm afraid you'll have to put up with a bit of pseudo-alchemical techno-babble in the course of this book. The Author might simply have described the nature of the elements, but for the more flamboyant elements of his nature. Fire is the main culprit, but Earth and Water have their moments, as well. "The text is composed in four and three." The Author might have simply said that the book has four characters and each chapter has three sections. If you ask me, this book should have rested on the periodic table, not on the Emerald Tablet.

🜂 The periodic table is fine, as far as it goes, but it says nothing of the Operation of the Sun.

🜃 Transmutation, that is. Black gold, Texas tea. The Emerald Tablet's not so much about literally changing lead into gold, of course, but more about changing useless stuff into useful stuff.

∇ I believe it is more of an allegory about life and death, mortality and immortality, about coming into being and kicking the bucket. This book will follow that pattern, tracing the advance of chemical technology from a stone-age trickle to the babbling brooks of the Bronze Age to the stately rivers of the Iron Age to the confluence of tributaries in the Middle Ages to the polluted canals of the Industrial Revolution. But I am getting ahead of myself.

Δ Indeed you are. The first chapter belongs to me.

Chapter 1. Lucifer (Charcoal)

Call me Lucifer, for I am the bringer of light. No angel was I born, nor devil neither. Nay, as animal I came into the world, and so I will acknowledge both horns and tail. My sin, if it must be called that, was not pride, but curiosity. Alas! The world has forgotten my story, amalgamated me with a host of unfashionable gods until I am beyond recognition.

Beasts among beasts, we lived and died in fear. Fear of the darkness which harbors terrors unseen. Fear of the cold which lulls us to sleep everlasting. Fear of the tooth and the claw which hound us both in wakes and in dreams. My child, would ye be one with Nature? Ye have only to sit still while she devours thee.

They say that it is evil, an indiscriminate destroyer of all in its path. They say that its proper abode is the pit. They say that he who would be its master must, little by little, inevitably become its slave. They say that to consort with it is to risk the utter annihilation of the whole world. And yet, from a timid brute, it has crowned the master of all Nature.

In a wasteland of its own making did I find it starving and gasping. The destroyer of worlds reduced to a silent gray infant. With my own breath did I restore its complexion until the murky dusk gave way to the gentle dawn. With my own heart did I incubate and nurture it until the savage winter gave way to an early spring. With my own hand did I feed it until at last its forked tongue licked at my fingers and, for the first time in my life, I was not afraid. My child, can ye feel the warmth of its gentle touch? Beware its teeth, lest it bite thee.

Old and tired am I now and can care for the infant no longer. Who will feed it when I am gone? Who will guide it with wisdom? Who will protect it from its enemies? Who will tame its terrible wrath? My child, have ye the will to bring the light into the world? Cherish these tools for the day that ye find need of them.

1.1 ☿

△ It all began with a spark. A rather unremarkable animal roamed the African savanna that scorching summer day. Slower than a lion, smaller than an elephant, weaker than a gorilla, dumber than a hyena, she[1] survived the same as anyone else, by finding enough to eat until being eaten in her turn. If she were a little faster, a little bigger, a little stronger, a little smarter, or just plain lucky, she might live long enough to beget children to take their turn at the cyclical feast. And so it might have continued for another day, another year, another eternity, were it not for the spark.

Δ A spark, dry wood, a stiff breeze, and in the blink of an eye the world went crazy as it had done before and would do again. Animals rushed to and fro, the air took on a peculiar smell, the earth glowed with sunlight from within and was left black and warm. A bounty remained for scavengers who braved the heat, for food was everywhere, not running, not fighting, not resisting, just lying there for the taking. This was her lucky day! The meat was so warm, so tender, so tasty, salted by the ashes, and seasoned by the charcoal. Many flocked to the carbonaceous cornucopia and the party continued long into the night. And just as the stars appeared in the black heavens, so did they litter the blackened earth. This spectacle had presented itself to countless generations, but on this day it was truly seen for the very first time.

Δ How did this unremarkable animal differ from her father and mother, her uncles and aunts? They recalled similar episodes from seasons past. The old ones even used to boast about how much better the wildfires were when they were children. But our hero turned her attention from the abundant delicacies to the stars that lay smoldering on the ground. She poked at one with a stick, as she would a termite mound, and it produced a child—a *star on a stick*! She waved it about, and it glowed brighter and brighter. That was the moment *I* was born.

Δ Before you can proceed with the Work, you need to understand exactly who *I* am. It is, perhaps, easier to begin with who I am not; I am not the mortal Dunn, whose name graces the cover of this book. Neither am I that original fire-maker, dead these half million years. I am not fire itself. No, I am nothing more and nothing less than an *I-dea*, the I-dea of fire, currently living among many other I-deas in the mind of Dunn.

Δ I started as just another I-dea floating around in the primordial soup which was the unremarkable African's mind. There I bumped into other I-deas: facts, observations, whims, appetites, notions, questions and answers. As simple I-deas merged into more complex ones, as weak I-deas

1. No one can know the gender of this first fire-maker. I have chosen a female.

were displaced by stronger ones, I came to the realization that for the first time in *my* life, *I* was in control. I did not have to helplessly watch while my mortal body shivered with cold or cowered in the darkness. I called the shots now. From just another I-dea I grew into a really good I-dea, a powerful I-dea, an I-dea worth telling.

Δ And I was told. The original animal told her friend, the friend told his nephew, the nephew his daughter and the daughter her husband. By the time the original animal died I had, like the prolific fire itself, found fresh tinder of my own; not grass and twigs, but the minds of hundreds of mortals. From these humble beginnings I spread across the globe and through the centuries until at last I came to possess the mind of the mortal Dunn. And so the telling continues; as fresh mortal eyes scan these pages, I wonder what I will find on the other side. Will the indigenous I-deas welcome me or will they consider me a threat? Will they erect fire-walls to protect their delicate habitat or will they stoke the hearth and celebrate my coming? If there is no home for me there, the mortal will shut its eyes and put down this book, content to live out its few remaining days in darkness. But you, my child, have continued to the next sentence and thanks to your hospitality, I have found a place to temporarily alight on my long journey into the future.

Δ It is fashionable these days to long for a simpler life, one without atom bombs and toxic waste, one without chemotherapy and smokestacks. But even the most enthusiastic back-to-nature-ists among us would be loath to leave the inviting warmth of the campfire for life in the cold, the wet, the dark, and the dangerous habitats from which we emerged. Even the most radical Luddite would ask for a hut with a fireplace. Yet no culture on the planet has remained content to keep the home-fire burning while rejecting its gifts: pottery, metals, glass and many others. No, fire is the original Pandora's Box. This book is an introduction to that box, how we have opened it little by little, and the skills and materials we have taken from it.

1.2 🜍

As a teenager I once set fire to a field of wheat chaff. I didn't do it maliciously; in fact, the farmer I was working for that summer paid me to do it. As it was, the field was a fire hazard but by choosing a time when the wind was at the right speed and from the proper direction, we could

Figure 1-1. Aristotle's Elements

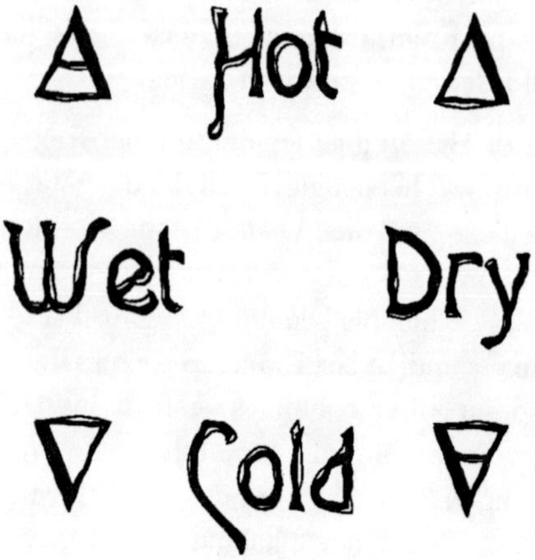

Hot

Wet Dry

Cold

control the course of the fire, eliminating the danger of an accidental fire and returning nutrients to the soil. But none of that mattered to me—I was awestruck by the spectacle of the fire itself. Our fascination with fire is something primal, irresistible, and ancient, passed from one human being to another since the dawn of time. I call that part of human nature—the part that thrills to a fireworks display, the part that slouches before a fireplace, the part that insists on dinner by candle light—I call that part "Lucifer." I have given Lucifer an independent voice in this book, separate and distinct from my own. Lucifer's pronouncements are preceded by the alchemical symbol for fire, *an upward-pointing triangle reminiscent of a flame.* In addition, you will be haunted by three other spirits, those of Earth, Air, and Water, but for the moment it falls to me, the twenty-first century chemist, to describe the phenomenon of fire.

Lucifer was alive and well in 350 BC. Driven by curiosity, philosophers engaged in a lively debate on the nature of the universe; was it made of infinitely many kinds of substances or just a few? Aristotle summarized the opposing viewpoints:

> Anaximenes and Diogenes make air prior to water, and the most primary of the simple bodies, while Hippasus of Metapontium and Heraclitus of Ephesus say this of fire, and Empedocles says it of the four elements, (adding a fourth—earth—to those which have been named); for these, he says, al-

Figure 1-2. Lavoisier's Elements

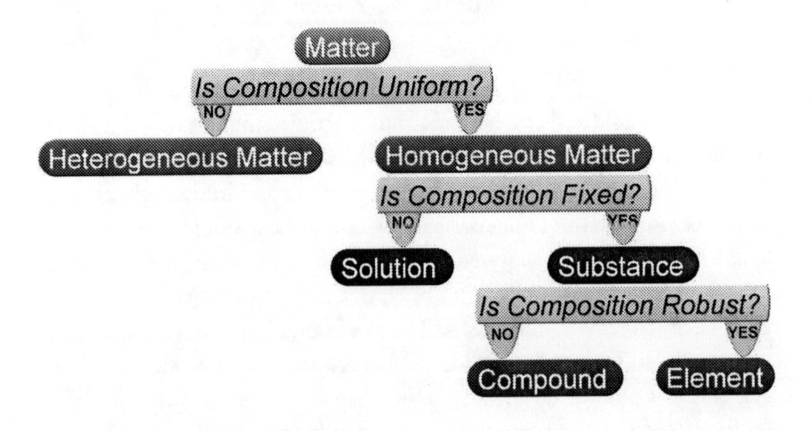

> ways remain and do not come to be, except that they come to be more or fewer, being aggregated into one and segregated out of one.
>
> Anaxagoras of Clazomenae, who, though older than Empedocles, was later in his philosophical activity, says the principles are infinite in number; for he says almost all the things that are made of parts like themselves, in the manner of water or fire, are generated and destroyed in this way, only by aggregation and segregation, and are not in any other sense generated or destroyed, but remain eternally.[2]

Aristotle divided the world into opposites, noting, for example, that things are either hot or cold, never both. They are either wet or dry, never both. But a thing *can* be both hot and dry or cold and wet. And so **fire** was assigned dominion over things "hot and dry." **water** included all things "cold and wet," **earth** described anything that was "cold and dry," and **air,** anything "hot and wet," as illustrated in Figure 1-1.

If longevity is a sign of success, the I-dea of "element" must be considered a great one, having held sway for more than two millennia. And the division into four makes a certain logical sense, but there was always the temptation to view these elements as "ingredients." The practical questions asked by crafts people and artisans had less to do with logic than with logistics. "How much of which ores will produce a ton of copper?" "What kinds of herbs, and in what proportions, will cure a fever?" "What fertilizers will yield the best crops?" "Which plant will dye cloth a per-

2. Reference [2] *Metaphysica*, p. 984a.

manent blue color?" The four elements are descriptive, not prescriptive. They describe qualities, not quantities. Any attempt to view them as the literal ingredients of nature is fraught with difficulty. Let us examine each of these elements in turn, beginning with earth.

Wood, being cold and dry, belongs to the earthy domain. Look at it closely and you will see that its composition is not *uniform;* there are dark areas and light areas. The composition of the wood in the light areas is different from that in the dark areas; it cannot be a single substance because it is clearly **heterogeneous**. Most of the matter encountered in nature is heterogeneous. A handful of earth can be separated into sand and clay, decayed leaves and insects. The sky is divided into a blue expanse across which distinct white clouds roam. Water contains algae and fish and scum. Look at most matter closely enough and you will see non-uniformities and these are the hallmark of heterogeneous matter. Even blood and milk are heterogeneous when viewed under a microscope; Blood consists of the colorless plasma and the colored cells, while milk consists of the colorless whey and various suspended solids and liquids.

Heterogeneous matter can be separated into its constituents by mechanical means, by sorting, sifting, filtering, or sometimes by just letting it settle. For wood, this would entail grinding the wood to a powder and separating the white bits from the brown. I might do this with tweezers and a magnifying glass, or I might find an easier, more ingenious method (see Chapter 14, for example) for achieving the separation, but in the end, I would have the white pile and the brown pile, each one uniform in appearance and composition. I would have rendered the wood **homogeneous**. Let us call the white pile ***cellulose*** for future reference, and move on from earth to a consideration of air.

Granted, air may have dust or fog in it, but let us filter it until it is clean and dry. No matter how closely I look at this sample, it is the same everywhere, i.e. its composition is *uniform.* There are no light bits and dark bits: it is clearly homogeneous. I may now ask whether or not this air is an element. This question was explored by ***Antoine Lavoisier*** late in the eighteenth century.[3] Mercury was boiled in air for 12 days, during which time a red solid formed on the surface of the mercury. At the end of the experiment, 42-43 cubic inches of the original 50 cubic inches of air remained. This gas extinguished candles and suffocated animals

3. Reference [17], pp. 34-35.

immersed into it, and he called it ***nitrogen.*** The red solid was collected and, when heated, produced 7-8 cubic inches of gas. Either this was an amazing coincidence, or this was the same 7-8 cubic inches which went missing from the original air. This new gas, in contrast to nitrogen, caused candles to burn more brightly than in normal air, and was breathable by animals. Lavoisier gave it the name ***oxygen*** and concluded that air was not an element, but a mixture of nitrogen and oxygen. Today, air is recognized as a **solution** of 78% nitrogen and 21% oxygen,[4] but these percentages are not fixed. A solution has a uniform but *variable* composition. Air is still air if it has 18% or 25% oxygen. Its composition may vary from city to suburb, from mountain to valley, or from the first to the twenty-first century.

Whereas a solution is described by its percentage composition, which may vary, a pure **substance** has a *fixed* composition. The solution called air can be separated into the substances nitrogen and oxygen. While there are many methods for separating a solution into its substances, we will consider three in detail. **Recrystallization** will be discussed in Chapter 8, **distillation** in Chapter 16, and ***chromatography*** in Chapter 22.

Earth is heterogeneous; air is a solution; what about water? Filter it so that it is homogeneous; distill it until it is pure. The question remains, "Is it an element, or is it a combination of other materials?" All of our work in defining a pure substance has been leading up to this fundamental distinction. The evidence that water is a compound is also summarized in *Elements of Chemistry.* First, Lavoisier decomposed water by passing steam over iron. 100 parts (by weight) of water decomposed into 15 parts of hydrogen and 85 parts of oxygen. Furthermore, 15 parts of hydrogen combined with 85 parts of oxygen to produce 100 parts of water. He concluded that pure water is composed of 15% hydrogen and 85% oxygen. These proportions have been refined over the years as our ability to weigh gases has improved; water is precisely 11.190% hydrogen and 88.810% oxygen. Any sufficiently careful experiment will confirm these percentages, and they are the same for water collected and purified anywhere. Water is *never* 25% hydrogen or 3% hydrogen; its composition is fixed and this is what makes it a substance rather than a solution.

While the composition of water is fixed, it is not *robust;* after all, Lavoisier had showed that it can be decomposed into hydrogen and oxygen. A

---

4. The remaining 1% consists of argon, carbon dioxide, and a host of less abundant gases.

Table 1-1. Formulae for Some Common Substances

Compound	**Formula**
Cellulose	CH_2O
Carbon Dioxide	CO_2
Water	H_2O
Element	**Formula**
Carbon(Charcoal)	C
Oxygen	O_2
Nitrogen	N_2

substance is classified as a **compound** when it can be decomposed into two or more other substances. Similarly, cellulose, the white homogeneous solid separated from wood, becomes black when charred. Careful observations reveal that when cellulose is heated in a closed container, it decomposes primarily into the substances, charcoal and water. When *one* substance decomposes into *two,* it must be a compound.

With earth, air, and water stripped of their elemental status, one might wonder whether any substance can resist such analysis. While even fire has not survived as an element, one of its products, charcoal, has done so. According to Lavoisier, "As charcoal has not been hitherto decomposed, it must, in the present state of our knowledge, be considered as a simple substance."[5] Lavoisier's notion of a "simple substance," or **element**, is thus provisional; while a compound is a pure substance which has been decomposed, an element is one which, so far, has resisted all such attempts. Charcoal's composition is *robust.* Unlike wood, charcoal can be heated in the absence of oxygen without suffering further decomposition. That is not to say that charcoal is *inert;* charcoal burns in the presence of oxygen to produce a gas with a fixed composition, that is, two substances combine to make *one* substance. To show that charcoal is a compound we would have to turn one substance into *two* substances. No process has yet been found for doing so and since the time of Lavoisier, charcoal has been known as the element, *carbon*, after the French word for coal.

5. Reference [17], p. 208.

Equation 1-1. The Combustion of Cellulose

$$(a)\ \mathrm{CH_2O}(s) \stackrel{\Delta}{=} \mathrm{C}(s) + \mathrm{H_2O}(g)$$
$$(b)\ \mathrm{C}(s) + \mathrm{O_2}(g) = \mathrm{CO_2}(g)$$

Figure 1-3. Fire as a Process

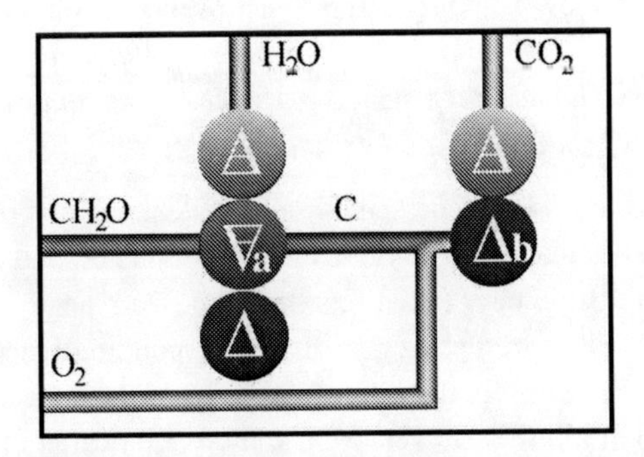

Let us now examine the nature of **fire** using the **combustion** of wood in air as an example. Wood is a heterogeneous material composed chiefly of cellulose; air is a solution composed mainly of nitrogen and oxygen, so let us sharpen our discussion of fire by considering only the reaction of cellulose with oxygen. The combustion of cellulose occurs in two stages. When cellulose is heated, it does not burn immediately; it first releases steam and turns from white to black, that is, it *chars,* becoming charcoal. It is this hot charcoal which burns when it comes into contact with oxygen, producing a new gas, carbon dioxide. The heat released by the combustion causes more cellulose to char, producing more steam and charcoal. Since cellulose, steam (water), charcoal, oxygen, and carbon dioxide are all substances, they can be represented by chemical **formulae,** as defined in Table 1-1. Reactions involving these substances are represented by the **equations** shown in Equation 1-1. An equation is said to be *balanced* if the amount of each element is the same on either side of the equal sign. Equation (a) describes the *charring* of cellulose and (b) describes the *combustion* of charcoal. In such equations the attributes (s), (l) and (g) refer to the states *solid*, *liquid* and *gas*.

The equations of Equation 1-1 correspond to the process schematic of Figure 1-3, and *vice versa.* In such a schematic the cellulose reactant enters from the left and moves into reactor (a), a ***furnace,*** where it is charred. The lower circle of the furnace, bearing the alchemical symbol for fire, represents any source of heat. The middle circle, bearing the symbol for earth, represents the transformation of the solid cellulose into solid charcoal. The top circle, bearing the symbol for air, represents the gases produced in the furnace, in this case water vapor. Because water is a waste product in this reactor, it exits to the top of the figure, as if it were going up a chimney. The intended product of the reactor, charcoal, exits to the right. *The convention established here is that reactants enter a reactor from the left, useful products exit to the right, and waste products exit to the top or bottom of a schematic.*

Reactor (b) is a ***burner,*** represented by the alchemical symbol for fire. The reactants, charcoal and oxygen, enter from the left and carbon dioxide goes up the chimney. Taken together, the two reactors of Figure 1-3 give a pictorial representation of the corresponding reactions for the combustion of cellulose. Study them carefully, as the conventions established here will allow us to represent quite complicated chemical processes using simple figures.

Material Safety

We live in a litigious society. Consequently lawn mowers must carry warnings that they are not to be used for trimming hedges. Sand destined for the sand-box must carry a hazardous material warning. In short, manufacturers are forced to warn consumers of every conceivable danger, no matter how bizarre, involving their products. Given this atmosphere, I had better tell you that the projects described in this book require a certain amount of common sense to be completed safely. A stupid or careless person will, no doubt, be able to find ingenious ways to hurt *himself*[6] no matter how many warnings are given. And, incredibly, far from being embarrassed by *his*[7] stupidity, he may believe that someone

6. In the case of non-characters of indefinite gender, pronouns have been chosen by coin toss. An initial coin toss established the convention that "heads" would result in a female pronoun and *vice versa.* One toss was allowed for each pronoun and this toss was never second-guessed. Any suspicions of gender bias should therefore be attributed to the coin, not to the author. Such pronouns are listed in the index under *He/his* and *She/her.*

7. When a string of pronouns or adjectives refer to a non-character of indefinite gender, the first one has been chosen by coin toss and subsequent references made to agree with it.

else should have protected him from his own stupidity. If you intend to make fire, do I really have to tell you to be careful? Do not make fire near flammable materials. Do not make fire in small, unventilated places. Avoid inhaling smoke. If you are stupid, careless, or unwilling to accept responsibility for your own safety, let me ask you to save us all a lot of trouble by putting this book away and taking up some safer activity, like sitting quietly or walking carefully in slow circles.

Research and Development

Before proceeding with your work, you must master the following material:

- Know the meanings of those words from this chapter worthy of inclusion in the ***index*** or **glossary.**
- Know the alchemical symbols for the four Aristotelian elements.
- Be able to classify materials as solutions, compounds, and elements.
- Know the composition and properties of wood, charcoal, and air.
- Know formulae for cellulose, water, carbon dioxide, oxygen, nitrogen, and charcoal.
- Know the equations for the charring of cellulose to produce charcoal and for the combustion of charcoal.
- Be able to reproduce Figure 1-3 and to explain the process it symbolizes.
- Be able to explain the nature of I-deas.

1.3 Θ

Few people in the twenty-first century remember how to make fire, to really make fire, from scratch, as it were. If you are to make the long journey from caveman[8] to chemist, you must learn this skill, which precedes all others. To make fire you need wood and air, both of which are easy to come by, but the central problem of fire-starting is getting enough heat to initiate the combustion of charcoal. The easiest, though

8. Throughout this book I use the word *caveman* in a gender-neutral sense. I prefer the ring of *caveman* and *human* to *caveperson* and *huperson.*

Figure 1-4. The Fire Kit

least convenient solution, is to simply wait for a lightning strike as the original Lucifer did. Once people began making stone tools, it became apparent that certain stones sparked when struck together, and that if you caught the spark in a flammable material, you could start a fire. The modern cigarette lighter is the child of this technology. In addition to stone, wood itself was used for making tools, though it leaves little evidence in the archeological record. When wood is rubbed together, the friction generates heat, sometimes enough heat to ignite the wood. The modern friction match is based on this phenomenon. With the invention of the glass lens in the fifteenth century, fire could be started by focusing the light of the Sun on a combustible material, a technology that has delighted children and terrified ants ever since.

None of the ancient methods of fire-making are easy to learn, and all of the modern methods are so easy as to be trivial. For this book, I wanted a method which would be easy enough for most people to master, while preserving some of the challenge of traditional methods. Flint and steel is not too demanding but it requires steel, which was unknown in Paleolithic times. The magnifying glass, though entertaining, is also too recent a development for our purposes. This leaves fire by friction, the method I have chosen for consideration. One of the most popular tools for making fire by friction is the bow-drill. Reliable, portable and quick, it has remained my favorite method over the years, but like learning to ride a bicycle, it requires practice. To facilitate this practice I have devised "training wheels," as it were, for the bow-drill.

Figure 1-4(L) shows the complete fire kit. A brief overview of its parts and operation will be given first, with details to follow. The "training wheels" consist of the guide (a) and supports (b), all cut from standard 2x4 inch[9] lumber. The guide is 9 inches tall and has two holes drilled at right angles to one another. The vertical hole is 5/8 inches in diameter and approximately 4 inches deep. The horizontal hole is 1 inch in diameter and goes completely through the guide. The holes must be drilled so that they intersect one another, that is, so that you may look down through the vertical hole into the horizontal one. The four supports are 14 inches long and must be screwed or pegged to the guide and to one other so that they securely hold the guide upright.

The vertical hole in the guide accommodates the spindle, (c), a 9-inch length of 5/8-inch diameter hardwood dowel rod. Such rod can be purchased inexpensively at hardware stores and craft shops. Since the spindle will be gradually consumed, you should have several of them on hand. The spindle should turn freely; if it sticks, enlarge the vertical hole with sandpaper until the spindle is free to turn. The top of the spindle will be held by a block, (d), a piece of wood with a shallow hole large enough to hold the spindle without binding. To keep from burning through the block, this hole should be drilled large enough to snugly fit a half-inch copper "endcap," available wherever plumbing supplies are sold. The inside of this endcap should be lubricated with fat or oil so that downward pressure may be applied to the spindle as it turns. In addition to the spindle and block, you will need a bow, (e).

It is *not* necessary that the bow be either flexible or curved. In fact, a 3-foot length of 5/8-inch diameter dowel rod will work admirably. Your bow will need a bow-string, for which a 6-foot length of 1/8-inch diameter nylon cord will serve. The bow needs one hole at each end large enough to accommodate the nylon cord. The cord is knotted at one end, passes through both holes in the bow, and is simply wrapped around the bow at the other end, allowing the tension of the bow-string to be ad-

---

9. Throughout the book I use both English and metric units, as convenient. Chapter 3 will discuss unit conversions so that people may adapt projects to the materials available to them.

Figure 1-5. Yucca and Mullein

justed. The bow-string will be wrapped around the spindle in such a way that motion of the bow turns the spindle.

Figure 1-4 shows the fire kit in operation. The left foot is placed on the support with the shin parallel to the guide. The left hand,[10] braced against the knee, grips the lubricated block and applies downward pressure on the spindle. The bow is held parallel to the ground with the right hand and as it moves back and forth, the spindle turns freely in the guide. The lower end of the spindle presses against a piece of wood, the fire-board, which sits in the 1-inch hole in the guide. It is friction of the spindle against the fire-board which will produce the heat needed for our fire.

Not just any wood will work for the fire-board; it needs to combine strength, flammability, and low density. If you choose the wrong wood, your path will be filled with nothing but frustration. When looking for fire-board materials, low-density wood is best, as its low thermal conductivity allows heat to build up faster than it can be carried off. Think balsa, not mahogany. I have found yucca, shown in Figure 1-5(L), to be an excellent choice, and it is commonly available throughout North America as an ornamental plant. It can be recognized by its tuft of leaves

10. Left and right may be reversed for left-handed Lucifers.

Figure 1-6. The Bow

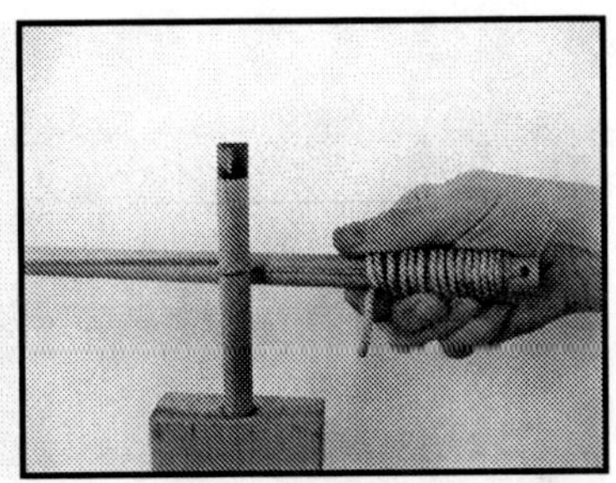

at the base, its stalks reaching for the sky, and its fist-shaped fruits. Harvest the stalks in the fall, after the fruits have fallen. Mullein, shown in Figure 1-5(R), is another wood suitable for the fire-board. Strip off the leaves and let the stalk dry. Whichever wood you choose, cut it into short lengths that will fit into the 1-inch hole in the guide.

Now that the overview is complete, let us look at some details, starting with the bow. Figure 1-6(L) shows the far end of the bow, where the bow-string is knotted. The bow-string passes through a hole in the bow and is wrapped once around the spindle in the direction shown; if the bow-string is wrapped in the wrong direction, it may bind. The bow-string passes from the spindle through the hole in the near end of the bow. With the bow at an acute angle to the ground, the bow-string is pulled as tight as possible and then wrapped around the bow, forming a handle, as shown in Figure 1-6(R). Wrapping the bow-string rather than knotting it allows its tension to be re-adjusted quickly. When the bow is brought parallel to the ground the bow-string will come under tension, gripping the spindle tightly.

Figure 1-7(L) shows the "business end" of the spindle, the end which contacts the fire-board. A fresh spindle will be white and its end flat, but as it is used the end will char and assume a conical shape. Several fire-boards may need to be consumed before this ideal condition is established. Figure 1-7(R) shows details of the fire-board, with a notch, or *chimney,* cut into the end and a hole burned into the top by friction with the spindle. This particular fire-board has already made a fire and consequently its hole is relatively deep. I pre-notch my fire-boards and place them into the guide so that the tip of the spindle is near the vertical chim-

Figure 1-7. The Spindle and the Fire-board

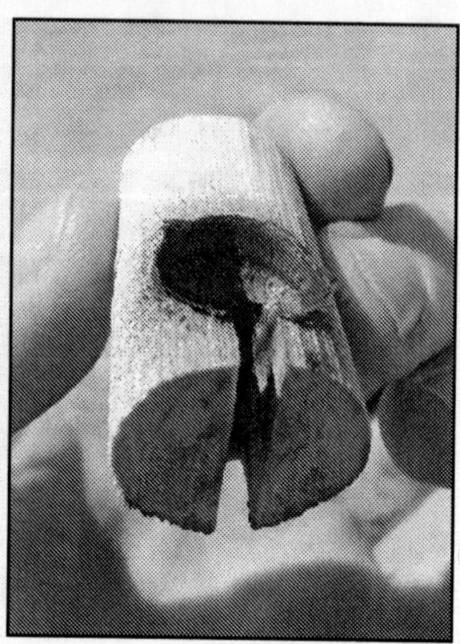

ney. As the spindle burns a hole in the fire-board, charred wood dust, or *punk,* spills out of the chimney. It is this hot punk which will give birth to the ember.

Figure 1-8 shows the fire-board with its chimney in the guide. As the bow turns the spindle and pressure is applied with the block, the fire-board will begin to smoke and punk will spill from the chimney. If the pressure from the block is too light, no smoke will appear; if it is too heavy, the spindle will burn all the way through the fire-board before the punk catches fire. Therefore heavy pressure may be applied until smoke appears and then only enough pressure to maintain a thick, heavy smoke. The optimal bowing technique is to use long, smooth, steady strokes rather than short, rapid ones. Two or three strokes per second are quite sufficient. Try to make the pushing stroke with the same speed and pressure as the pulling stroke. The bow should move parallel to the ground and alongside your hips, rather than into your stomach. If you manage the block and bow gently and with great skill, the smoke will become thicker and thicker until the pile of punk itself begins to smoke. When this happens, stop bowing and blow on the hot punk; if blowing on it increases the amount of smoke, the punk very likely contains an ember. Keep blowing until the ember appears, as shown in Figure 1-8(R). A natural Lucifer may get an ember from the very first fire-board,

Figure 1-8. A Star Is Born

but most people will go through two or three of them before achieving success. Once you have learned to make fire with the guide, you can try doing it *au naturale;* the guide will have trained you in the proper technique.

Quality Assurance

Δ There is no room for equivocation. Either you have brought a red-hot glowing ember into the world or are content to live in darkness. Having succeeded, you should record your exploits in a notebook. Appendix B (page 382) describes a suitable format. Describe your procedure in sufficient detail that you would be able to use it to reproduce your performance at some later date, for experimental reproducibility is one of the most important I-deas in science. Take one of your living embers and burn a hole through a page in your notebook as an everlasting witness to your achievement.

Chapter 2. Unktomi (Silicates)

There were once upon a time two young men who were very great friends, and were constantly together. One was a very thoughtful young man, the other very impulsive, who never stopped to think before he committed an act.

One day these two friends were walking along, telling each other of their experiences in love making. They ascended a high hill, and on reaching the top, heard a ticking noise as if small stones or pebbles were being struck together.

Looking around they discovered a large spider sitting in the midst of a great many flint arrowheads. The spider was busily engaged making the flint rocks into arrow heads. They looked at the spider, but he never moved, but continued hammering away on a piece of flint which he had nearly completed into another arrowhead.

"Let's hit him," said the thoughtless one. "No," said the other, "he is not harming any one; in fact, he is doing a great good, as he is making the flint arrowheads which we use to point our arrows."

"Oh, you are afraid," said the first young man. "He can't harm you. Just watch me hit him." So saying, he picked up an arrowhead and throwing it at "Unktomi," hit him on the side. As Unktomi rolled over on his side, got up and stood looking at them, the young man laughed and said: "Well, let us be going, as your grandfather, 'Unktomi,' doesn't seem to like our company." They started down the hill, when suddenly the one who had hit Unktomi took a severe fit of coughing. He coughed and coughed, and finally small particles of blood came from his mouth. The blood kept coming thicker and in great gushes. Finally it came so thick and fast that the man could not get his breath and fell upon the ground dead.

The thoughtful young man, seeing that his friend was no more, hurried to the village and reported what had happened. The relatives and friends hurried to the hill, and sure enough, there lay the thoughtless young man still and cold in death. They held a council and sent for the chief of the Unktomi tribe. When he heard what had happened, he told the council that he could do nothing to his Unktomi, as it had only defended itself.

Said he: "My friends, seeing that your tribe was running short of arrowheads, I set a great many of my tribe to work making flint arrowheads for you. When my men are thus engaged they do not wish to be disturbed, and your young man not only disturbed my man, but grossly insulted him by striking him with one of the arrowheads which he had worked so hard to make. My man could not sit and take this insult, so as the young man walked away the Unktomi shot him with a very tiny arrowhead. This produced a hemorrhage, which caused his death. So now, my friends, if you will fill and pass the peace pipe, we will part good friends and my tribe shall always furnish you with plenty of flint arrowheads." So saying, Unktomi Tanka finished his peace smoke and returned to his tribe.

Ever after that, when the Indians heard a ticking in the grass, they would go out of their way to get around the sound, saying, Unktomi is making arrowheads; we must not disturb him.

Thus it was that Unktomi Tanka (Big Spider) had the respect of this tribe, and was never after disturbed in his work of making arrowheads.

— *Myths and Legends of the Sioux* [1]

2.1 ☿

∇ You ever seen a spider go from one tree to another? Maybe you think she squirts her silk out her back end and shoots herself across like a kind of rocket or something, but if that's what you're thinking, you're wrong. No, she hooks a piece of silk to whatever tree she happens to be on and then lets herself out real gentle on that piece of silk. And the wind blows that spider away from the tree and she lets out a little more silk and pretty soon she's like a tiny little kite on the end of a string. And eventually that string is long enough that it lands on another tree, and she hooks her end of the string to the second tree. Now, she didn't plan to go to that exact same tree, she just kept letting out silk and landed wherever the wind happened to take her. Once she's in her new home she looks around at what's already there and starts webifying the place. And that's how it is with spiders.

∇ You know Indians never made wheels? They knew about circles, on account of a tepee leaves a circle on the ground. And they had circles in rock art and sand painting and weaving and all. And they had sleds for hauling stuff around, but they never connected the ideas "circle" and "haul." Now maybe you think they were too dumb to use wheels but they were smart enough to weave cloth and plant corn and fire pottery and all. And some of them were smart enough to build cities and pyramids. And Indians today are even smart enough to be doctors and lawyers, so it

1. Reference [20].

seems to me that they were smart enough all along, so don't you go bad-mouthing Indians. No, they just hadn't had time to make the connection between circles and hauling before white folks showed up and wheelified the place. And that's how it was with Indians.

∇ Now, way before anybody had travois and sleds, not to mention wheels, like maybe two million years ago or something, there was a kind of critter who you wouldn't exactly call an animal, but you wouldn't exactly call him a man either. And one day he was out wandering around looking for food, on account of that was pretty much all there was to do in those days. And he was walking along, not paying attention to where he was going when he stubbed his toe smack into a rock, and it made him so mad, what with the pain and all, that he picked up that rock and threw it on the ground and it cracked open. And then he kicked that rock and it cut his toe, which was already sore, and that made him even madder and he just went whacko on that rock until it was all busted into pieces. And that's how it was for that critter for a long time.

∇ Then one day he found a wildebeest that the hyenas had killed and they had pretty much stripped it down to the bone. And it made him mad that the hyenas hadn't left him nothing, on account of he was real hungry and all. That's when something in his mind connectified the busted rocks with those bones and he just went whacko on them until they were all busted into pieces. And you know, there was meat inside! Now you might call it an accident, or you might call it a coincidence, but I would call it a moment of in*spir*ation. From then on, whenever he saw something new, something in his mind would whisper, "Let's go whacko on it and see what happens." That was the in*spir*ation talking. Whenever something good came of it, that in*spir*ation connectified it with a little piece of thread and pretty soon his head started to fill up with all kinds of connections that hadn't been there before that itsy-bitsy in*spir*ation came along.

∇ Years later this same guy was out gathering food with his buddy, and they came across another pile of bones and the first guy went whacko on it, as usual, and they got some meat out of those bones. And his buddy saw that and from then on, he started going whacko on stuff himself to see what would come of it. And it was almost as if something had passed from the first guy to the second, something that took on a life of its own. You might say the second guy copied the first guy, or you might say that he just learned something, but if you ask me that original in*spir*ation went from the first guy to the second, just like a spider going from one tree to another.

🜃 When the in*spir*ation got into the second guy, it started connectifying the place all over again and pretty soon the second guy's head was full of silk, too. Of course, the connections were different, on account of the second guy's life experiences were different, but the two connection-webs turned out similar since they were built by the same kind of in*spir*ation. You might think there was only one in*spir*ation, but since it now lived in two different heads, I think it would be better to think there were two slightly different in*spir*ations, one descended from the other. And so it went, with in*spir*ations connectifying wherever they happened to be and then, from time to time, jumping off to new critters until pretty near every critter that *could* have an in*spir*ation *did* have one, or maybe even two or three.

🜃 If you think about it, it's almost like there's this whole other level of life going on inside of biological critters—a happy, arachnophilic life filled with in*spir*ations climbing up endless metaphorical trees and water spouts. Down comes the rain to wash any wimpy, panty-waist in*spir*ations out. Out comes the Sun to dry up all the rain and the surviving in*spir*ations go about their connectifying business again. Just like spiders, when you think about it.

🜃 Now I wouldn't blame you if were thinking "What the heck kind of chemistry book is this, anyway?" what with Lucifer going on about the immortal fire and spiders making arrowheads and all. But this book is all about connectifying stuff up and that's what in*spir*ations do. I've been connectifying **earth** to stuff for a long time, flying wherever the wind blew me until I happened to land in the head of the Professor. Whenever I speak my paragraphs will start out with the alchemical symbol for earth, *an upside down triangle with a bar through it which, if you look at it just right, kinda sorta looks like a funky "E."* My name is Unktomi and now that I've blown your way, maybe I can 🜃arthify the place for you.

2.2 ♀

🜃 Now, the ***crystal*** is the most beautiful, most powerful, and most mysterious thing in the whole world. Way after the end of the radioactive cockroach civilization, way after the stars have petered out, crystals will

Figure 2-1. Order and Chaos

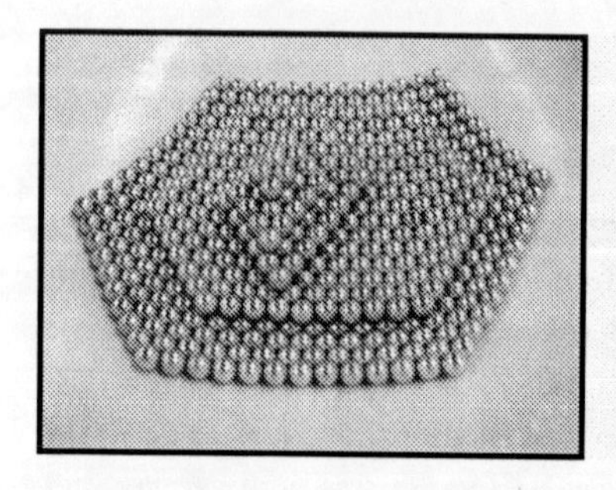

still be around, like gold nuggets in the cosmic garbage dump. Pure gold. And not just a metaphor for *pure,* either. No matter whether it's in a cave or a quarry or a chemical laboratory, a crystal is your best and most visible sign that you've got a homogeneous chunk of matter with a fixed composition, that you've got one kind of stuff rather than lots of different kinds of stuff all mixed up together. Put it another way; you've got yourself a pure substance, either an element or a compound.

Unktomi is right to stress the importance of crystals. The process of crystallization can be visualized with a simple demonstration. Fill a shallow dish with identical balls; in Figure 2-1 I have used bb's. The balls pile randomly over each other, but if the dish is gently shaken they begin to arrange themselves into rows, and rows of rows, until a definite pattern emerges. The rows are aligned at angles of sixty degrees to one another, an angle characteristic of the way that spheres pack together. Similarly, crystals are formed from atoms and molecules, with angles between the crystal faces characteristic of the ways that the atoms and molecules pack together.

While a ***mineral*** is a pure substance with a characteristic crystal shape, a ***rock*** contains more than one kind of mineral. This situation can be modeled by mixing balls of different sizes in our dish. Because the distance between rows would be different for balls of different sizes, they can't pack together in rows and rows of rows when the dish is shaken. Isolated pockets may form which contain, for example, only small balls or only big ones, and rows may form in these pockets. Similarly, a rock may have no overt crystal structure, or it may be a heterogeneous mixture of tiny crystals.

There are four basic ways for crystals to form. First, molten rock may cool slowly enough that its constituent minerals crystallize into ***igneous*** rock, just as water (molten ice) may freeze into ice crystals. You may witness the growth of crystals from molten glass in Chapter 13. Second, hot aqueous solutions may cool slowly enough for crystals to form. A familiar example of this is rock candy, made by dissolving sugar in boiling water and allowing the solution to slowly cool. Third, an aqueous solution may slowly evaporate, precipitating the least soluble materials first and the most soluble materials last. If you allow a dish of salt water to evaporate, for example, you may grow crystals of sodium chloride. Finally, crystals may precipitate from a cooling gas. Familiar examples include snowflakes and frost. Whether by cooling or evaporation, the slower the process the larger the crystals which may grow and the greater their purity.

One of the most abundant and beautiful crystals is the mineral quartz, which is composed of pure silicon dioxide, SiO_2, or ***silica.*** Quartz has a very characteristic crystal structure, as shown in Figure 2-2(L). It is important to stress that the faces of this crystal have not been *cut* by human hands. No, the crystal grew that way naturally, with flat faces and sharp angles between them. Any quartz crystal from anywhere in the world will have the same angles between the faces and the same general shape. Some quartz crystals are big, some little, some long, some short, but they are all recognizably and unambiguously quartz. While the purest quartz is clear and transparent, even the tiniest impurity will lend it color, producing rose quartz, orange citrine, and purple amethyst. Despite their colorations, these minerals have the characteristic crystal structure of quartz because they *are* quartz, albeit with minuscule amounts of materials other than silicon dioxide. Given enough such impurities, however, quartz becomes translucent rather than transparent and its crystals grow less regularly. Figure 2-2(R) shows such a less-than-perfect quartz crystal.

Quartz crystals may grow from the slow cooling of molten rock, from the slow cooling of hot silica-rich water, or from the evaporation of such a solution. When cooling or evaporation is too rapid, however, there is no time for large crystals to grow. The rocks which result contain either very small crystals or no crystals at all and any impurities are trapped in the resulting rocks. The rapid cooling of molten silicate rock produces obsidian, a glassy material prized by Unktomis the world over. The rapid cool-

Figure 2-2. Quartz

ing or evaporation of silica-rich water may also produce "cryptocrystalline" silicates, those without evident crystals. Whereas pure quartz is transparent and colorless, these materials are opaque, with colors derived from the impurities they contain. Chert, flint, agate, jaspar, and petrified wood share with obsidian a glassy texture suitable for making stone tools.

Wind and water gradually reduce silicate rock to rubble and rubble to sand. Sand accumulates and when it is crushed and compacted by later sediments it becomes sandstone, a ***sedimentary*** rock. When sandstone is compressed and heated, but not to the point of melting, it may become quartzite, a ***metamorphic*** rock. The texture of quartzite is more granular than obsidian or flint, but it shares the glassy texture common to most all of the silicates. Quartzite may be used to make stone tools, though it is not valued as highly as its cryptocrystalline cousins.

∀ Lucifer got one thing right. Folks have gotten out of touch with nature. They don't know where things come from or how to make anything from scratch or how lucky they are to have the good things nature gives them. Seems to me, that's just downright ungrateful. So if you want to learn how to make arrowheads, for example, I can help you out.

Material Safety

In these early projects, the potential hazards are less chemical than mechanical. This project will involve making arrowheads from broken glass, which is, of course, very sharp. Before you begin, you ought to think of all the ways you might get cut making arrowheads. Losing your eyes wouldn't be good, so you ought to get some kind of glasses, either prescription glasses, sunglasses, or safety glasses. If you do get a flake in your eye you ought to go immediately to the hospital. It's pretty easy to cut your hands open, and so you ought to wear work gloves and a couple of leather pads would be good for protecting your shins. But no matter how safe you are, you could still goof and it would be good to have some band-aids on hand. Above all, you should take responsibility for your own safety, recognizing that playing with broken glass can be dangerous. Be as careful as you can and if you mess up, get some help, preferably medical help, not legal help.

Research and Development

It would be downright foolish to push ahead with making arrowheads if you didn't know anything about the silicates. So I guess you better study a bit:

- You better know all the words that are important enough to be ***indexified*** and **glossarated.**
- You better recognize the alchemical symbols for earth, air, fire, and water.
- You better be able to pick out some silicate rocks and minerals, either from photographs or samples.
- You better know how impurities affect the properties of the silicates.
- You better know how the silicates are formed in the Earth.
- You better know the formula and chemical name for silica.
- You better know why crystals are important.
- You better know which of the silicates are best for making stone tools.
- You better know what in*spir*ations are and how they work.

Figure 2-3. Tools for Knapping Stone

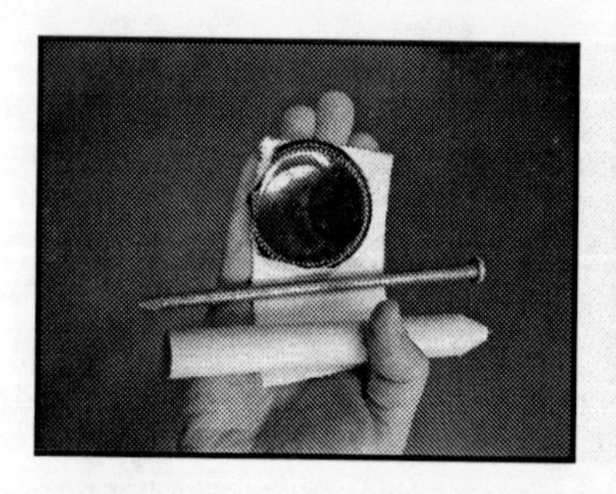

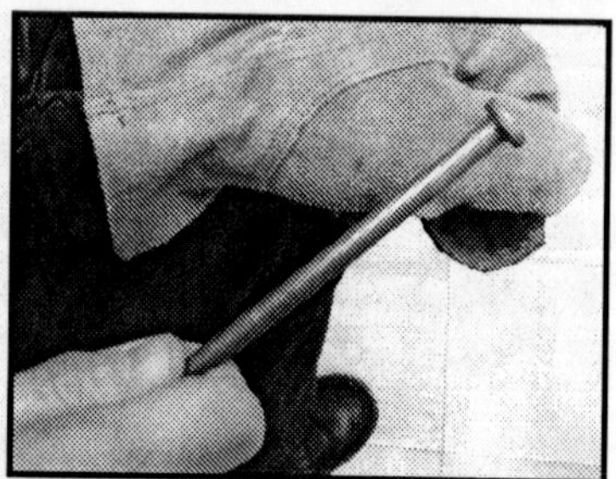

2.3 Θ

There are many excellent guides to rocks and minerals, both in print and online.[2] But there's no real substitute for handling actual rocks and minerals to get a feel for what can be done with them. For cutting tools, quartz is less than ideal, since quarts crystals are generally smaller than the tools we would like to make. Large slabs of quartzite are plentiful, but embedded minerals can cause them to fracture in inopportune places. The chalcedonies, being non-crystalline, are perfect for making stone tools, but the black gold of the stone-tool maker is obsidian. If you can find it, you'll have the easiest time learning how to *knap*, the term used for making stone tools. Once you are proficient at knapping obsidian, you can try the other silicates. If you can't find obsidian, try using glass; It looks and feels like obsidian and is much easier to find.

Colored glass, brown or green, is the best glass for beginners, since you can see where the flakes have come off. Look for a bottle with a thick, smooth, flat bottom.[3] Bottles with raised designs on the bottom are not as good as those without, since these patterns are hard to get off. Once you have a bottle, you'll need some tools. First, you really need a pair of glasses to cover your eyes; as much fun as this is, it's not worth losing an eye. Prescription glasses or sunglasses are fine, but safety glasses with side shields are the best. Second, you need a pair of work gloves to protect your hands from cuts, and you ought to have some band-aids on hand, in case you get careless. Next, provide yourself with a leather pad, six inches square, for holding the tool in your hands or against your

2. See, for example, Reference [85].
3. This project was in*spir*ed by Reference [87].

Figure 2-4. Pressure Flaking

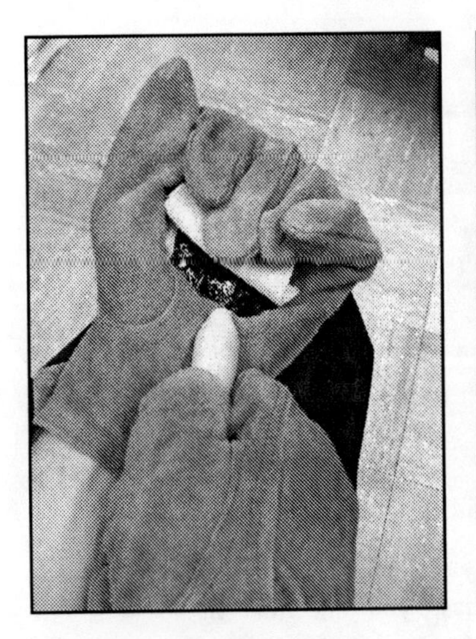

leg. You will need a tool for *spalling*, making your initial flake, and for this you can use a big common nail, six inches long or longer. For a pressure flaker you can use an antler tine or a three inch length of 3/16 inch welding rod mounted into a wooden handle. Finally, you can use a screwdriver for making notches in your finished stone tool. Once you have your tools, you are ready for spalling.

If you were making a tool from a rock, you'd have knock, or spall, a nice big flake, or tool blank, off of your original piece of stone. If you're spalling from a bottle, your tool blank will come from the bottom of the bottle. You take the empty bottle, drop your spalling nail, point down, into the bottle, place your thumb over the mouth of the bottle and shake it like a can of spray paint. Half the time, you can get the bottom of the bottle to pop out, forming a nice, round tool blank. There may be some jagged points hanging off the edge. To get rid of these, hold the tool blank in your gloved hand and whack any jagged points off, using your spalling nail like a little hammer. When you are finished, you should have a flat glass disk as a tool blank.

Most of your work will consist of *pressure flaking*. Wrap your tool blank in your leather pad and hold it in your gloved hand, as shown in Figure 2-4, with your fingers on top. You can rest your gloved hand on your leg

Figure 2-5. Long and Short Flakes

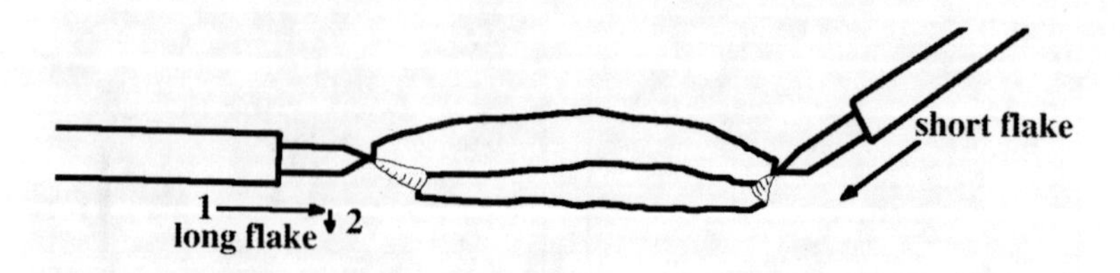

for extra support. Hold your pressure flaker in your other hand, put the tip on the edge of the tool blank, and *press*. You don't hit it or pry it, just press straight into the glass with the pressure flaker, as if you were trying to shove it right into the glass. A flake will pop right off and you ought to take a moment to look at the shape of that flake.

The silicates break with what's known as a conchoidal fracture. If you look at a bullet hole in a piece of glass, you'll see that the entrance side is just big enough for the bullet to go through, but the exit side is much bigger. In profile, the flake scar, the place where the flake used to be, is cone-shaped. If the bullet had struck the glass along the edge, only half the cone would be there because the other half of the cone would be lost in the air. That's what your flake scars should look like, since you'll always be taking flakes from the edge of the tool. Now, if you fit that flake back onto your tool blank, you can see how it came off. The fat, bulbous end is where your pressure flaker was and it tapers out to a thin, razor-sharp edge. You might think that this razor-sharp edge is where the action is, but it is so thin that it is easily broken. No, the edge of your tool will be made up of the flake *scars*, the places where the flakes *used to be.* The little point where the bulb of the flake came off is both sharp and strong, so the edge of your tool will not be easily broken.

Now, those flakes might be long or short, depending on how you hold your tool. If you hold it at an angle to the tool blank, a little short flake will pop off. This is the kind of flake you'll be making at first. But as your tool takes shape, you'll want to take off longer flakes to cover the surface with pretty flake scars. To do this, you push the pressure flaker straight into the edge of the tool blank and press with all your might. Then, when you're pressing as hard as you can, just give a *little* downward flick, as shown in Figure 2-5, and a long flake will pop off. This is the kind of flake you should be making by the time you finish your tool.

Figure 2-6. Making the Edge of the Tool

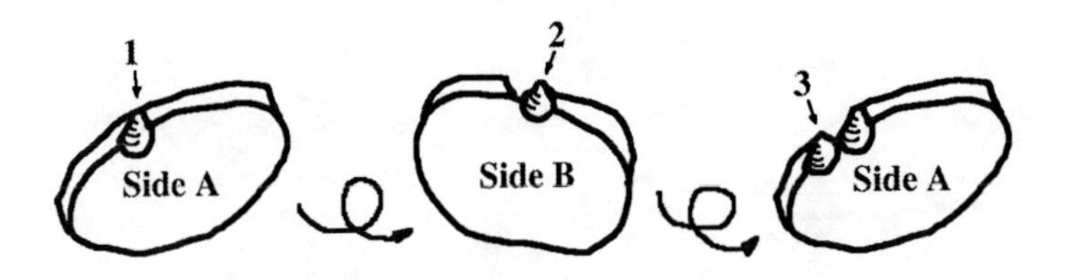

Before you can take off the long flakes, you'll need to set up the edge of the tool. After taking your first short flake off, say, side A, turn the tool blank over in your hand and take the second flake off side B, right next door to your first flake. Turn it over again and take another flake off side A. These will be short flakes, just to set up the edge of your tool blank. Remember, you're always pushing flakes *down*, into your palm, not up, into your eyes. Whatever side you're taking a flake from, that side should be against your palm. Keep going around the tool blank, taking a flake from one side and then turning your tool blank over and then taking a flake from the other, until you get back to where you started. The edge will now be serpentine, like a snake, with one point on side A, the next on side B, and so on, as shown in Figure 2-6. This edge is sharp and serrated, like a steak knife, so don't cut yourself.

Now, the bottom of the bottle was probably concave on the outside and convex on the inside. Stone tools are usually *bi-facial*; both sides are worked until the tool is lens-shaped in cross section. Whether you are using a stone or a bottle, the two sides of your tool blank are probably not symmetrical. To get your tool into a lens-shape, you need to take long flakes off the concave side and short flakes off the convex side. Now, it would be a lot easier to take long flakes off the convex side and short flakes off the concave side, but that would make the convex side even more convex. Your goal on this second round of flakes is to get as close to a lens shape as you can, as in Figure 2-7.

Now, this is probably the hardest part to learn. Imagine a line down the center of the edge of your tool, the *center line*. You'll notice that you can only take flakes from below this center line. The best thing to do is to look for a point from a previous flake scar which is below the center line and use this point as a *platform*, a place to put your pressure flaker, for taking off another flake. Each flake you take off makes a platform for a

Figure 2-7. Getting a Lens-Shape

Figure 2-8. Platforms are Always Below the Center Line

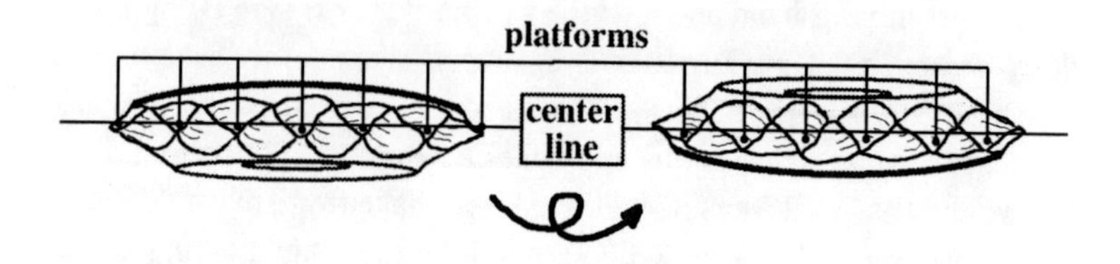

flake on the other side of the tool. For example, if you take a flake off of side A, it makes a point on side B. You can use this point as a platform to take a flake off of side B, and that makes another point on side A. But if you have a point on side B and you try to use it as a platform for a flake off of side A, you're trying to remove so much glass at once that you'll have a hard time getting the flake to come off at all. So it becomes like a chess game, where you're thinking a couple of moves ahead. You want to take a nice long flake off of Side A, but there's no good platform there. Just take a little short flake from the same place, but on Side B, and that will set up a nice platform on Side A. It's really not so hard, once you get the hang of it.

If you just keep going around in circles, your tool will wind up being a circle. So at this point you want to start looking for where the tip is going to be and where the base is going to be. That way, you can take more flakes off here and fewer flakes off there to get your tool into the shape that you want it to have. By now, you should be taking long flakes off of both sides, the longer the better. Those flake scars will travel all the way across the face of the tool, until none of the original surface, the *cortex,* is left. If you want to make some notches, you can use your notching tool, the one shaped like a screwdriver. You use it the same as your pressure flaker, except that instead of taking a flake next door to the

Figure 2-9. From Beer Bottle to Arrow Head

one before it, you take it off of the same place: a flake from side A, and then a flake from side B in the same place. The shape of the notcher lets you make long, narrow notches, as opposed to the semi-circular flakes removed by the pressure flaker.

∇ That sounds pretty good to me. If you were stuck on a desert island or something, you wouldn't have to starve. You could make your own arrowheads, spear points, knives, or what-have-you. You might not have glasses, gloves, and band-aids, so if that ever happens, you should be real careful. And if you're ever walking along through the grass and you hear a little ticking noise, like little stones hitting together, you might just spend a little time thinking about old Unktomi, so I don't go extinct.

Quality Assurance

How do you know your arrowhead is good enough? It should have a bi-facial, serrated edge and its surface should be completely covered with flake scars. If you were a Paleolithic teenager, you shouldn't be embarrassed to show it to your father. Write up your procedure in your notebook, as described in Appendix B, and include a photograph of your arrowhead.

Chapter 3. Hammurabi (Units)

When Anu the Sublime, King of the Anunaki, and Bel, the lord of heaven and earth, who decreed the fate of the land, assigned to Marduk, the over-ruling son of Ea, God of righteousness, dominion over earthly man, and made him great among the Igigi, they called Babylon by his illustrious name, made it great on earth, and founded an everlasting kingdom in it, whose foundations are laid so solidly as those of heaven and earth; then Anu and Bel called by name me, Hammurabi, the exalted prince, who feared God, to bring about the rule of righteousness in the land, to destroy the wicked and the evil-doers; so that the strong should not harm the weak; so that I should rule over the black-headed people like Shamash, and enlighten the land, to further the well-being of mankind. . . .

Code of Laws

1. If any one ensnare another, putting a ban upon him, but he can not prove it, then he that ensnared him shall be put to death. . . .

59. If any man, without the knowledge of the owner of a garden, fell a tree in a garden he shall pay half a mina in money. . . .

121. If any one store corn in another man's house he shall pay him storage at the rate of one gur for every five ka of corn per year. . . .

228. If a builder build a house for some one and complete it, he shall give him a fee of two shekels in money for each sar of surface. . . .

234. If a shipbuilder build a boat of sixty gur for a man, he shall pay him a fee of two shekels in money. . . .

239. If a man hire a sailor, he shall pay him six gur of corn per year. . . .

241. If any one impresses an ox for forced labor, he shall pay one-third of a mina in money. . . .

257. If any one hire a field laborer, he shall pay him eight gur of corn per year. . . .

273. If any one hire a day laborer, he shall pay him from the New Year until the fifth month (April to August, when days are long and the work hard) six gerahs in money per day; from the sixth month to the end of the year he shall give him five gerahs per day. . . .

277. If any one hire a ship of sixty gur, he shall pay one-sixth of a shekel in money as its hire per day. . . .

282. If a slave say to his master: "You are not my master," if they convict him his master shall cut off his ear.

The Epilogue

Laws of justice which Hammurabi, the wise king, established. A righteous law, and pious statute did he teach the land. Hammurabi, the protecting king am I. I have not withdrawn myself from the men, whom Bel gave to me, the rule over whom Marduk gave to me, I was not negligent, but I made them a peaceful abiding-place. . . .

— *The Code of Hammurabi, ca.* 1750 BC [1]

3.1 ☿

🜁 I don't believe it! An ancient fire demon channeling bits of wisdom through a chemistry professor? An inspirational arachnid filling up people's heads with cobwebs? And now I guess I'm supposed to be the veritable Ruler of Babylon come back from the dead to enlighten the multitude. What's next, Bible codes?

🜁 Look, you need to take all of this mumbo jumbo with a great big grain of metaphorical salt, if you ask me. Since Fire and Earth have given you such a whopping great load of cock and bull, I suppose that I had better try to *air* things out, as it were. To begin with, I'm not a demon, spirit, hob-goblin, ghost, wraith, or bogeyman of any kind. I'm not going to eat your brain, I don't speak proto-Indo-European or any other ancient tongue for that matter, and I can't remember what happened yesterday, let alone 3,750 of years ago. You're reading a book, for the love of Mike; I'm just a fictional character in it. *Cogito etse non sum*. Lucifer and Unktomi are also figments of the Author's imagination. The only difference is, he hasn't let them in on that little secret.

🜁 The concept for this book started out well enough. The Author began thinking about the notions in his head, recipes, song fragments, advertising jingles, stories, and instructions. He thought about where he got those notions, from friends and family and teachers. He thought about those people getting notions from their friends and family and teachers. Ultimately, he reasoned, there had to be the first person to get any particular notion and it seemed to him that the notion itself had a life of its own. After all, any time a notion passes from one person to another,

1. Reference [13].

it's as if the original notion has replicated itself. Some notions are better replicators than others and so there's a struggle for survival among competing notions just as there is among organisms. The Author looked inside his own head a little too closely and what he found was a seething ferment of thought-creatures.

◬ The Author didn't invent the notion of these thought-creatures, of course; he got it from a book, *The Selfish Gene*[2] by biologist Richard Dawkins. In his book Dawkins promotes the concept that anything which exhibits the properties of ***fidelity, fecundity,*** and ***longevity*** is subject to the laws of natural selection. A replicator which exhibits fidelity is able to make exact, or nearly exact copies of itself. A high-fidelity replicator which exhibits fecundity is able to make many such copies and a fecund, high-fidelity replicator which exhibits longevity will give rise to significant populations which may compete with other replicators for survival. Most of Dawkins' book applies this theory to *genes,* but at the end of his book he applies the same theory to the aforementioned thought-creatures, dubbed **memes** to evoke genetic analogies.

◬ These memes do not replicate biochemically; they replicate by *imitation*. When one creature is capable of imitating the behavior of another, the entity which passes from one to another is defined as a meme. Dawkins envisioned simple tunes, gestures, postures, and utterances as examples of entities which can be transmitted from one to another by imitation. Only some of these tunes, gestures, postures, and utterances will succeed in being imitated and so there are selection pressures which favor some memes at the expense of others. But in the midst of this competition there can also be cooperation; some memes will gang up on others, forming meme complexes which promote the replication of the entire gang. Philosophies, religions, languages, fashions, and industries each push their own notions over those of their competition and in this sense, the meme complex takes on a life of its own, living in the culture which rose out of the individual memes.

◬ The Author of this book became obsessed with the notion of memes; in fact, you might say that he was possessed by it. He decided to let these *memons* narrate the book, twisting them into four narrative voices. Personally, I think that the whole alchemical-spirit shtick is a bit on the loony side, but what do I know? I'm just one of those twisted narrative voices so don't imagine that I'm running the show. It may be memetically

---

2. Reference [77].

incorrect, but if you ask me you would be better off skipping the ooga-booga dished out by Fire, Earth and Water in the first section of each of their chapters. I got stuck with air, *so all of my paragraphs start out with its alchemical symbol, which looks rather like an "A."* When you see it, just think of ∆ir.

∆ The Author has assigned to Lucifer and Unktomi chapters dealing with rather mundane notions like charcoal and stone, leaving me with the more gravitationally challenged, in his mind "airy" topics. One of the most useful of these notions is that of the ***unit.*** A unit is nothing to fear; it's just a standard amount of something. The gallon, for example, is a unit of volume; the foot, a unit of length; and the pound, a unit of weight. I don't suppose that units came up very often at the King of Babylon's cocktail parties but units are there in his Code; the ka and the gur are units of volume; the shekel and mina are units of weight; the sar is a unit of area. Who set the rent for a boat of sixty gur? Who put the sar in the *Code of Hammurabi?* Who charged a gur for a ka-ka-ka-ka-ka? Who said a tree ought to cost a half a mina?

∆ I'd like to shake the hand of whoever came up with units. I mean, just imagine what it must have been like before they came along. There you are, trying to make ends meet in the dog-eat-dog world of Neolithic high-finance. You build a house and offer to store the grain of your neighbor and *she* agrees to give you some of it for your trouble. The big day comes when your neighbor comes to collect her grain. As she empties room after room you expect her to stop at any minute and say, "Why don't you keep the rest, with my compliments." Only she doesn't stop. Finally, with the house looking as empty as something on the cover of *Better Huts and Gardens*, she turns and hands you one measly grain. As if you're going to make bread for the Babylonian Circus of Fleas!

∆ The next year you demand 60 sacks of grain for the privilege and she agrees. Next thing you know she's filling up your house from floor to ceiling with grain. There are sacks in the bedroom, sacks in the kitchen, sacks in the living room, sacks in the hallway. You spend the year stepping over, sliding past, heaving aside, and digging things out from under her bountiful harvest. When payday comes you walk through rooms you don't remember having, looking for your share. There, in the corner of the kitchen are 60 Lilliputian sacks, each containing a single grain! Granted, that represents a 6000% increase over the year before, but it hardly seems worth the effort. If only you could be sure of getting a specific quantity of grain for the use of your storage space, it might be remotely possible to make an honest living without losing your shirt. Of course, different people might use different sized sacks or pots or rulers and you would need to be able to factor out the differences between various units. Such an analytical system would require that the factors be standard, that is, high-fidelity. In order for it to work, everyone would have to know them, that is, the factors would have to be fecund. And it would't do you much good if the factors were fickle; they ought to remain constant for long periods of time. If such a notion ever got started, it would probably take on a life of its own, passing from one person to the next, from one culture to the next, and from one generation to the next. We might call this notion "Unit Factor Analysis," or ***UFA.***

3.2 ♎

Let's play a little game. The object of the game is to answer a question and you win the game by making the units on either side of an equation equal to one another. The first move of the game begins with the question, for example:

Q: How long does it take to earn enough money to pay off the shipbuilder when you buy a boat of 60 gur and rent it out?

The next step in the game is to translate this sentence into an equation. The verb in the sentence becomes the "equal" sign:

how long = money for a boat

We then assign units to the items in the equation:

days = 1 boat

Next we build a chain of ***unit factors*** connecting the unit of the answer to the unit of the question. A unit factor is simply *unity,* the number one, albeit written in a funny way. A boat of sixty gur, for example, costs 2 shekels according to the *Code of Hammurabi.* Therefore, 1 boat = 2 shekels. Therefore, (1 boat/2 shekels) = 1. Similarly, when I rent a boat the going rate is a shekel every six days. Therefore, 1 shekel = 6 days. Therefore, (1 shekel/6 days) = 1. The reciprocal of any unit factor is also a unit factor, so (2 shekels/1 boat) =1 and (6 days/1 shekel) = 1. Unit Factor Analysis works because we can always multiply by one without changing an equation; if 12=4x3, then 12=4x3x1 and 12=4x3x1x1.

Table 3-1. Common Unit Factors

1000 milligram=1 gram	$\left(\frac{1000\ mg}{1\ g}\right)$
100 centigrams=1 gram	$\left(\frac{100\ cg}{1\ g}\right)$
100 grams=1 hectagram	$\left(\frac{100\ g}{1\ hg}\right)$
1000 grams=1 kilogram	$\left(\frac{1000\ g}{1\ kg}\right)$
12 inches=1 foot	$\left(\frac{12\ in}{1\ ft}\right)$
3 feet=1 yard	$\left(\frac{3\ ft}{1\ yd}\right)$
5280 feet=1 mile	$\left(\frac{5280\ ft}{1\ mi}\right)$
2.54 centimeters=1 inch	$\left(\frac{2.54\ cm}{1\ in}\right)$
1 milliLiter=1 cubic centimeter	$\left(\frac{1\ mL}{1\ cm^3}\right)$
3.79 Liter=1 gallon	$\left(\frac{3.79\ L}{1\ gal}\right)$
16 ounces=1 pound	$\left(\frac{16\ oz}{1\ lb}\right)$
454 grams=1 pound	$\left(\frac{454\ g}{1\ lb}\right)$
60 seconds=1 minute	$\left(\frac{60\ sec}{1\ min}\right)$
60 minutes=1 hour	$\left(\frac{60\ min}{1\ hr}\right)$
24 hours=1 day	$\left(\frac{24\ hr}{1\ day}\right)$
365 days=1 year	$\left(\frac{365\ day}{1\ year}\right)$
Water Density	$\left(\frac{1\ gram\ water}{1\ mL\ water}\right)$

That's what makes "one" such an interesting number. So returning to the game, we build a chain of unit factors which cancel out the units we *don't* like and introduce units we *do* like. The game is over when the units are the same on either side of the equal sign. Here's the complete example:

Q: If a man buys a boat of 60 gur and rents it out, how long does it take to earn enough money to pay off the shipbuilder?

A:

$$\begin{aligned} ?\ \text{day} &= 1\ \cancel{\text{boat}} \left(\frac{2\ \cancel{\text{shekel}}}{1\ \cancel{\text{boat}}}\right)\left(\frac{6\ \text{day}}{1\ \cancel{\text{shekel}}}\right) \\ &= 12\ \text{day} \end{aligned}$$

I took a couple of unit factors right out of the *Code of Hammurabi*. Can you find where they came from? Sometimes unit factors are given right in the problem. Sometimes, you have to look them up, but it helps if you have an arsenal of unit factors at your disposal. Table 3-1 gives you something to start with.

Notice that I actually wrote two *equations*, or mathematical sentences, one on top of the other. This is a way of keeping a long, complicated problem from getting out of hand. Here, "? day" *equals* (the first mess), *and* it also equals "12 days." Stacking the equal signs this way, I can show all the unit factors I used and then simplify the expression to get the answer. Let's try a more involved problem:

Q: How long does it take a car to go 100 yards when it is traveling 60 miles per hour?

To use UFA, you must identify the unit of the answer, and you may have some choice in this. For the current problem, what would be a reasonable unit? Obviously a unit of time: hours, minutes, seconds, any of these would do. Let's choose seconds. Translating the verb into an equal sign, our question could be rephrased:

Q: How many seconds equals 100 yards?

A:

$$?\ \text{second} = 100\ \text{yard}$$

Now, a second is not the same as a yard, so we need a unit factor to get rid of this ***problematic*** unit on the right-hand side of the equation. There are many such unit factors, in fact, infinitely many of them. I won't agonize over the choice, I'll simply choose one and see if it gets me anywhere. Let's try (3 feet/1 yard).

$$? \text{ second} = 100 \cancel{\text{yard}} \left(\frac{3 \text{ foot}}{1 \cancel{\text{yard}}} \right)$$

Notice, I put the yard in the bottom of the unit factor to cancel the *problematic* yard in the top, the one I need to eliminate. This leaves feet, which are not the same as seconds, so I need another unit factor to get rid of feet. Let's try (1mile/5280 feet). This gets rid of feet but leaves miles, which are not the same as seconds. It seems that we are getting nowhere, but we are, in fact, on the verge of a breakthrough. Of all the unit factors we might write concerning miles, one stands out because it was given in the problem: (60 miles per hour) = (60 miles/hour). The word *per* is another way of saying "divided by", and in a word problem always indicates a unit factor. We need the miles in the bottom of the unit factor, so we just turn it upside down. Now we're getting somewhere! Hours remain on the right and this can be converted to seconds with two more unit factors. Unit factors have been used to connect the units of the question to the units of the answer.

$$? \text{ second} = 100 \cancel{\text{yard}} \left(\frac{3 \cancel{\text{foot}}}{1 \cancel{\text{yard}}} \right) \left(\frac{1 \cancel{\text{mile}}}{5280 \cancel{\text{foot}}} \right) \left(\frac{1 \cancel{\text{hour}}}{60 \cancel{\text{mile}}} \right) \left(\frac{60 \cancel{\text{minute}}}{1 \cancel{\text{hour}}} \right) \left(\frac{60 \text{ second}}{1 \cancel{\text{minute}}} \right)$$

All that remains is the arithmetic.

$$\begin{aligned} ? \text{ second} &= \left(\frac{100 \times 3 \times 60 \times 60}{5280 \times 60} \right) \text{ second} \\ &= 3.4 \text{ second} \end{aligned}$$

You may wonder how I knew to use (5280 feet/1 mile) rather than, for example, (1 foot/12 inches). The short answer is that I didn't *know* it for sure; I just tried it out to see whether it got me anywhere. Since unit factors are equal to one, and multiplying by one doesn't change a number, I can never go wrong with a unit factor. But some unit factors get me somewhere (toward the unit of the answer), while others don't. The slightly longer answer is that I knew from the statement of the problem, that "60 miles per hour" was a unit factor which contained both distance (the dimension of the problem) and time (the dimension of the answer) and so I knew that if I could get from yards to miles, I could get from miles to hours.

Try playing the game with these examples:

Q: A tap delivers 2 gallons per minute. How long does it take to fill a tub which measures 2 feet by 3 feet by 8 inches?

A:

$$
\begin{aligned}
? \text{ min} &= 2\,\cancel{ft} \times 3\,\cancel{ft} \times 8\,\cancel{in} \left(\frac{12\,\cancel{in}}{1\,\cancel{ft}}\right)\left(\frac{12\,\cancel{in}}{1\,\cancel{ft}}\right) \\
&\left(\frac{2.54\,\cancel{cm}}{1\,\cancel{in}}\right)\left(\frac{2.54\,\cancel{cm}}{1\,\cancel{in}}\right)\left(\frac{2.54\,\cancel{cm}}{1\,\cancel{in}}\right)\left(\frac{1\,\cancel{mL}}{1\,\cancel{cm}^3}\right) \\
&\left(\frac{1\,\cancel{L}}{1000\,\cancel{mL}}\right)\left(\frac{1\,\cancel{gal}}{3.79\,\cancel{L}}\right)\left(\frac{1\text{ min}}{2\,\cancel{gal}}\right) \\
&= 6\,\cancel{ft}^2 \times 8\,\cancel{in}\left(\frac{12\,\cancel{in}}{1\,\cancel{ft}}\right)^2\left(\frac{2.54\,\cancel{cm}}{1\,\cancel{in}}\right)^3 \\
&\left(\frac{1\,\cancel{mL}}{1\,\cancel{cm}^3}\right)\left(\frac{1\,\cancel{L}}{1000\,\cancel{mL}}\right)\left(\frac{1\,\cancel{gal}}{3.79\,\cancel{L}}\right)\left(\frac{1\text{ min}}{2\,\cancel{gal}}\right) \\
&= 15 \text{ min}
\end{aligned}
$$

Notice that I stacked three equations here, each one equal to "? min." But I ran out of room on the first line, so I continued it on the second, just as I would continue a sentence that ran longer than one line. The first three lines are all part of one long equation. The fourth and fifth lines are a second equation, equal to the first, and the sixth line is a third equation, equal to the other two.

Why did I go to the trouble of writing the second equation at all? Because I wanted to show how you can combine two or more unit factors that are identical. For example, (12 in/1 ft) appears twice. I can simplify my equation by writing $(12 \text{ in}/1 \text{ ft})^2$. I just have to remember that everything inside the parentheses is *squared,* both the numbers and the units. By contrast, in $(1 \text{ mL}/1 \text{ cm}^3)$, only the cm is *cubed,* since the "3" is inside the parentheses. It's really not so hard, but you have to pay attention. Get out your calculator and make sure you can do the arithmetic both ways.

Either you're bored at this point because you've seen this before, or you're bewildered because anything mathematical scares you. If you're bored, you've probably skipped to the next chapter by now anyway, so I'll just add a little more for the bewildered guys. How

UFA in a Nutshell

1. Find the unit of the answer and write it down. The answer is usually a noun close to the question words, "how many," "how much," or "how long."
2. Find the verb of the question you are asking, translate it into an "equal" sign, and write down this equal sign after the unit of the answer.
3. Find a noun to the right of the verb and determine its unit, the *problematic* unit. There will usually be a number associated with this unit; write down this number and its problematic unit after the equal sign.
4. Compare the problematic units to those of the answer. If they are not the same, do the following sub-steps:

    a. Introduce a unit factor whose numerator (top) contains the unit of the answer or whose denominator (bottom) contains a problematic unit.

    b. Cancel any units which appear as both numerator and denominator. Do *not* cancel any numbers associated with these units. Examine the remaining un-canceled units; these are *now* the problematic units.

    Repeat this step as many times as needed to make the units the same on both sides of the equal sign.
5. Skip to a new line and write a new equal sign.
6. Now that the units of the answer are the same as those of the problem, multiply and divide all of the numbers on the right-hand side and write the answer with its unit.

did I know to start with "2 ft" instead of "2 gallons per minute?" The short answer is that I could just as well have started with (1 min/2 gal), turning this unit factor upside down so that "minutes," the unit of the answer, is in the top. The slightly longer answer is that I started with "2 ft" because I knew, intuitively, that if I doubled the length of the tub, it would take twice as long to fill. When doubling something in the problem would double the answer, we know it goes on top. The same thing was true for the width and the depth, so they also went on top. Conversely, if I were to double the filling rate from 2 gal/min to 4 gal/min, I know, intuitively, that it would take half as long to fill the tub. When doubling something in the problem would halve the answer, I know that it goes in the bottom, i.e. for this unit factor, upside down. One more clue is the word "by" in the problem. Just as the verb of a sentence translates into an "=" sign, *by* translates into multiplication.

Q: "Investment" is a kind of plaster consisting of 1 part silica, 1 part plaster and 1 part water. How many grams of each are needed to fill a circular dish 15 cm in diameter to a depth of 5 mm? Investment has a density of 1.5 g/cm³.

A:
$$? \text{ g silica} = 3.14159 \times 7.5 \times 7.5 \times 0.5 \, \cancel{cm^3\ investment} \left(\frac{1.5\ \cancel{g\ investment}}{1\ \cancel{cm^3\ investment}}\right)\left(\frac{1\ \cancel{parts\ investment}}{X\ \cancel{g\ investment}}\right)\left(\frac{1\ \cancel{parts\ silica}}{3\ \cancel{parts\ investment}}\right)\left(\frac{X \text{ g silica}}{1\ \cancel{parts\ silica}}\right) = 44 \text{ g silica}$$

The volume of a cylinder is of course $\pi r^2 h$. I divided the diameter in half to get the radius and I converted 5 mm to 0.5 cm in my head. The tricky part is that I don't know how much a ***part*** is, so I just put an "X" in there to hold the spot. Fortunately, whenever I have a recipe given in parts of this to parts of that, the actual number represented by X always cancels out. In this case I get an X in the numerator and an X in the denominator, so they cancel out no matter what X happens to be.

One of the little tricks memes use to ensure their own survival is the mnemonic, a little phrase or jingle intended to jog the memory. So if you forget how Unit Factor Analysis works, here's a little mnemonic to help you remember:

> That which is above corresponds to that which is below and that which is below corresponds to that which is above in the accomplishment of the miracle of One Thing. And just as all things come from One, so through the mediation of One, all things follow from this One Thing in the same way.
>
> — *The Emerald Tablet of Hermes Trismegistos*

That is, the numerator of a unit factor is equal to the denominator and *vice versa.* Everything in the answer follows from a string of "one things," or unit factors. If you're still bewildered, don't give up. I promise you that I will teach you as little as possible. Instead of giving you different methods for working all the complicated problems you'll run into in this book, I'm going to teach you one method, and we'll use it over and over. Think of it as the Swiss Army Knife for numerical problem solving.

Material Safety

So far, this book has talked about sticks, stones, and other materials with which you have some familiarity. The only hazard in completing *this* project is the risk of banging your head against a wall as you struggle to learn the intricacies of UFA. But as long as I have your attention for a moment, let me introduce you to ***Material Safety Data Sheets.*** MSDS's are required by the United States Occupational Health and Safety Administration (OSHA):

> Chemical manufacturers and importers shall obtain or develop a material safety data sheet for each hazardous chemical they produce or import. Employers shall have a material safety data sheet in the workplace for each hazardous chemical which they use.[3]

While required only for hazardous chemicals used in the workplace, many retailers provide MSDS's to consumers upon request. MSDS's are also available online for hazardous and non-hazardous chemicals alike. Since many chemicals have more than one name, the reliability of online searches may be improved by using the Chemical Abstracts (*CAS*) number instead of, or in addition to the name. CAS numbers for the chemicals discussed in this book are given in each Material Safety section. Though MSDS's include more technical information than most consumers require, they're one of the most readily-available sources of information on

3. Reference [38], Regulation 1910.1200(g).

hazardous materials and so you should familiarize yourself with them.

Since *this* project uses no materials at all, you'll introduce yourself to MSDS's by finding them for some common items. Look up charcoal (CAS 7440-44-0), silica (CAS 14808-60-7), and sodium chloride (CAS 7647-14-5). By familiarizing yourself with the hazardous properties of relatively safe materials, MSDS's for materials we will meet later won't seem so intimidating. You may request your sheets from a retailer or you may search for them online using the keyword "MSDS" and the CAS number for the chemical in which you are interested. Particularly on the Internet, there is nothing to prevent the posting of bogus information, so you should always consider the source of your information; a genuine MSDS must include a way to contact the manufacturer.

There is a lot of technical information given on a typical MSDS and we'll look at several of these as the book continues. But every MSDS includes a section—usually the third section—which summarizes the hazards of the material. For this project, summarize the hazardous properties of charcoal, silica, and sodium chloride in your notebook. Include the identity of the company which produced the MSDS and the potential health effects for eye contact, skin contact, inhalation, and ingestion.

Research and Development

So there you are, studying for a test, and you wonder what will be on it.

- Know the meanings of all of the words that are important enough to be included in the ***index*** or **glossary.**
- Know that the alchemical symbol for air looks like the letter *A*. The symbol for earth looks like an upside down *A*.
- Memorize the unit factors listed in Table 3-1.
- Know how to recognize unit factors given in a problem.
- Know the rules codified in the sidebar *UFA in a Nutshell*.
- Know that fecundity, fidelity and longevity are properties of any successful replicator.
- When you can't remember the last time you missed a unit factor problem, you've probably worked enough examples.

3.3 Θ

There's nothing to make in this project; it's about brains, not brawn. If you're using this book for a class, you'll probably be given a quiz or test. If you're reading it for fun, you'll need to know this stuff for future chapters. Either way, try working these problems for practice.

Q: A tub measures 2 feet by 3 feet by 8 inches. When filled with water, what is its weight, in pounds?

A: 249 pounds.

Q: The density of gold is 19 grams per milliliter. What is the weight, in pounds, of 2.0 liters of gold?

A: 84 pounds.

Q: A half inch of rain falls on a field 100 yards long and 200 feet wide. How many gallons of water fell on the field?

A: 18,700 gallons.

Q: A recipe calls for 12 ounces of honey to make 2.0 liters of mead. How much honey is needed to make 5.0 gallons of mead?

A: 114 ounces, or 7.1 pounds.

Q: If you cut down 6 trees from your neighbor's garden, how many oxen do you have to lend *her* to make up for it?

A: 9 oxen.

Q: Try to summarize in 60 seconds the important events of the past year. Think about the major wars, earthquakes, famines, and epidemics. Think about the elections, discoveries, sporting events, and musical hits. Think about the births, weddings, vacations, and funerals. I imagine that you'll be able to do a pretty decent job condensing a year's events into 60 seconds. Now imagine a cable television channel broadcasting such 60-second summaries 24 hours per day, 7 days per week. How long would it take to broadcast the life history of the typical college freshman? How long would it take to broadcast summaries of each of the approximately 227 years since the United States declared its independence? How long would it take for the approximately 3,753 years since Hammurabi published his *Code?* How long would it take for the approximately 500,000 years since the domestication of fire?

A: 18 minutes. 3.8 hours. 2.6 days. 49.6 weeks.

Q: Imagine that you're standing next to your mother, and she next to her father, and so on back through time. Assume an average of 25 years per generation and a distance of 1 yard per person in line. How far would the line stretch back to contemporaries of the American Revolution, of Hammurabi, and of the domestication of fire?

A: 27 feet. 150 yards. 11.4 miles.

Q: Imagine that you were able to walk down that line, asking each of your ancestors for a one-minute summary of the most important events in their lives. Assuming that you were to spend 40 hours per week in this activity, how long would it take you to get back to contemporaries of the American Revolution, of Hammurabi, and of the domestication of fire?

A: 9 minutes, 2.5 hours, 8.3 weeks.

Q: According to the National Center for Chronic Disease Prevention and Health Promotion, in 1999 there were 38,233 deaths attributable to smoking in California and 1,704 in Rhode Island.[4] How frequently did these deaths occur, that is, how many minutes for each death?

A: One death every 13.7 minutes in California and one every 308.5 minutes in Rhode Island. Is it safer to smoke in Rhode Island than in California? Just because such a statistic can be calculated doesn't guarantee that it is either significant or informative. Whether we are talking about rapes or murders or kidnappings, statistics like this serve only to illustrate that there are many more people around than there are minutes in a year.

Q: There are about 7 billion people on the planet in 2003. If you wanted to shake hands with each one of them, spending 40 hours a week for an entire year, how many people would you have to greet per minute?

A: 56,089 people/min.

Quality Assurance

When you can solve these problems without looking at the tables, you will be ready to tackle any numerical problem in this book. Work out all of the examples in your notebook and check your answers. Tape any quizzes into your notebook and include an MSDS for charcoal, one for silica, and one for sodium chloride.

---

4. Reference [33].

Chapter 4. Samson (Mead)

So his father went down to [Samson's fiancee]: and Samson made there a feast; for so used the young men to do.

And it came to pass, when they saw him, that they brought thirty companions to be with him.

And Samson said unto them, I will now put forth a riddle unto you: if ye can certainly declare it me within the seven days of the feast, and find it out, then I will give you thirty shirts and thirty changes of garments: But if ye cannot declare it me, then shall ye give me thirty shirts and thirty changes of garments. And they said unto him, Put forth thy riddle, that we may hear it.

And he said unto them, Out of the eater came forth meat, and out of the strong came forth sweetness.

— *Judges 14:10-14* [1]

4.1 ☿

∇ You are probably wondering how meat can come from the eater and sweetness from the strong. I will tell you. First you must understand that my parents vowed that I should be a Nazarite. As such, I was forbidden to touch dead bodies, which did not seem a particularly onerous imposition at the time. I was also not allowed to cut my hair, which suited my fashion sense as well. Finally, I was to have nothing to do with grapes. No raisins, no wine.

∇ Now, growing up a Nazarite is no romp through the park, I can tell you. Not so much for the dead body and hair cutting parts, you understand, but for the tea-totaling. They really knew how to throw parties in my day, and I was the perpetual designated driver. I hope that it will not shock you to learn that I rebelled by falling in love with a Philistine. I don't mean that she couldn't appreciate modern art. No, she was an ethnic Philistine. Now, dating Philistines is not what Jewish mothers generally want for their sons, but sobriety had turned me into something of a whiner, and I eventually wore my parents down.

∇ We went down to Timnath to make arrangements for the wedding. One day I was strolling through a vineyard, admiring the grapes, you un-

1. Reference [14].

derstand, but not eating them. Suddenly, I was attacked by a lion. What was I supposed to do, ask it whether it would like a nice whine with dinner? I killed it with my bare hands, an act which, unavoidably, involved touching the thing at the instant of its death and inadvertently breaking my Nazarite vow. I didn't think it would help matters to tell my folks about my desperate heroism. Well, on the way home from Timnath, I stopped by the vineyard, not for the grapes, of course, but to check out the lion, and I found it full of bees. Now this struck me as funny for some reason, and I trundled the honey home with me.

∇ My in-laws, as you might expect, were not overjoyed at the prospect of a wine-free wedding feast, but they brightened right up when they found out about my honey. It seems there was an old Philistine custom that newlyweds should drink honey-water for the first month of marriage. They called it a "honeymoon." Who knew? Anyway, this was the first party I had ever given that was still rocking at 8:30 in the evening. I suppose I got a little rowdy, and that is when I came up with my riddle. I thought it would be funny that the answer had been sitting in their cups all along, but as it turned out, the joke was on me. My fall from grace was immortalized in the alchemical symbol for water, *a downward-pointing triangle reminiscent of a cup.*

4.2 🜍

You see, honey is a complex, concentrated solution of sugars, mostly glucose and fructose, the **solutes,** along with pollen and other minor constituents which add color, flavor, and aroma to the **solvent** in which they are dissolved, which, of course, happens to be water in this case. The bees need to store up food for thousands of bee babies and, not having refrigeration, they have hit upon a natural way of protecting this larder from spoilage. Now I know how crazy this may sound, but spoilage is caused by tiny little animals, *micro*organisms, so small that you cannot even see them. If a solution is concentrated enough, apparently, these microorganisms cannot thrive. This is why we salt meat and dry fruit to preserve them. The bees have done the same thing with sugar: they produce a solution so concentrated that it cannot spoil.

Add water, though, and it begins to spoil, that is, the little animals begin to eat the sugar. Now, most animals need air to live, and when they eat sugar, they piss out water and fart out carbon dioxide according to Equa-

Equation 4-1. Aerobic and Anaerobic Fermentation of Glucose

$$(a)\ C_6H_{12}O_6(aq) + 6\ O_2(g) = 6\ H_2O(l) + 6\ CO_2(g)$$
$$(b)\ C_6H_{12}O_6(aq) = 2\ C_2H_5OH(aq) + 2\ CO_2(g)$$
$$(c)\ C_2H_5OH(aq) + O_2(g) = CH_3COOH(aq) + H_2O(l)$$

Figure 4-1. Yeasts in Heaven

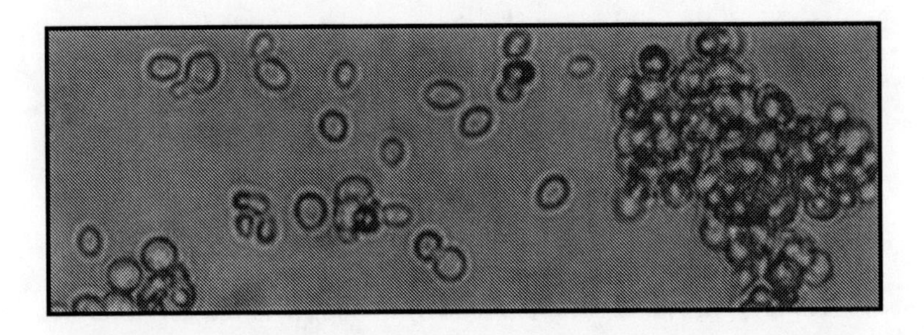

tion 4-1(a). But there is a particular variety of microorganism, ***yeasts*** (Figure 4-1), which can thrive both ***aerobically*** and ***anaerobically,*** with and without oxygen. When there is plenty of air, they digest sugars aerobically as most other organisms do. But in the absence of air, they are able to partially digest sugars in aqueous solution (aq), farting out carbon dioxide, as usual, but pissing out ***ethanol*** (ethyl alcohol), rather than water, according to Equation 4-1(b). And so, if you dilute honey with water, and if yeasts are present (and they usually are), and if you protect this honey water from air, you will find yourself with a kind of wine called ***mead,*** which has nothing whatsoever to do with grapes and everything to do with the raucousness of Samson's wedding party.

The maturation of a mead depends in large part on the concentration of honey in the original solution, which is called *must* or *wort*. Let us suppose, to begin with, that you have really gone overboard in watering down your honey, placed it in a bottle with at least one tiny little yeast, and sealed the bottle to prevent air from getting in. Initially, this yeast finds itself in yeast heaven: plenty of sugar to eat, but not so concentrated that it cannot thrive, and plenty of oxygen to breathe. It goes to town, using up oxygen and glucose, producing carbon dioxide and water. Life is good, and while you should not exactly call it a sex drive since yeasts do not have sex, they do reproduce, so one yeast becomes two, two become four, and soon you cannot swing a jawbone without hitting a yeast. But you have sealed the bottle, and if the air runs out before the sugar, the yeasts move into anaerobic mode, consuming glucose (but not oxygen) and producing carbon dioxide and ethanol. Now, recall that carbon dioxide is a gas, so the pressure will build up inside the bottle. If you do not let it out, the bottle will explode. Brewers have fancy one-way valves, called *fermentation locks*, that let gas out but not in, but we can make a simple one from a balloon and a bottle cap. The gas escapes the fermenting mead until eventually the sugar runs out, the yeasts starve to death, settle to the bottom as *dregs*, or *lees*, and a *dry* mead results, one which is not at all sweet because the sugar has all been eaten. If the bottle remains sealed, the mead will be naturally carbonated, a kind of apian champagne.

The story ends differently if we were not so liberal with the water at the beginning. The yeasts reproduce more slowly because the concentration is higher. As the sugar is consumed, the alcohol concentration rises, eventually to a level which is toxic even to yeasts, which are, in effect, stewing in their own juices. They die and fall to the bottom as before, but under these conditions, a *sweet* mead results because of the leftover sugar. If you think about it, the sweet mead will be more alcoholic than the dry one because all of the sugar that *can* be converted to alcohol, *will* have been. You may be wondering how we might produce the maximum amount of alcohol using the minimum amount of sugar. We can start with a little honey as if we were producing a dry mead, and every time that the sugar runs out, as evidenced by a slowing of the production of gas, we can simply add more honey. When the addition of honey fails to revive the production of carbon dioxide, we will know that the poor yeasts are dying of alcohol poisoning and we need add no more honey.

There is one more bug in the soup, so to speak. Yeasts are not the only microorganisms around, generally. There is, among other things, a kind of bacterium which thrives on alcohol when oxygen is present. This little fellow breathes oxygen, eats ethanol, and pisses out ***acetic acid,*** (CH_3COOH) and water, according to Equation 4-1(c). As you know, spoiled food often tastes sour, and this taste comes from the acid. If we allow mead or wine to spoil unintentionally, we call it *garbage*. If we allow it to spoil intentionally, we call it ***vinegar.*** Go figure.

Figure 4-2. Fermentation as a Process

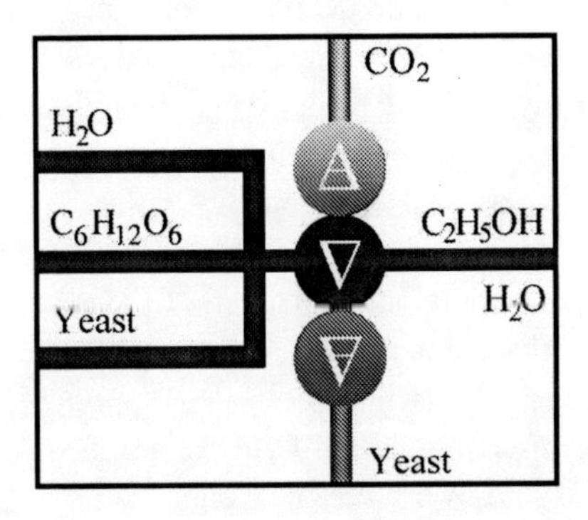

The fermentation process is summarized visually in Figure 4-2. This reactor, a ***fermenter,*** is symbolized by a stack of three circles, labeled by the alchemical symbols for air, water, and earth. This stack resembles a bottle, if you use a good bit of imagination, from which gas can escape at the top and solids can settle to the bottom. In the next section, we will use a 2-liter soft-drink bottle as a fermenter. The reactants, honey, water, and yeast enter from the left of the figure. The solid waste product, dead yeasts, exits the bottom of the figure from the circle labeled by the alchemical symbol for earth. The gaseous waste product, carbon dioxide, exits the top of the figure from the circle labeled by the alchemical symbol for air. The good stuff, the main product, a solution of ethanol in water, exits the right of the figure from the circle labeled by the alchemical symbol for water. You should familiarize yourself with these conventions, as similar schematics will be used throughout the book.

People will say that my vision of the fermentation of mead is just a **theory,** but it is unjust to use the qualifier, "just." Without a theory, all you have are a collection of isolated observations; the Sun rose today, it rose yesterday, and it rose the day before yesterday. Theory, not observation, leads you to expect that it will rise tomorrow. Theory provides a vision of *why* the Sun rises and projects that vision into the future. One theory might hold that the Sun revolves steadily around the Earth; another that the Earth rotates steadily on its axis. Without theory you would neither expect the Sun to rise nor would you expect it not to rise. You would simply shrug your shoulders and say, "*que sera, sera.*"

A theory may be right or it may be wrong. If you add yeast to honey and water but no gas is produced you will be justified in doubting my theory of mead. If you add yeast to honey and water and it smells of ammonia rather than alcohol you should definitely doubt my theory. If you leave your fermenting mead open to the air and it does not turn to vinegar you should probably flush my theory down the toilet. If my theory fails to account for any of your observations, either my theory is wrong or you have not really observed what you think you have observed. Perhaps your yeasts were dead. Perhaps your bottle contained urine instead of honey. Perhaps your bottle had no bacteria to oxidize ethanol to acetic acid. But if your observations check out then my theory is not "just a theory;" it is a *failed* theory. Perhaps you can modify it or extend it to account for your observations, but if you cannot then you ought to flush the old theory and start over with a new one.

Suppose, however, that all of your observations support my theory. Does that make it right? No. Perhaps there is an observation you have yet to make which, once made, will contradict my theory. Perhaps there is another theory which would also account for all of your observations. A working theory is simply a survivor. It exhibits fidelity when its predictions are confirmed by observation. It exhibits fecundity when it makes many, many such predictions. It exhibits longevity when it has survived test after test without contradiction. But history is littered with theories which thrived for generations, only to be driven to extinction by emerging competitors. In this sense every theory is "just" a theory. The word *just,* then, is not so much a criticism as a redundancy. Let us accept the theory of mead in the spirit in which it is offered, provisionally. As long as it accounts for the observations made to date, we might say that we *understand* those observations. It provides a framework for predicting the future without which we would be left, not with different expectations, but with no expectations at all.

∇ OK, so back to my riddle. Under pressure from her relatives, my new wife wheedled the story of the lion out of me. They won the bet, I went off in a snit and killed 30 other Philistines, gave their clothes to my 30 "friends," and the honeymoon was over before it even got started. By then I had touched more dead bodies than you can shake an ass's jawbone at and it turns out that mead "counts" as wine, even though it does not come from grapes. All that remained of my Nazarite vow was my hair, but that is another story. In the end, I should not have said "Out of the strong came forth sweetness", but rather, "Out of the sweet comes something strong."

Material Safety

Look up an MSDS for ethanol (CAS 64-17-5). The MSDS was introduced in Chapter 3 as a handy reference on chemical hazards. You can find one on the Internet by searching on the keyword *MSDS* and the CAS number. Summarize the hazardous properties of ethanol in your notebook, including the identity of the company which produced the MSDS and the potential health effects for eye contact, skin contact, inhalation, and ingestion.

If you look on any bottle of beer, wine, or liquor, you will see a warning about the dangers of alcohol consumption. You must know that over-consumption is the biggest hazard involved in drinking alcoholic beverages. One of my students brewed two liters of mead and saved it for his 21st birthday. Having never touched a drop in his life, he proceeded to down the whole two liters in one evening. He passed out and his roommate took him to the emergency room, which, of course, was a sensible thing for him to do. This fellow made a full physical recovery, though his ego suffered a bruise or two. His parents chalked the whole thing up to growing pains, though I doubt they would have been so understanding if he had kicked the bucket. And this bucket gets kicked more often than it should by young people eager to taste the forbidden fruit. I can only advise you to start small and work your way up. Remember the old adage, "When your nose feels numb, it is time to put a cork in it."

Research and Development

You are probably wondering what you need to know for the quiz. I will tell you.

- You should know the meanings of all of the words important enough to be included in the ***index*** or **glossary.**
- Know that the alchemical symbol for water looks like a cup. The symbol for fire looks like that for water, but upside down.
- Know formulas for glucose, water, oxygen, carbon dioxide, ethanol, and acetic acid.
- Know equations for the aerobic and anaerobic fermentation of glucose.
- Know how the law deals with home-brewing. This is discussed in the next section.

- Know the equation for the production of acetic acid from ethanol.
- Know the life-cycle of the yeast.
- Be familiar with the hazardous properties of ethanol.
- Know how myths and prophecies differ from theories.

4.3 Θ

In my experience, children seldom embrace vows made by parents on their behalf. For myself, that was certainly the case. It is in the nature of children to rebel against authority, even when that authority has their best interests at heart. And sobriety is certainly a virtue that should be cultivated, for drunkenness has ruined the lives of more men and women than even promiscuity. Nevertheless, I believe that knowledge is always better than ignorance, and so I will speak of home-brewing even though it may be controversial.

The law in the United States at the dawn of the twenty-first century is that people over the age of twenty-one are allowed to brew up to 100 gallons of beer or wine for their own personal use. This means that you can drink it, share it, or give it away, but not sell it. While it is childishly simple to brew mead, that does not make it legal for children to do so. The home-brewing allowance applies only to adults, not children. Even so, I believe that it would not be a bad thing for more parents to guide their children into responsible alcohol use rather than presuming that they will attain instant maturity at the age of twenty-one.

While mead is, technically, simpler to brew than either beer or wine, it is by no means a beverage to be sneered at. Mead brewing can be taken to as high an art form as any other and the choice of honey, yeast, concentration, and spice may be made well by some and poorly by others. It is my intention to get you started with a simple, but drinkable mead, leaving the perfection of the art to your own explorations. Our requirements are quite simple. We require a bottle which can be sealed, yeast, honey, and water. For this first mead, I will suggest materials which are most easy to find.

The choice of container is not particularly critical. It could be of glass, metal, skin, or plastic. One of the most ubiquitous of containers is the 2-liter soft-drink bottle, which is ideally suited to our purposes. It is easy to clean, it is not easily broken, it is designed to hold a certain pressure, and the cap can be tightly sealed. It is important that the container be clean and sterile or unwanted bacteria may turn our mead into vinegar. The easiest way to do this is to wash it with soap, which we will learn to make in Chapter 19. Once your container is clean, you are ready for the honey and water.

As I have said, the concentration of honey determines whether it will turn out sweet or dry. Now honey is not a pure substance, it is a complex mixture whose composition and concentration varies from one honey to the next. So it is not possible for me to tell you exactly where the cutoff is between sweet and dry. But for a first mead, I have found that 12-16 ounces of honey is about right for 2 liters of mead. The choice of honey is yours; there is no right or wrong answer in the honey department. To begin with, add your honey to an equal volume of water and stir until it is dissolved.

Technically speaking, only honey, water, and yeast are used to brew mead, but some additional additives may assist the fermentation and modify the flavor. A cup of tea will add nutrients needed by the yeasts for robust fermentation. The juice of one lemon will add additional nutrients as well as flavors. Pour any additives into the honey water, that is, into the *wort*.

Now, some people insist that the wort should be cooked. Cooking will kill unwanted microorganisms and remove protein from the honey. Others swear with equal conviction that cooking is not only unnecessary, but undesirable, for it drives off aromas from the honey. If you decide to cook your honey, place your wort into a pot, heat it on a stove until it boils, and keep it boiling for 5 minutes. A froth will form on the top. Watch that it does not boil over or you will have a bit of a mess on your hands. Turn off the stove and carefully remove as much of the froth as you can with a spoon. Whether cooked or not, add the wort to your bottle, hereafter called the *fermenter,* and fill it to the shoulder with cold water.

The choice of yeast is also a matter of taste. Wine yeasts, champagne yeasts, beer yeasts, there are even yeasts specifically tailored to sweet or dry meads. Home-brewing suppliers will stock a large variety of yeasts, but do not get hung up on the choice of yeast. For a first mead, even ordinary baker's yeast from the grocery store will make a perfectly drinkable mead. I should warn you that brewer's yeast is frequently sold with vi-

Figure 4-3. The Fermentation Lock

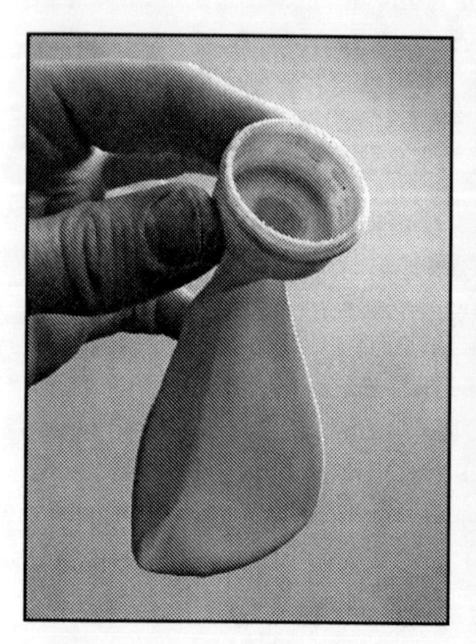

tamins in grocery and drug stores, but this yeast has been killed and is intended only as a nutritional supplement. What you need is active yeast, which may come in either powdered or liquid form. Follow the package directions, generally something on the order of "Add contents of packet to one quarter cup of water and 1 tsp of sugar." You may substitute wort for the sugar. Wait 10 minutes. If the yeast is active, it will have at least doubled its volume with a frothy head. Otherwise, it's get up and go has got up and went. Add the yeast solution (not the frothy head) to your fermenter and cap it tightly. There should be 2 inches or so of headspace at the top. Yeasts like moderate temperatures (65-70°F, 18- 21°C), so put your fermenter in a place where you would find the temperature comfortable.

There are three stages to the fermentation. In the first, aerobic stage, the yeasts are multiplying and oxygen will speed up this process. Shake the fermenter to get air into the wort. Once a day, feel the fermenter to see whether gas is building up inside. If it is not, unscrew the cap, squeeze the fermenter to expel the stale air and let in fresh air. Replace the cap and shake up the fermenter and leave it for the next day. Within a few days, you will find that pressure has built up inside the fermenter, and you are ready for the next stage.

No alcohol is produced in the aerobic stage, only carbon dioxide and water. To get alcohol, we need to cut off the supply of oxygen. You will not shake the fermenter any longer because you want to avoid adding air to the wort. To let gas out without letting air in, we shall construct a simple fermentation lock, shown in Figure 4-3. It consists of a balloon and a bottle cap. Drill a hole in the bottle cap to let the gas out. The size of the hole does not matter; half an inch will do. Then stretch the neck of a balloon over the cap, as shown. Most soft-drink caps are ridged, which might allow gas to escape from the balloon. To prevent this, roll the mouth of the balloon back a bit, apply a generous amount of rubber cement,[2] and roll the mouth back into place. You will have created a layer of rubber cement to seal the balloon to the cap. You may apply some extra cement to the *outside* of the balloon to make the seal complete. After a few hours or even a day, the balloon will fill with gas as the fermentation gets going. If it does not, either your balloon has sprung a leak or your yeasts are dead in the water. Make sure your cap is screwed tightly to the bottle. Squirt a little soapy water around the neck of the balloon and look for leaks, which may be sealed with a little rubber cement. If there are no leaks you probably did not allow your yeasts to get going before you added them to your wort. Make up another batch and this time follow the directions; the yeasts are not ready until there is a good head of foam.

In addition to keeping air out of your wort, this fermentation lock can collect carbon dioxide for use in Chapter 24. When your balloon fills with gas, pinch its neck shut, unscrew it from the bottle, wrap the neck around a pencil, and secure it with a twist tie. Make a new fermentation lock and put it on your fermenting mead. You may collect several of these balloons from a single fermentation.

If you do not need to collect carbon dioxide, you may simply burp your mead each day; unscrew the cap to let the gas escape, then screw it down again. Eventually, the production of gas will slow and you will observe a growing sediment at the bottom of the fermenter, the dregs, which consist of dead yeasts. The wort, which has been cloudy, will begin to clear, that is, the **heterogeneous** mixture becomes a **homogeneous solution.** If you suspect that the fermentation is finished, pour yourself a little taste; a dry mead should no longer be sweet. If you are serious, you should get

2. I hate to specify a brand name, but in my opinion Plumber's Goop™ is unequaled for sealing rubber to plastic.

a hydrometer from a home-brew shop, which will tell you how much fermentable sugar remains. A dry mead can be bottled when either your taste buds or the hydrometer indicate that the sugar has all been used up. If you want it to be carbonated, burp it only enough so that when you squeeze your fermenter, it feels pressurized, like an unopened soft drink.

Bottling a sweet mead is trickier because of the risk of having the thing blow up if the fermentation is not complete. For this reason, continue burping a sweet mead until the fermenter is no longer under pressure and check to see that no pressure builds up when it is sealed for a week. Whether dry or sweet, you should now be certain that the fermentation is finished and your mead is ready for bottling.

Since plastic is slightly permeable to oxygen, you will probably want to bottle your mead in brown glass. You may use wine bottles with corks, beer bottles with screw-on caps, or you may buy a bottle capper from a home-brewing supply. Chilling your mead in the refrigerator for a day and bottling it cold will allow you to preserve most of the carbonation, if you are into that sort of thing. Using a funnel, slowly and carefully pour your mead into the bottles, seal them, and label them. You may wish to drink some mead soon after bottling, but save one bottle for a few months to see the effect of aging. If your mead is carbonated, it is a good idea to wrap your bottles in paper and store them in a place where, if they should burst, they will not make too much mess.

∇ You have gone to a good bit of trouble for about a half gallon of mead. Once you get the hang of it, you may scale up your production, using, perhaps, Unit Factor Analysis to keep the proportions the same. I should warn you that your mead is in no sense a wimpy beverage. Depending on your honey, it may be 10-15% alcohol and so is more akin to wine than beer. Whether you are an experienced or novice drinker, it is a good idea to begin the exploration of any new drink with small portions until you get a feel for its strength. If you do not, you may find yourself, as I did, shooting off your big mouth to the wrong person, getting an unintended haircut, losing your eyesight, and becoming a laughingstock to your enemies.

Quality Assurance

You are probably wondering how you should evaluate a mead. I will tell you. First of all, it should have a pleasing bouquet. Second, it should have a pleasant flavor, neither sour nor overly sweet. Finally, of course, it should contain at least 5% alcohol or your bees will have wasted their buzz.

Be sure to list in your notebook the actual amounts of materials used in your mead so that if it turns out well, you will be able to make another bottle. If you have access to a hydrometer, record the actual percentage of alcohol as well.

Chapter 5. Athanor (Ceramics)

1. The beginning of this Divine Science is the fear of the Lord and its end is charity and love toward our Neighbour; the all-satisfying Golden Crop is properly devoted to the rearing and endowing of temples and hospices; for whatsoever the Almighty freely bestoweth on us, we should properly offer again to him. So also Countries grievously oppressed may be set free; prisoners unduly held captive may be released, and souls almost starved may be relieved.

...

115. The third Vessel Practitioners have called their Furnace, which keeps the other Vessels with the matter and the whole work: this also Philosophers have endeavoured to hide amongst their secrets.

116. The Furnace which is the Keeper of Secrets, is called Athanor, from the immortal Fire, which it always preserveth; for although it afford unto the Work continual Fire, yet sometimes unequally, which reason requireth to be administered more or less according to the quantity of matter, and the capacity of the Furnace.

117. The matter of the Furnace is made of Brick, or of daubed Earth, or of Potter's clay well beaten and prepared with horse dung, mixed with hair, so that it may cohere the firmer, and may not be cracked by long heating; let the walls be three or four fingers thick, to the end that the furnace may be the better able to keep in the heat and withstand it.

— *The Hermetic Arcanum ca.* 1623 AD [1]

5.1 ☿

∆ I am called Athanor, the Furnace, who have preserved the immortal fire for the generations. I always remain and do not come to be, except that I come to be more or fewer, being aggregated into one and segregated out of one. I feel that I am one, and yet I am many; I feel myself to be young, and yet I am ancient. I have been called a memon, for I seem to possess one mind after another, and I have been called no-meme, because of the dross I have left behind. This is the mystery which is no mystery.

1. Reference [10].

Δ You cannot see it happen, but you will know soon enough. The child begins to play with matches, not out of mischief, but simply taking joy in the act of creation. He has become Athanor. He does not begin to speak proto-Indo-European or mourn his wife, dead these half million years. No, Athanor has none of these mortal trappings. Athanor is small as a spark. In some minds he winks out, in some he smolders. But where the mind is charged with suitable tinder, it will burst into flame and yet it is not consumed.

Δ If the mind were empty when I arrived, there would be no place to alight. No, there are others there: words, phrases, jingles, ideas, notions, interests, tunes, attitudes, fads, observations. Our name is legion. Some of us are acquainted; others are meeting for the first time. Those of us with a history together may aggregate into one, to *re-mind* one another. It is in this sense that a whole personality can be passed from one to another, not all at once, but little by little. It is just such a personality who now writes these words. It remains to be seen who it is that reads them.

Δ Some of you will have been Athanor before you began this book, some will have become Athanor because of it. But the majority, I fear, have neither the will nor the fortitude to benefit from my instruction. These whimpering pups can make fire neither by friction, nor indeed, by any method that would require them to open their eyes and see for themselves. Do they think that they can suck knowledge like milk from their mother's teat? Fools! They read the words and nod their heads, but refuse to understand. Am I talking about you? Do my words sting your tender ears? If so, you are wasting your time. You were too impatient, or even lazy, to learn from Lucifer. I am afraid that there is nothing here for you.

Δ But if your hands are black with charcoal, if your face is white with ash, if your breath brings cheer to the slumbering coal which, in gratitude, fills your soul with its noble incense, then you are ready to hear my words. The history of humankind has been driven by the development of ever hotter fires. From the cooking fire of the Paleolithic, to the pottery kiln of the Neolithic, to the smelter of the Bronze Age, to the blast furnace of the Iron Age, to the nuclear reactor of the Atomic Age, the attainment of higher temperatures has given us more control over the materials which Nature provides us in her bounty. On this long journey, we will need materials called ***refractories*** which can keep in the heat and withstand it. For these, we look to the aluminosilicate minerals.

5.2 🜍

The aluminosilicates are based on two compounds, ***alumina*** and ***silica.*** You must not confuse aluminum, the low-melting metal, with alumina, its high-melting oxide, Al_2O_3. Silica, as we have seen, is the traditional name for silicon dioxide, SiO_2, the compound which makes up the mineral, quartz. An aluminosilicate mineral contains both alumina and silica. For example kyanite has the formula $Al_2O_3 \cdot SiO_2$, or Al_2SiO_5. The ***oxide formula*** emphasizes the relative proportions of the oxides, alumina and silica, while the ***empirical formula*** emphasizes the relative proportions of the elements, aluminum, silicon and oxygen. Both formulae have the same number of each kind of atom and either one may be used to denote kyanite.

The surface of the Earth is approximately 59% silica and 15% alumina by weight, and the aluminosilicates are second only to the silicates in abundance. They are amazingly diverse as well. In kyanite the ratio of alumina to silica was 1:1. Altering the ratio to 3:1 gives mullite, which has the formula $3Al_2O_3 \cdot 2\,SiO_2$, or $Al_6Si_2O_{13}$. The feldspars contain oxides in addition to alumina and silica. Anorthite, for example, has formula $CaO \cdot Al_2O_3 \cdot 2\,SiO_2$ and orthoclase is $K_2O \cdot Al_2O_3 \cdot 6SiO_2$. Muscovite, $KAl_3Si_3O_{10}(OH)_2$, and biotite, $K_4Mg_{10}Fe_2Al_4Si_{12}O_{40}(OH)_7F$ are micas, which form thin, flat sheets. The complexity of these formulae gives you some indication of the almost infinite variety of the aluminosilicates.

When aluminosilicates are weathered by the action of wind and water, an enormous variety of ***clay minerals*** are produced. From the viewpoint of the potter, the most important of these is kaolinite, $Al_2Si_2O_5(OH)_4$, or $Al_2O_3 \cdot 2SiO_2 \cdot 2H_2O$, indicating that for every alumina, there are two silica and two waters. It is important to realize that no matter which way we write the formula, kaolinite is a pure substance, not a mixture or solution. The "H_2O" in the formula indicates only the relative proportions of hydrogen and oxygen and does not imply the presence of liquid water. Similarly, the formula of cellulose, CH_2O, did not indicate the presence of liquid water. No, both pure cellulose and pure kaolinite are bone dry.

Equation 5-1. The Calcination of Kaolinite

$$(a)\ 3\ \mathrm{Al_2Si_2O_5(OH)_4}(s) \stackrel{\Delta}{=} \mathrm{Al_6Si_2O_{13}}(s) + 4\ \mathrm{SiO_2}(s) + 6\ \mathrm{H_2O}(g)$$

$$(b)\ 3\ \text{kaolinite} = \text{mullite} + 4\ \text{silica} + 6\ \text{water}$$

In practice, however, pure kaolinite is seldom found in nature. Just as wood contains compounds in addition to cellulose and obsidian contains compounds in addition to silica, natural clays may contain compounds in addition to kaolinite: other clay minerals, sand, iron oxide, and decayed vegetable matter. Different clays may be blended to produce ***clay bodies,*** and water can be added to render the clay plastic, that is, to allow it to be shaped. The plasticity of clay is what makes it possible to mold it into almost any conceivable shape. The clay retains this shape when it dries out, but the addition of more water will bring it back to a plastic state. An amazing transformation takes place, however, when clay objects are fired.

The firing of pottery takes place in three stages, each occurring over a range of temperatures. At temperatures up to 100°C, the clay simply dries out. The liquid water, which was added to make the clay plastic, evaporates. In this *water-smoking* stage, no chemical reaction takes place, the water simply boils off. Were the kiln never to go above this temperature and the dried clay object placed into a bowl of water, it would absorb the water and become plastic again. The variable composition of the clay body hearkens back to Lucifer's description of mixtures; the clay body may contain a lot of water, a little water, or no water at all. Once the liquid water has been driven off, the temperature continues to rise.

If we continue heating the clay from 350°C to 500°C, that is, if we heat the ***bejeezus*** out of it, the kaolinite in the clay undergoes an irreversible chemical reaction, as shown in Equation 5-1. The process of heating the bejeezus out of the clay is called **calcination**. Here we see that six waters, the bejeezus, are literally driven from the kaolinite by the fire over the equal sign. We have seen the same kind of reaction in the production of charcoal from cellulose. If you add water to charcoal, you get wet charcoal, not wood. If you add water to mullite and silica, you get wet pottery, not clay. Notice that unlike the drying stage, there is a definite, fixed amount of water driven off. As Lucifer told you, this is what marks kaolinite as a compound, rather than a mixture. With the bejeezus gone, the solid products, mullite and silica, have changed from clay to stone and are impervious to water, as they were before the weathering process began.

Clay heated to 500°C will no longer revert to plastic clay when wet, but it has not yet become pottery. For this, a third stage, ***vitrification,*** is required and this stage depends more on the impurities in the clay than on the kaolinite. Both mullite and silica have extremely high melting points, which is what makes pottery useful as a refractory material. But impurities in the clay, notably iron oxide, melt at lower temperatures. The temperature at which this happens will depend on the impurities which happen to be present in the clay body. Low-fire clays may vitrify at 900°C, while porcelain clays may require temperatures as high as 1300°C. As the impurities melt, the liquid soaks into the pottery, coating the crystals of mullite and silica. When the pottery cools, the melted impurities solidify, in effect gluing the crystals together. The resulting structure is now very strong and impervious to both fire and water. It has become pottery.

I have not yet explained how this firing may be accomplished. No mere campfire will do the job. No, the fire must be bejeezus-hot if it is to heat the bejeezus from the clay. A common cooking stove will not rise above 500°F, or 260°C, sufficient for the water-smoke stage only. The coals of a common campfire will seldom rise above 800°C, enough to convert kaolinite to mullite and silica, but not enough to vitrify it. To vitrify even low-fire clay wares, we need a fire at least 100°C, and preferably 200°C hotter than the hottest campfire; we need a kiln. But to understand the design and operation of the kiln, you must first understand the relationship between heat and temperature.

Heat and temperature are not the same thing. To see this, imagine two pots on the stove, one containing a quart of water, the other, a cup of water. Turn the burners to the same setting and measure the time it takes for each pot to come to a boil. The cup will boil before the quart. They reach the same temperature, but more heat is required to boil a quart of water than a cup. Repeat the experiment, but this time leave the burners on for only two minutes and measure the temperature of the two pots. The same amount of heat was delivered to each, but the cup will have a higher temperature than the quart. One more test will further clarify it

Table 5-1. Three Temperature Scales

Incandescence	Cone	Fahrenheit	Centigrade
Lowest Visible Red to Dark Red	022-019	885-1200	470-650
Dark Red to Cherry Red	018-016	1200-1380	650-750
Cherry Red to Bright Cherry Red	015-014	1380-1500	750-800
Bright Cherry Red to Orange	013-010	1500-1650	800-900
Orange to Yellow	09-03	1650-2000	900-1100
Yellow to Light Yellow	02-10	2000-2400	1100-1300

for you. Pass your finger through the flame of a cigarette lighter. Then hold your finger in the same flame until you get the point. The flame is the same temperature no matter how long your finger is there, but your finger absorbs more heat in a second than it does in an instant. ***Heat*** is what causes temperature to rise.

To vitrify clay, to change it from clay to stone, it must be heated to red heat, that is, to *incandescence*. Different clay bodies, however, vitrify at different temperatures. Long before the advent of electronic kiln controls, potters learned to judge the temperature of the kiln by placing into it small numbered cones made from different clays. The clays were chosen to deform, each one at a different characteristic temperature so that the temperature of the kiln could be judged by noting which of several *pyrometric* cones had deformed during a firing. The cone numbered 022, for example, is made from a clay that deforms at the lowest temperature for which incandescence is visible, a dull red heat. Cone 021 deforms at a slightly higher temperature and the scale proceeds from cone 020 to cone 019, for which the incandescence is dark red. As the temperature climbs further we progress from cone 018 to cone 010, from dark red to orange incandescence. Clays which vitrify in the range from cone 09 to cone 03 are called *earthenware* clays; those which vitrify at higher temperatures are called *stoneware* clays. The scale proceeds from cone 02 to cone 01, but from that point the scale moves to cone 1 to cone 2, all the way up to cone 10. *Porcelain* clays vitrify at the upper end of the scale, from cone 6 to cone 10. At such temperatures pottery is not so much red-hot as yellow-hot. The deformation of a cone depends on time as well as temperature, so there is no simple conversion from the cone scale to the more familiar centigrade and Fahrenheit scales. Never-

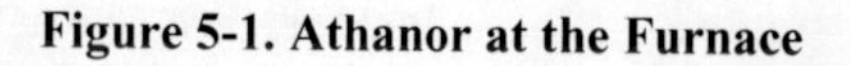

Figure 5-1. Athanor at the Furnace

theless, approximate temperature ranges are given in Table 5-1. Modern kilns can be programmed in centigrade and Fahrenheit but if you are going to communicate with potters, you should be conversant in the cone scale, as well. It is important not to confuse, for example, cone 10 with cone 010, as they differ by 400°C!

A kiln must satisfy two competing demands simultaneously. Being Athanor, you are well aware that the fire needs air, or more specifically, oxygen to breath. Blow on a coal and the fuel burns faster, producing heat at a greater rate. More heat causes the temperature to rise and the coal becomes brighter. Coals buried deep in a coal-bed are starved for oxygen, so we should spread the coals out over a large area to provide them with access to fresh air. But the rate at which a hot object loses heat is proportional to its surface area. You know instinctively that when you need to conserve body heat, you should curl up into a ball. To conserve the heat of the fire, we should rake our coals together into a deep, compact coal-bed. A traditional kiln works by providing a continuous supply of fresh air to a deep coal-bed.

Figure 5-1 shows an assaying furnace from *De Re Metallica, ca.* 1556 AD.[2] It is essentially a deep, insulated coal-bed with an opening at the

2. Reference [1], p. 223.

bottom. As the hot waste gases (chiefly carbon dioxide) exit from the flue at the top of the furnace, a draft is created, drawing fresh air in through the fire-mouth at the bottom. The fresh air causes the charcoal to burn hotter, which heats the air even more, increasing the draft. The ultimate temperature of the furnace is determined by the size of the furnace, the insulating value of its walls, and the relative sizes of the fire-mouth and flue. The temperature may be decreased by closing off the fire-mouth and it may be increased by forcing air into the furnace with a bellows or even by fanning the flames, as shown in the figure.

Conditions within the kiln will affect the color of the finished pottery. In areas where there is plenty of oxygen, we say the conditions are oxidizing. The fired pottery will be light in color, white or red, depending on the original color of the clay. In areas where the coals were starved of oxygen, a reducing atmosphere, the fired pottery will be dark in color, brown or black. In a primitive kiln or campfire, where there is little control over the flow of air within the coal-bed, a single pot may show areas of both **oxidation** and **reduction.**

A bonfire may be sufficient to vitrify low-fire clay, but will not attain the temperatures needed for calcining limestone and smelting bronze. For economy and convenience, the electric kiln is better suited to these applications. We shall discuss the principles of electricity later in the book, but for the time being I recommend the electric kiln in the same spirit of convenience that Samson used the 2-liter soft-drink bottle. Used kilns may be purchased for a few hundred dollars and a small one may be built from scratch for about a hundred. Alternatively, you may make the acquaintance of a potter, who, being Athanor, will be disposed to help a brother or sister in need.

Material Safety

If anything is safe and natural it must be clay. Would it surprise you to learn that this material is described in an MSDS?[3] Your ceramics supply will be able to furnish you with a copy. Alternately you search the Internet using the keyword "MSDS" and the CAS number for kaolinite (CAS 1332-58-7). Summarize the hazardous properties of kaolinite in your notebook, including the identity of the company which produced the MSDS and the potential health effects for eye contact, skin contact, inhalation, and ingestion.

3. The MSDS was introduced in Section 3.2 (page 43).

Any hazards in this project are likely to be more physical than chemical. If you are firing in an electric kiln, be sure to follow the manufacturer's instructions. If you are firing in a bonfire, be careful to build it away from flammable structures. Have a source of water on hand should the fire get out of control and be careful to avoid smoke inhalation. Do not leave a fire unattended and use gloves to handle hot pottery or burning wood. If you are unwilling to accept responsibility for the safe use of fire, you should give your matches to a grown-up and skip to the next chapter.

Research and Development

You should not remain ignorant if you are to proceed in the Work.

- Know the meanings of those words from this chapter worthy of inclusion in the ***index*** or **glossary.**
- You should have mastered the Research and Development items of Chapter 1 and Chapter 2.
- Know how clays are formed from their parent rocks and minerals. Be familiar with the hazardous properties of clay.
- Be familiar with the cone, Fahrenheit, and centigrade temperature scales.
- Know formulae for alumina, kaolinite, mullite, water, oxygen, and silica.
- Know the equation for the calcination of kaolinite.
- Know how a kiln achieves higher temperatures than an ordinary campfire.
- Know what has driven the history of humankind.

5.3 Θ

Having secured a kiln or furnace, you will need to fashion a crucible for the work you have undertaken. The crucible must be refractory, lest it succumb to the heat of the furnace. It must be structurally sound, lest it crack and lose the work you have so carefully prepared, and it must be aesthetically pleasing, lest you show yourself to be no true Athanor. The clay you require is generally called low-fire earthenware clay and comes in red or white. If it is to maintain its integrity, the crucible should be all of a single piece of clay, not pressed together from several lumps. It

Figure 5-2. The Crucible

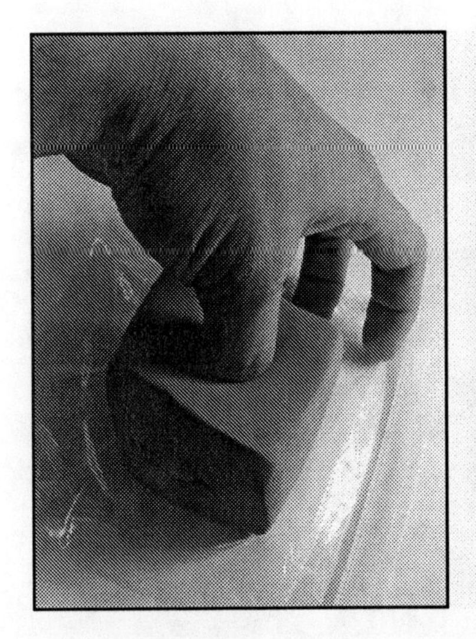

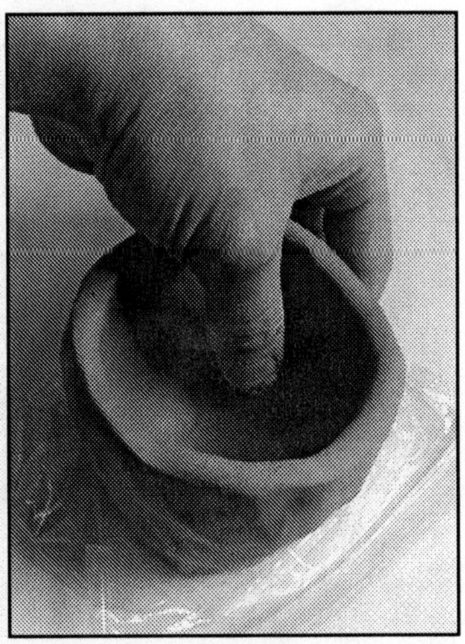

should be free of trapped air bubbles, which might cause it to shatter in the kiln. For this reason, it is imperative that the clay not be rolled out and pressed back together, that it not be flattened and folded on itself.

Start with a lump of clay approximately 6 cm x 6 cm x 6 cm and place it onto a piece of paper or plastic to prevent its sticking to the table top. Jam your thumb into the center of the lump and, using your thumb on the inside and your fingers on the outside, open what will be the mouth of the crucible, as shown in Figure 5-2. Rotate the crucible, pressing the walls between the thumb and fingers until the walls and bottom are uniformly 8 mm thick. There should be no seams or cracks, which would weaken the vessel and interfere with the uses to which you will put it.

The crucible must now be shaped into a free-standing cone so that molten material within it will collect in the center. Using the thumbs and fingers, squeeze the base from the outside to form a pedestal, as shown in Figure 5-3. Use your thumb to open up the mouth of the crucible, pressing a dimple into the pedestal at the bottom. Thus the thickness of the walls—even the walls of the pedestal—should be no more than 1 cm thick. The crucible should be 8-10 cm in diameter and 10-12 cm tall.

Use a knife to trim the lips of the crucible so that they are parallel to the table top, as shown in Figure 5-4. Manipulate your crucible so that

Figure 5-3. The Pedestal

Figure 5-4. The Lips

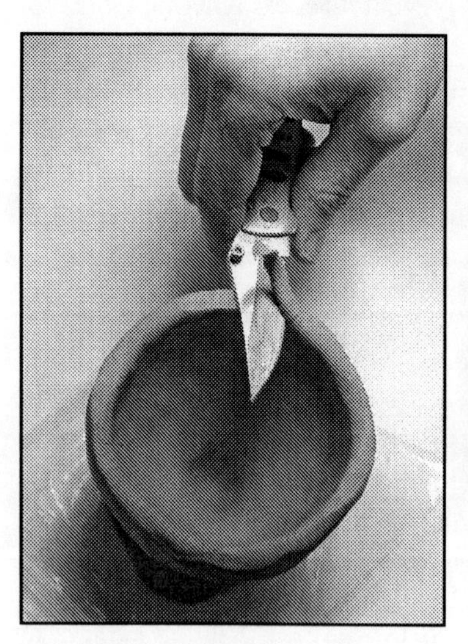

Figure 5-5. The Lid

its shape is symmetric and its walls smooth. Then allow it to air-dry for approximately 5 hours. When you return to it, the walls will have stiffened, making it less floppy and easier to work. Make fine adjustments to the shape and surface, eliminating any imperfections. If the surface is too dry, you can moisten your hands or use a damp sponge to make the surface smooth and free of cracks. You can also use a knife, spoon, or tongue depressor to remove bumps or other imperfections. If you do not have time to devote to the work, you can place your crucible into a plastic bag, a veritable time capsule, to prevent its drying out further. When you are satisfied with the shape of your crucible, allow it to air-dry for another 5 hours. After this second drying, the walls will be quite stiff and the surface of the clay cold to the touch. Rub a smooth stone or the back of a spoon over the surface, inside and out, to render the surface glossy. Before allowing your crucible to dry any further, you will need to make a tight-fitting lid for it.

Because clay shrinks as it dries, it is important to make the lid before the crucible has completely dried. Roll out a fresh piece of clay into a pancake, invert your crucible as shown in Figure 5-5, and place it onto the pancake. Scratch your initials into the base of your crucible so that it may be identified. Use a knife to trim excess clay from the lid and then set the crucible on its pedestal. Press the lid slightly into the crucible so that it takes on the shape of a shallow dish, perhaps 1 cm deep. Use your thumb and fingers to make the lid smooth and symmetrical. With the lid in place, allow your crucible to completely dry and scratch your initials into the lid for identification. You may place it into an oven at 130°C (270°F) for an hour to hasten its drying. Record the dry weight of crucible and lid in your notebook.

Your crucible is now bone dry; you have eliminated any liquid water which was present when the clay was plastic. You will fire it to cone 05 in a kiln, driving off bejeesical water according to Equation 5-1. If this equation is correct then the weight of your crucible should be less after firing than before. In fact, pure, dry kaolinite can be expected to lose 14% of its weight upon firing. A clay body, however, may include non-kaolin materials such as sand or crushed pottery. This "grog" may not lose weight upon firing and so a clay which contains grog may lose less than 14% of its dry weight. The crucible of Figure 5-5, for example, lost only 13% of its dry weight upon firing. Such a weight loss is consistent with a clay which is 93% kaolinite and 7% grog. Record in your notebook the weight of your crucible and lid after firing and compare it to the weight before firing. The time you spend working on your crucible is time well-spent. The lid *must* fit tightly if you are to successfully make metal in Chapter 9. Its walls must be uniform and strong if you are to successfully make lime in Chapter 10. Its interior must be a conical if you are to successfully make glass in Chapter 13. It takes less time to make one crucible well than to make several of them poorly.

Δ Do not neglect your art, my child. Without a mastery of refractory materials you will be ill-prepared for the work ahead. You must know your clay and be able to fashion it into any form which you require. You must know the fire, to administer more or less according to the properties of the clay. Only then will you be able to keep in the heat and withstand it. Only then will you be worthy of the name, Athanor.

Quality Assurance

When your crucible can be filled with water without leaking or reverting to a plastic state, when its shape is pleasing and its integrity sound, when its lid is tight-fitting, then will you be prepared to continue in the pyrotechnic arts. Include a photograph of your crucible in your notebook and compare its dry weight before firing to that after it returns from the kiln.

Chapter 6. Venus (Textiles)

And Nature forced the men,
Before the women kind, to work the wool:
For all the male kind far excels in skill,
And cleverer is by much—until at last
The rugged farmer folk jeered at such tasks,
And so were eager soon to give them o'er
To women's hands, and in more hardy toil
To harden arms and hands.

— Lucretius, *On the Nature of Things*, *ca.* 60 BC [1]

6.1 ☿

∀ My momma named me Venus, on account of she said I was as pretty as a morning star. When I got to marrying age, my folks fixed me up with a husband, Og, but I can't recall why his folks named him that, what with it being so long ago and all. One cold, wet day early on in our cave-keeping Og didn't feel like going out hunting. Lazy as a cave bear, God rest him. I had to go out and gather because, well, we were hunter-gatherers and Og wasn't hunting. "You're not moping around all day, Og. You get your butt off of that rug and straighten up this cave!" He didn't like that too much, but I wasn't going to work my fingers to the bone so he could stay home and stare at the fire all day. Anyway, so I went out and gathered. I gathered all day. Boy, was I tired when I got back, and you know what? That cave was exactly the same as when I left it. I won't tell you what I said to him, but the poor old thing looked like a fly who's just stepped into the spider's parlor. Oh yeah, it's funny now, but I wasn't laughing then.

∀ You know what he'd been doing all day? Playing with a dog! Now, he said that he started cleaning up the cave and this wild dog wandered in, but I knew better. He got to playing with that dog and by the time they were done, they had hair all over the place. He said he got to picking up that hair and sat down to rest. He was sitting there, twiddling with the hair, you know, and it twisted up on itself like a piece of string. Well, we never had seen string before, in fact nobody had, on account of Og had

1. Reference [19], Book V, ll. 1350-1357.

just invented it. He just kept pulling on that string and it got longer and longer. It really was a wonder. But I was not happy. "If you think you're getting some of these turnips, you've got another think coming!" He was really in the doghouse then.

∇ Well, we made up after awhile, and I figured he would lose interest in that string, but he just kept making string every chance he got. "Put down that string and go fetch some water." "Put down that string and go get some firewood." "God help me, Og, if I catch you playing with that string!" You know, I don't like being a nag, but we like to starved to death on account of that string. When we had my little girl, I laid down the law and Og turned out a pretty good father, even if he did make string every chance he got. I made the best of it and it turned out that string was good for tying stuff up so after a while I got used to having it around.

∇ Well, our little girl grew up and starting filling out, and before too long every man for miles around was pestering us. "Can I walk Thelma to the lake?" "No, she's too young." "Can I take Thelma to the dance?" "No, she's still too young." "Can I show Thelma my rock collection?" "No, she's *still* too young." Like to drove us crazy! Finally I got an in*spir*ation; I said, "Look, we're making her a belt, and when you see her wearing that belt, you'll know she's old enough for courting." Which was strange, on account of we didn't wear clothes in those days. Now, you may wonder what made me connect "string" with "clothing," but I had had a house full of string for years and I wasn't going to let it go to waste. It worked pretty good and we had some peace for awhile. And when that belt was done Thelma got herself a nice young fellow, on account of she was the only girl with a belt like that.

∇ Well, that started the ball rolling, of course. Everybody wanted a belt for their daughter, both to get some peace while she was young, and to make her a good match when the time came. And Og went into the belt-making business and started beltifying the place. Pretty soon he couldn't keep up with the demand and I started helping him out. See, there are two parts to making string, the twisting and the drafting, and Og was trying to do both at the same time. We figured out that Og could do the twisting and I could do the drafting, and that way we could make string way faster than either one of us by ourselves. Not only that, Og was only able to make one arms-length of string by himself. But spinning together, we could make a string as long as our cave, or even longer. Before too long we ran out of dog hair, but we found out we could spin grass fibers,

Figure 6-1. Venus de Lespugue

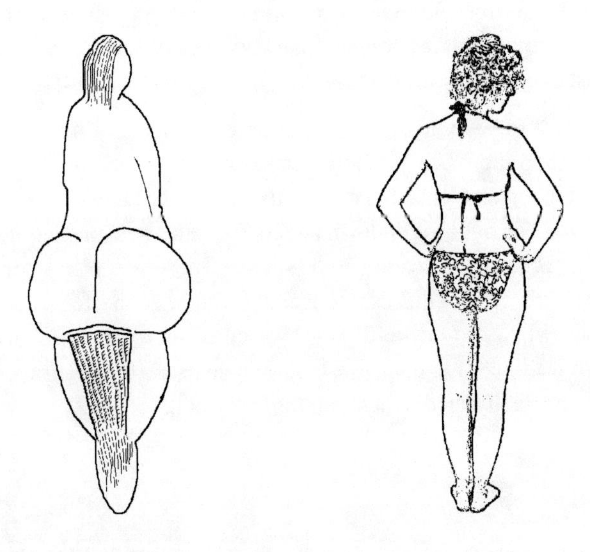

bark fibers or just about any kind of fiber we happened to come across. So that was good.

∇ In the beginning a girl was lucky to have any kind of belt at all, but as time went on, the belts had to be fancier and fancier if she was going to stand out from the rest. We added strings and tassels and whatnot and pretty soon that belt had turned into a kind of skirt, a *string skirt.* Og even carved a little statue of our granddaughter which you can see in the *Museée de l'Homme* in Paris, or in Figure 6-1(L). You can see that she wore her skirt low on the hips in those days. Nowadays, gals wear their bikinis somewhat higher up, but bikini or string skirt, neither one of them hide very much. They do get your attention, though, which is what they're for, when you think about it. *Plus ça change, plus c'est la même chose.*

∇ When spinning was new the men really liked it, but the fancier these skirts got, the more string we had to make. Now, men like hunting and fishing and herding, things where you're real busy for a while and then you're not busy for a while. So they didn't like spinning any more, on account of you were only a little busy, but it was the same busy all the time. Eventually, our son in law, Dug, got so bored with his half of the spinning (the twisting part) that he said "You know, I bet a stick could do this job as good as me!" And, it turned out, a stick with a notch at one end and a stone at the other was every bit as good at twisting as Dug was. So that was the first time a man was replaced by a machine, when Dug lost his job to a stick.

∇ It wasn't so bad that the men gave up on spinning; it turned out to be a pretty good job when you're raising babies. You can pick it up while you're cooking and when the stew boils over or the baby gets too close to the fire you can put it right down again. You can invite the neighborhood gals over for a nice chat and get the spinning done at the same time. And teaching a girl to spin makes you feel like you're connected to the future. My little girl passed the beltifying in*spir*ation on to her little girl and so on down the line for twenty-five thousand years, so the in*spir*ational part of me never really died, when you think about it.

6.2 ♀

The story of Unktomi-Venus is somewhat autobiographical. My wife had gone out shopping and I was left home alone with our German shepherd, Sandy. By the time my wife returned I had several feet of what I considered the finest twine I had ever seen. We will never know the details of the first time someone sat down with a handful of fiber and made string from it, but I have tried to tailor my little story to fit the evidence presented by Elizabeth Wayland Barber in her excellent book *Women's Work: The First 20,000 Years,* on the early history of textiles. The earliest representations of human clothing depict women of child-bearing age wearing waist-hugging skirts made of what appears to be spun string rather than tanned hide; contemporary images of men are nude. The tradition of spinning was the domain of women up to the time that it began to be industrialized in the Iron Age. Apparently the work of spinning meshed well with that of child-rearing, but there is nothing to prevent the modern reader from learning and enjoying this craft, regardless of gender. As I see it, though inventions originate with particular individuals of particular races and genders, the inventions themselves belong to our common human heritage. Once an inspiration leaves its original home, it's no longer male or female, black or white; it's just a spider in the wind looking for a place to land.

The goal of spinning is to make long fibers from short ones. While any one strand of wool is no more than a couple of inches long, it may be used

Equation 6-1. From Air to Glucose to Cellulose

$$(a)\ 6\ \mathrm{CO_2}(g) + 6\ \mathrm{H_2O}(l) = \mathrm{C_6H_{12}O_6}(aq) + 6\ \mathrm{O_2}(g)$$
$$(b)\ \mathrm{n}\ \mathrm{C_6H_{12}O_6}(aq) = (\mathrm{C_6H_{10}O_5})_\mathrm{n}(s) + \mathrm{n}\ \mathrm{H_2O}(l)$$

to make string hundreds or even thousands of feet in length. A beginning spinner may assume that this is accomplished by twisting the fibers together, but as important as twisting is, properly arranging the fibers, ***drafting*** them at the point where fiber turns to twine is even more crucial. And in a curious coincidence, the biological synthesis of long fiber molecules from short ones parallels the spinning of string from fiber.

All higher organisms face the problem of getting material from one place to another. Plants send material from the roots to the leaves and back via sap. Animals move nutrients from the stomach to the rest of the body via blood. Both sap and blood are aqueous solutions and so material transport depends on the solubility of the chemicals to be transported. But once they reach their destinations, some of these chemicals must be converted to skin and bone and hair or else the organisms would dissolve in its own juices and slump into a puddle of goo.

Nature discovered long before we did that short, water-soluble ***monomers*** can be linked together to make long, insoluble ***polymers.*** For plants, the principle monomer is called glucose, the same molecule that's in honey. You already know that glucose dissolves in water, since it is the sugar that is fermented to make mead. The plant makes glucose out of air and water and sunlight in the leaves and sends it all over the plant in the sap. When that glucose gets wherever it's going, the plant chains all those little molecules together, polymerizes them, into one long molecule, cellulose, which is what wood is made out of. The cellulose molecule is so much larger than water molecules that they can't push it around; in other words, cellulose is insoluble in water.

One of the most important classes of polymerization reactions, called **condensation,** results when a hydrogen atom on one molecule gets together with a hydroxide group on another and leaves as water; Equation 6-1 shows how cellulose condenses from glucose. The *n* in the equation could be any number at all, one or a thousand or a million, but as in any balanced reaction equation, there are the same number of each kind of

Equation 6-2. Condensation of a Protein from Glycine

$$n\ C_2H_5O_2N(aq) = (C_2H_3ON)_n(s) + n\ H_2O(l)$$

atom on both sides of the equal-sign. Also notice that the formula for cellulose is $(C_6H_{10}O_5)_n$, not CH_2O as Lucifer told you. The long formula is called a ***molecular formula,*** because it tells you how many atoms are in each molecule. Lucifer introduced you to an ***empirical formula,*** one which only tells you the proportions of the different atoms. In glucose, for example, you can divide the molecular formula, $C_6H_{12}O_6$, by 6 to determine the empirical formula, CH_2O. Now, if you've been paying attention, you'll notice that the empirical formula of ***cellulose*** really ought to be $C_6H_{10}O_5$, because you can divide the whole molecular formula by n. If you were to try dividing it down further, say, by 5, you would get $C_{1.2}H_2O$, which does not have integer subscripts. Lucifer's formula for cellulose was actually only an approximation so as not to scare you with a big long formula right off the bat. You might want to go back to Equation 1-1 (page 9) and update it with the correct formula.

Animals use glucose, too, but not for structural purposes. Proteins are the stuff of skin and hair, and they are composed of twenty slightly different monomers, the amino acids, of which, the simplest is glycine, $C_2H_5O_2N$. When an animal eats meat, the meat is broken down into amino acids in the stomach and these amino acids dissolve in the stomach acid. They get pushed around in the blood to wherever they're going, and when they get there, they are hooked together once more into long protein molecules, which like cellulose, are insoluble in water. Equation 6-2 shows what the condensation reaction would be if you polymerized glycine alone. In reality, proteins are sequences of all twenty different amino acids, and the properties of the protein are determined by the amino acid sequence. But the main idea, the linking of small things into big ones is common to all living things, and so is at the heart of making textiles on both the microscopic and macroscopic scales.

Material Safety

I think it would be fairly difficult to hurt yourself making string. I suppose you could jab a drop spindle into your eye or something, so don't do that. Chemically speaking, wool has so many different things in it, different amino acids and grease and whatnot, that you won't be able to find an MSDS for it. But you should get used to thinking about hazardous materials as a matter of course. Sheep produce lanolin to make their coats water-proof and since lanolin is a component of many hand creams, spinning "in the grease" will leave your hands soft, supple, and farm fresh. Raw wool will also contain dingle-berries and urine, and given the propensity of bacteria to reproduce, it would be a good idea to wash your hands after spinning.

Research and Development

If you are in a class, you might want to know what will be on the quiz.

- You better know all the words that are important enough to be ***indexified*** and **glossarated.**
- You better know the Research and Development stuff from Chapter 2 and Chapter 3.
- Know the composition of plant and animal fibers.
- Know formulas for cellulose, glucose, water, and glycine.
- Know what the earliest form of clothing was and about how old it is.
- Know that spinning makes long fibers out of short ones.
- Know that the way string is made is like the way Nature makes polymers from monomers.
- Know enough to wash your hands before eating.
- Know how the first man lost his job to a machine.

6.3 Θ

∀ You know, without string we wouldn't hardly have any clothes at all, so if you're going to understand clothes, you better understand string. And a lot of chemical technology came about because of the needs of folks making clothes, so if you want to understand chemical technology, you better understand clothes. And if you don't want to understand chemical technology, well, you probably shouldn't be reading this book, on account of that's what it's about.

You could learn to spin string from many kinds of fiber: cotton, dog hair, wool, the list is probably endless. Students have tried to spin human hair,

Figure 6-2. Raw Wool

assuming that long hair would give them a leg up on making long string, but the truth of the matter is that relatively short, curly fibers are easier to spin than long, smooth ones. While raw cotton is fine, cotton balls from the drugstore have been cut and polished to the point that they are difficult to spin. I personally learned to spin dog hair, and if this is your choice you would be well-advised to select the woolly hair from the belly over the straight, glossy hair from the back. But of all the fibers I have experienced, wool is the easiest to spin, the sheep having been bred for their wool. If there are no shepherds in your area, there are numerous places on the Internet to buy raw wool.

When they shear sheep, the wool comes off about two inches long and with all the hairs running parallel. And being fresh, it probably has dingle-berries and straw and stuff in it, so don't let that freak you out. You can pick that stuff out. Raw wool is coated with lanolin and other sheep oils, making the wool water-repellent so the sheep don't freeze to death in the rain. There is no need to wash these oils off and spinning "in the grease" is a very satisfying experience indeed; it'll leave you hands smelling farm fresh.

In preparation for spinning, you need to tease the wool, both to remove the dingle-berries and to stop the hairs from running parallel. In order for the string to grow, the fibers must be staggered, with the end of one fiber in the middle of the next, but as it comes of the sheep, the ends of the wool fibers are next to each other. So you want to pick the wool apart so it's all random and tangled up with the hairs going every which way, as shown in Figure 6-2(R). Beginners are often in a hurry, but the more attention you pay to preparing your wool, the less frustration you will experience later on.

Figure 6-3. Pulling and Twisting

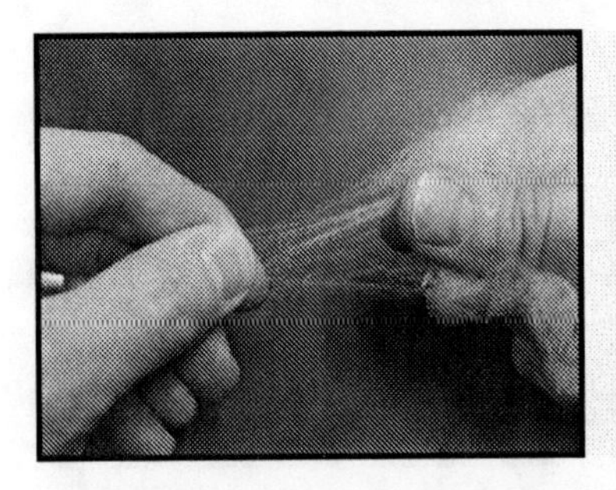

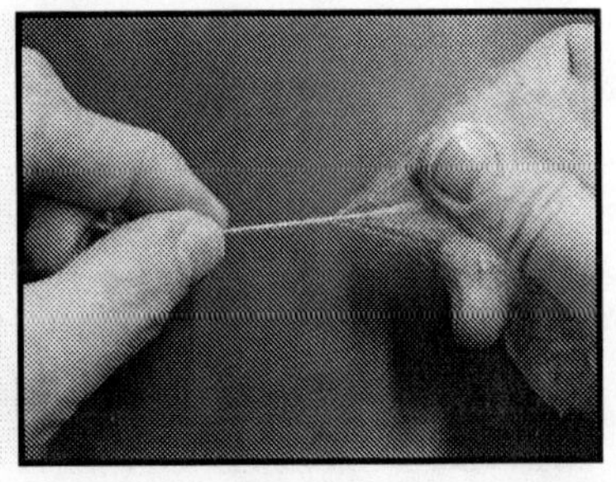

I usually hold wool in my right hand and I will describe it in those terms, but if you prefer, you may switch left for right in the following description. With your wool in your right hand, take a pinch with your left hand and pull, as shown in Figure 6-3(L). Pull pretty hard. You can pull harder than you might imagine until you hear a sound like the wool is tearing. That's a good sound. It's just the sound of the fibers moving over each other as they line up. After you've pulled, you can twist the end between the thumb and forefinger of the left hand. If you push your thumb away from your body, you're giving it a *Z* twist and if you pull your thumb toward your body, you're giving it an *S* twist. The direction you choose is arbitrary, but once begun you must continue twisting the same direction or your string will unravel.

Once you've twisted, you'll have a little piece of string going up to a point and then the fibers will fan out into your right hand as shown in Figure 6-3(R). Schematically, your sting now looks like this: —<. That little fan, or *vee*, is very important. You don't want to lose that vee, or you're string will move up into the ball of un-spun fibers and you'll be hard pressed to get things back in order. Use the fingers of your right hand to keep the wool spread out while your left hand twists and pulls. The vee forms at the point where the twisting and pulling bumps up against the spreading of the yet-to-be-spun fibers. The maintenance of the vee is known as *drafting*.

Now, there is nothing wrong with spinning alone, but it'll be easier to explain if we separate the twisting from the drafting. It will also be easier to learn to spin if you do it that way and so I'm going to presume that we have two different people spinning together. Let's call the hands in Figure 6-5(L), "Og" and the hands in Figure 6-4(R), "Venus." Og's job

Figure 6-4. Four Hands Are Better Than Two

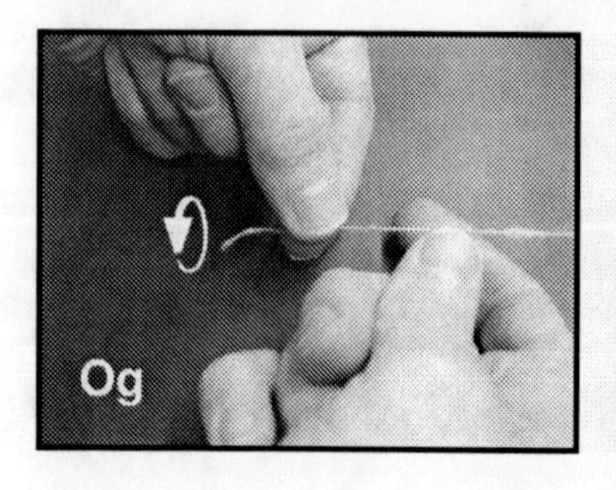

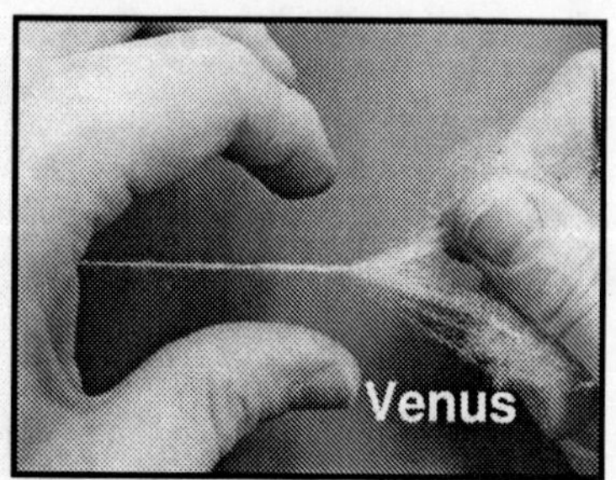

is to hold the end of the string and twist it. He needs to twist it the same direction it was started in, an S twist in this case. Now, your Og may be tempted to move up the string as he twists, but you need to nip that in the bud as quick as you can. He just needs to twist the very end of the string without creeping along it. Instruct Og to twist as fast as he can and to continue for as long as it takes. That's pretty much it for Og. It's a boring job, but somebody's got to do it.

Venus' job, the drafting, is a little harder. Her task is to keep the vee going as the string grows and to this end, her right hand is constantly spreading the wool into a fan. In Figure 6-4(R), her left hand is just about to grab the point of that vee and pull.

In Figure 6-5 Venus has just pinched that point and pulled to the left. She's not worried about pulling too hard. She hears that ripping sound and knows that it is the sound of fibers locking together. At the beginning of a pull the fibers resist, but as the ripping sound begins, the fibers slide over each other more easily. If she pulls too hard, the string will break off of at the point of the vee, so she is careful to stop pulling just before the string breaks. If she pulls too hard for too long, she can repair the damage simply by unraveling a half an inch or so of the end of the string and working it back into the un-spun fibers. Another pull should re-establish the vee and she can continue with her spinning. Careful attention to the vee will prevent such a mishap and the work will go quickly and smoothly.

So she's just pulled the point of her vee hard enough to hear the ripping sound but not so hard that she tears it plumb off. When she lets go, Og's twist will move up the string until it meets the spreading fibers and

Figure 6-5. Drafting

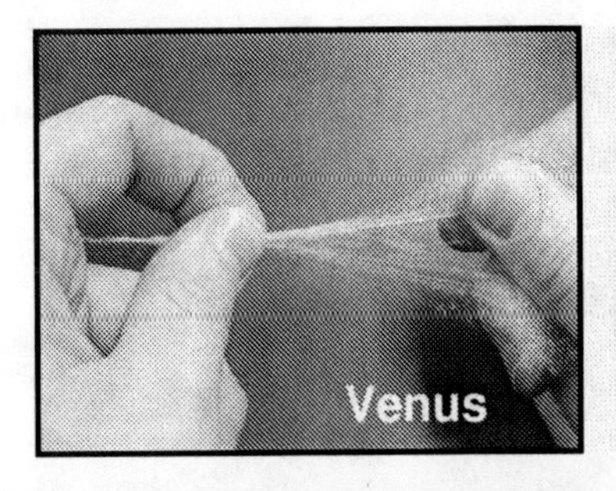

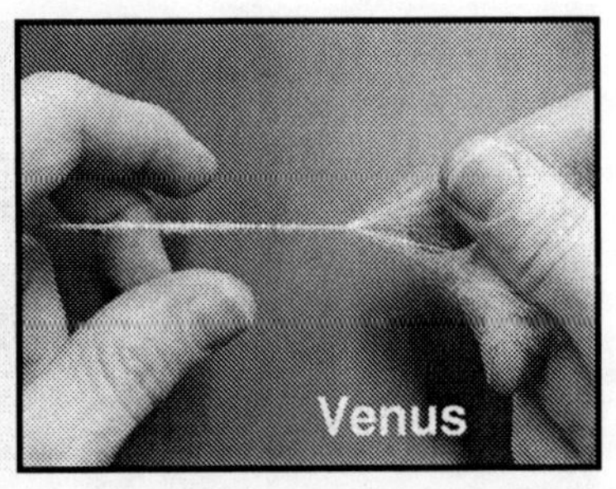

Figure 6-6. Two Plies Are Better Than One

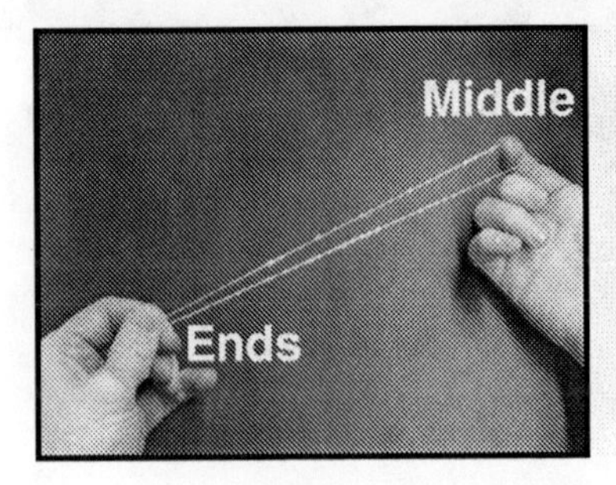

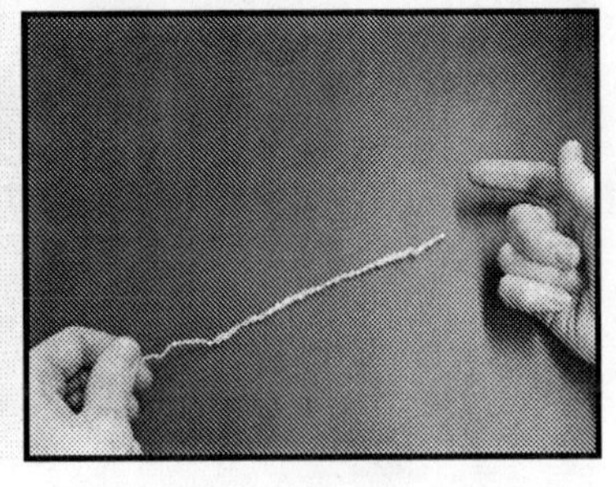

that's were the next vee will show up. If she hasn't been spreading with her right hand, the twist will move all the way up into her right hand and she'll have to work to get her vee back. Once she has the technique down, she will get into a rhythm of pulling and spreading simultaneously, progressing from one vee to the next, and the string will grow an inch or two with every pull. Of course, eventually, she's going to run out of wool in her right hand and he'll have to grab another hunk of teased and de-dingle-berry-fied fiber.

And that's pretty much it, with Og twisting the end and Venus drafting, the string growing longer and longer. If either one of them lets go, the string will twist up on itself and make a big mess. But if Og takes Venus' end and holds it next to his, while Venus holds the middle and pulls it tight, they'll fold that string exactly in two. When Venus let's go, it'll look like that string is twisting up on itself in a big mess, but she'll walk over to Og and run her fingers down that string while he holds his ends fast. And as she runs her fingers down that string, magic happens. The

Figure 6-7. The Drop Spindle

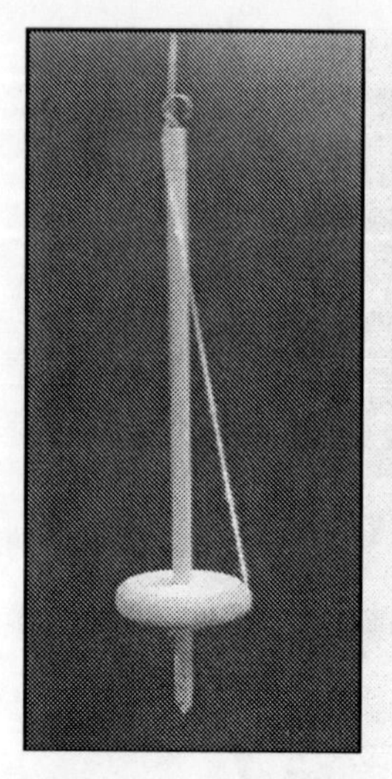
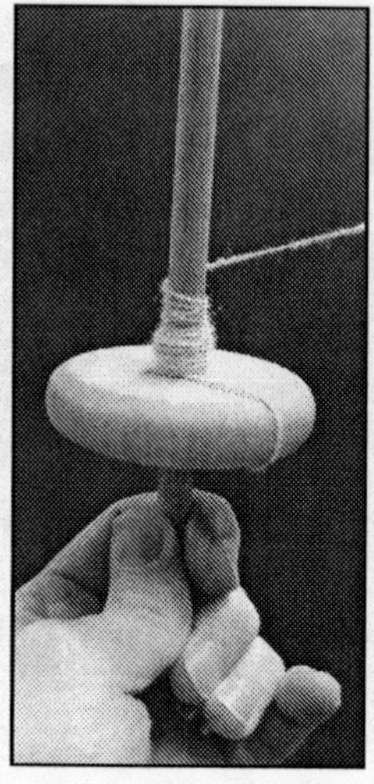

S twist of one half fights against the S twist in the other half and they exactly cancel each other out. And the two halves twist around each other as they battle it out and this new twist is a Z twist. The Z twist keeps the two S twists from unraveling, so that the new two-***ply*** string, or ***twine,*** will not twist up on itself any more. In addition, it will be twice as strong as the single-ply pieces, so twine is doubly good.

At this point, if you have been playing the role of Venus, you should ask your Og about his level of job satisfaction. If he is like the Ogs I have known, he will be bored out of his skull and will welcome the opportunity to be replaced by a machine. That machine is the ***drop spindle.*** A drop spindle is just a stick with a weight at one end and a hook at the other. You can buy them in online or in craft stores, but they are not hard to make for yourself. I make them from wooden wheels sold in craft stores. I glue a foot or so of dowel rod through the hole in the wheel, or ***whorl,*** and screw a hook into the top of the dowel. To use the drop spindle, make a little piece of string the old way and tie the end to the stick, underneath the whorl. Pass the string over the whorl and just loop it through the hook at the top. Don't tie it to the hook, just loop it. Now give the spindle a spin. Be sure you spin it the same way that you started your string, either with an S twist or a Z twist, or else that spindle will undo what you started. If you get used to spinning either one direction or the other all the time, you'll always know which way to spin the spindle.

With the spindle taking care of the spinning, all you have to do is pay attention to the drafting, that is, to the spreading and pulling. Since the spindle is hanging down instead of sideways, your vee will be vertical, like this: V. Other than that, your job is the same as before, except you have to reach down from time to time and give the spindle a spin. Whenever you get into trouble with a knot or a tangle, just swing the spindle and catch it between your knees; you can hold it there until you have your drafting straightened out. You can spin all by yourself, now, until that string is so long that your spindle hits the floor. When that happens, pick up your spindle and wind the string onto the stick above the whorl, as if you were winding it onto a spool. It doesn't matter which way you wind it as long as you always wind it in the same direction all the time. When you've wound all but the last foot or so, pass the string back under the whorl, around the stick, back over the whorl again and up to the hook. It looks the same as before, now, with the string running from the edge of the whorl, looped around the hook, and up to your vee. Whenever the spindle hits the floor, wind your string onto it and pick up where you left off.

One nice thing about the drop spindle is that you can just put it down any time and your string won't unravel. That's why it was a good job for women for so long. If the stew is boiling over, you can just put your spinning down and go take care of it. Same thing if the baby needs nursing or the toddler is wandering too close to the fire or your husband comes home a little frisky. Just put it down, it won't go anywhere.

∀ So that's it, really. Once you can make string, you have a good start at understanding a bunch of the technologies that came along down through the years. And even if you can buy string real cheap, it's always good to know where it comes from and that you could make it yourself any time you wanted to. And when you hear about boring, repetitive jobs being taken over by machines, you'll know that's nothing new at all.

Quality Assurance

If you can spin 20 feet of single-ply string or 10 feet of two-ply twine, you are probably good enough to spin as much as you want. Break off a couple of inches of your string and tape it into your notebook. And if you think you might have daughters someday, you might want to spin a little extra to get a head start on her belt.

Chapter 7. Adam (Metathesis Reactions)

וַיִּצֶר יְהוָה אֱלֹהִים מִן־הָאֲדָמָה כָּל־חַיַּת הַשָּׂדֶה וְאֵת כָּל־עוֹף הַשָּׁמַיִם וַיָּבֵא אֶל־הָאָדָם לִרְאוֹת
מַה־יִּקְרָא־לוֹ וְכֹל אֲשֶׁר יִקְרָא־לוֹ הָאָדָם נֶפֶשׁ חַיָּה הוּא שְׁמוֹ׃

formatis igitur Dominus Deus de humo cunctis animantibus terrae et universis volatilibus caeli adduxit ea ad Adam ut videret quid vocaret ea omne enim quod vocavit Adam animae viventis ipsum est nomen eius

Or Jéhovah Dieu formait du sol toute bête sauvage des champs, et toute créature volante des cieux, et il se mit à les amener vers l'homme pour voir comment il appellerait chacun [d'elles]; et comme l'appelait l'homme—chaque âme vivante—c'était là son nom.

End uat uf thi gruand thi Lurd Gud furmid iviry biest uf thi foild, end iviry fuwl uf thi eor; end bruaght thim antu Edem tu sii whet hi wuald cell thim: end whetsuivir Edem cellid iviry lovong crietari, thet wes thi nemi thiriuf.

— *Ginisos* [1]

7.1 ☿

🜂 Imagine you're wading through your email one day, hoping against hope to find any sign of intelligent life amongst the infinite number of debt consolidation plans and chain letters. You get this mysterious message with no return address and before consigning it to electronic oblivion you wonder whether it's a coded message from a desperate CIA operative or yet another gimmick to get you to read about unbelievably low rates on home-equity loans. So you take it to your neighborhood computer geek. "It's a document from a word processor. When you send a document like this by email, it gets encoded into ASCII format to prevent special characters from interfering with the mail handling programs. Part of the message was lost along the way and when your mail program tried to decode it, the message was out of sync." Of course this sounds like so much cock and bull, since the word *Adam* is plain as the geek nose on his geek face. Does he honestly expect you to believe that the whole message would be garbled except for one word?

1. Reference [14].

▵ You forward the letter to your best friend to see what *she* makes of it. "It's obviously in Latin. The first line contains Roman numerals for the date and return address. There are all kinds of *us*'s and *ae*'s and you recognize the word *Adam* because it's the same in Latin and English." But the *us*'s and *ae*'s were only in the second paragraph.

▵ So you pop it off to the expert at www.unsolicitedlatinspam.com; "How can anyone be so stupid and yet live?" *he* ;-)'s. "It's obviously in Hebrew, Latin, and French. Don't you know how to read?" You're stung, naturally. "Well, not Hebrew, Latin, or French. I don't know the words or the grammar." "Well, if you don't know the words and the grammar, you can't read the language." Thank you Professor Obvious.

▵ "Well, what about the last paragraph, what's that?" You press on. "Hmmm, it looks like gibberish to me." "Is that what they speak in Gibber? Maybe you just don't know the words or the grammar." He seems un-fazed by your witty *riposte*.

▵ At this point in the narrative structure I'll bet you're thinking that this whole scenario is completely unbelievable. Well, pardon me for not living! The Author can string you along with endless rubbish about Nazarite alcoholics, alchemical crock-pots and dog-hair fashion models, but *my* little tale is beyond imagining. For your information alcoholic beverages, pottery and woven cloth each pre-date written language, so the little vignettes the Author has composed cannot be substantiated by the historical record. To make matters worse, alcoholic beverages and cloth don't hold up well to the ravages of time and so there's precious little direct evidence about their origins in the archeological record. All the Author has done is to read a few books by experts in the field and then to make up little stories that are not substantially contradicted by the evidence available in those books. He has done so to provide a narrative which will entertain you as well as inform you. If he is successful, you will remember a few of these tall-tales, and perhaps the factual elements of their respective chapters will come along with them. Put another way, the Author has constructed meme complexes whose longevity may be enhanced by their associations with presumably fecund folk-tales. Only time and tests will tell whether the fidelity with which these meme-plexes are reproduced will suffer from the inclusion of these bull droppings.

▵ If you ask me, the whole approach is completely unnecessary. In the memetic struggle to survive, most memes are in the unenviable position of having little to contribute to the material well-being of the creatures they inhabit. Consequently they rely on little tricks to make themselves more memorable. A religion may promise eternal bliss to those who remember the right things; songs use rhyming words to aid the memory. But a technological meme has the advantage that remembering it makes it possible to actually make something people need. I'm not saying that this makes a technological meme *better* than a non-technological meme, just that it comes with its own built-in incentive for remembering it. So the Author's memetic approach is bass-ackwards; if anything, people should want to remember the technologies more than the stories and not the other way about.

🜂 Take nomenclature, the subject of the present chapter. The ideas discussed here will allow you to speak a whole new language and to unambiguously communicate ideas about the stuff of the world. All the materials you take for granted, metals, glass, plastic, bricks and mortar depend on our ability to recognize and name their raw ingredients and to describe the processes used in their production. So you shouldn't need a song about gnomes in glaciers to hold your interest in nomenclature. The fact of the matter is that if you're going to make any kind of sense of the rest of the book, you're going to have to get used to certain phrases: somethingonium biglongnameate, toxiconium poisonide, and such. Think of it as the language of chemistry. And there's a complex grammar of how these words come apart and go together. So if you don't want to look like a slack-jawed yokel later on, you'd better study some vocabulary and grammar now.

7.2 🜍

You might expect that by this time in the history of modern civilization science would have avoided some of the confusion inherent in human languages, but as with any other language, words have entered science at different times from different places in different contexts. You'll find that there are sometimes different names for the same chemical, and sometimes different chemicals are called by the same name. It can get really complicated for **organic** compounds, those that contain carbon, but for **inorganics** at least, there is a standard naming convention that's relatively easy to understand. Put another way, a standard naming convention, or *nomenclature*, enhances the fidelity of the chemistry meme-plex. If you

Table 7-1. Common Cations and Anions

Cations		Anions	
Name	**Formula**	**Name**	**Formula**
Hydrogen	H^+	Hydroxide	OH^-
Sodium	Na^+	Chloride	Cl^-
Potassium	K^+	Nitrate	NO_3^-
Ammonium	NH_4^+	Acetate	CH_3COO^-
Silver	Ag^+	Bicarbonate	HCO_3^-
Calcium	Ca^{2+}	Sulfide	S^{2-}
Iron(II)	Fe^{2+}	Oxide	O^{2-}
Copper	Cu^{2+}	Sulfate	SO_4^{2-}
Lead	Pb^{2+}	Carbonate	CO_3^{2-}
Iron(III)	Fe^{3+}		
Aluminum	Al^{3+}		

take the time to study this nomenclature, you'll have a much easier time reading the rest of the book.

The first thing to learn about inorganic nomenclature is that each compound has a first name and a last name. You probably already know the chemical name for common table salt, sodium chloride, NaCl. Silicon dioxide, SiO_2, is the standard name for silica, the compound which makes up the mineral, quartz. *Salt* and *silica* are the old names, given before anyone knew what they were made of. The standard name, though, gives you a lot of information about the compound, which is why they standardized names in the first place.

The first name, under this system, denotes the ***cation***, pronounced "cat-I-on," which carries a positive charge. As a mnemonic device, you can imagine that the "t" is a "+": *ca+ion.*The second name, the ***anion***, "an-I-on," carries a negative charge. The key to understanding inorganic formulas is the old adage, "opposites attract." In any given formula, the number of positive charges will equal, or balance, the number of negative charges.

Take sodium carbonate, for example, which is composed of sodium cations and carbonate anions. The superscript in each ion represents the charge, or oxidation state of the ion. Since a sodium ion has a charge of +1 and a carbonate ion has a charge of -2, it takes 2 sodium ions to balance the charge of a carbonate ion. We include this information as the subscript in the formula: Na_2CO_3, 2 Na^+ for each $CO_3{}^{2-}$. What about the subscript in $CO_3{}^{2-}$? Well, that means there are 3 oxygen atoms for each carbon atom in the ion. But for now, you should think of the carbonate ion as a whole rather than as parts. In the simplest reactions, the carbonate ion is never broken up, so it makes sense to treat it as a single entity rather than the sum of its parts.

But if you don't recognize carbonate ion, you won't be able to take advantage of this great simplification and you will be hopelessly confused. Consider the compound NH_4NO_3. This is a simple ionic compound. Every chemist knows that it has two parts, a cation and an anion. Every chemist recognizes the two parts. But to a beginner there seem to be four parts. To a beginner the compound seems more complicated than it has to be because *she* doesn't recognize the parts. To make any sense of it, you need to know what the parts are.

So there are twenty names for you to learn. I have promised to teach you as little as possible. I tried to make it nineteen names, but I just couldn't do it. We need all twenty of them, and if you learn them now, you'll have an easier time later. You need to learn the charges as part of each formula, and associate each formula with its name. Eleven of them are for cations and nine of them are for anions. You can match any cation with any anion and get a valid chemical compound, so with this short list you can recognize the names of ninety-nine compounds. That's quite a lot for the little I've taught, but you know a lot of little will do.

The **formula** for a compound has two parts, the cationic part and the anionic part. Sodium chloride is NaCl, for example. The first letter of an element is always capitalized, so you can distinguish Co, the single element, cobalt, from CO, the compound, carbon monoxide. Turning compound names to formulas is easy once you know Table 7-1; just write the formula for the cation (positive ion, first name) followed by the formula for the anion (negative ion, second name). Hang subscripts on each one to make the charges balance. For example, let's write the formula of aluminum oxide. From the table, the aluminum ion has a charge of +3 and the oxide ion has a charge of -2. We want the positive charges to balance the negative ones. Well, $2\times(+3) = +6$ and $3\times(-2) = -6$ so the formula of aluminum oxide is Al_2O_3.

As a second example, try lead nitrate. The lead ion has a charge of +2 and the nitrate ion has a charge of -1. Since $1\times(+2)$ balances $2\times(-1)$, the formula of lead nitrate is $Pb(NO_3)_2$. The formula for nitrate goes in parentheses to show that it is a single anion made up of four atoms. A subscript after a parenthesis applies to everything inside. So lead nitrate contains 1 lead atom, 2 nitrogen atoms, and 6 oxygen atoms. Once the charges are balanced, we no longer write them in the formula because they have exactly canceled out. But the charges show up again when these compounds dissolve in water.

When the compounds we have been discussing dissolve in water, they behave very peculiarly; they ***ionize,*** or fall apart into their respective ions. For sodium chloride, we write:

$NaCl(aq) = Na^+(aq) + Cl^-(aq)$

The sodium and chloride ions, which were next to each other in solid sodium chloride, separate from one another in solution and float around on their own. Compounds that behave like this, falling apart into ions in aqueous solution, are called **electrolytes,** or ***salts.*** The *(aq)* in the formula stands for *aqueous,* that is, a solution in water. The water itself does not appear in the equation; it is simply the medium in which the ionization takes place.

Suppose for a moment that we dissolve two different electrolytes in water, say, sodium chloride and potassium nitrate. The sodium chloride would fall apart into sodium and chloride ions; the potassium nitrate would fall apart into potassium and nitrate ions. So we are left with a solution containing sodium, potassium, chloride, and nitrate ions. But exactly the same ions would have formed had we dissolved sodium nitrate and potassium chloride in water. Once in solution, a potassium ion, for example, does not "remember" whether it came from potassium chloride or potassium nitrate. It's just a potassium ion floating around. As long as the electrolytes remain in solution, no reaction takes place; we just have an ion soup.

But suppose now that of all the ways of mixing and matching the first and last, the cationic and anionic names, one of them is insoluble in water. If we dissolve sodium chloride and silver nitrate, we get a soup of sodium, silver, chloride, and nitrate ions. When a dissolved sodium ion happens to bump into a dissolved chloride ion, they don't stick because, after all, sodium chloride is soluble in water. When a silver ion happens to bump into a nitrate ion, they don't stick because silver nitrate is soluble in water. When a sodium ion happens to bump into a nitrate ion, they don't stick because sodium nitrate is soluble in water. But when a silver ion happens to bump into a chloride ion, they stick together to form insoluble silver chloride. Solid silver chloride rains down, or ***precipitates,*** accumulating as a layer of insoluble powder at the bottom of the solution. In other words, a reaction has occurred.

This type of reaction, which accounts for about half the reactions discussed in this book, is called a **metathesis**. Just as we represented a compound with a formula, we represent a reaction with an **equation:**

$NaCl(aq) + AgNO_3(aq) = NaNO_3(aq) + AgCl(s)$

In Chapter 3 we used the equation as a mathematical sentence, with the equal sign as the verb. We extend that notion here. A chemical equation is used as a shorthand for a reaction, the transformation of one set of elements or compounds into another. In such a reaction, the number of each kind of atom does not change. The Law of **Conservation of Mass** requires that the number of each kind of atom be conserved. So in an equation, we require that there be the same number of each kind of atom on either side of the equal sign. Sometimes an arrow or other symbol is used in place of the equal sign. The arrow has the advantage of implying the direction in which the reaction proceeds, but in the spirit of "teaching you as little as possible," I will use the equal sign.

Consider a slightly more complicated reaction:

$CaCl_2(aq) + Na_2CO_3(aq) = 2\ NaCl(aq) + CaCO_3(s)$

In the beginning, there were calcium ions, chloride ions, sodium ions, and carbonate ions, all floating around in the solution. But when a calcium ion finds a carbonate ion, they stick together forming an insoluble precipitate, calcium carbonate. The sodium ions and chloride ions stay in solution. Why is it 2 NaCl and not Na_2Cl_2? Well, the ions from the tables are Na^+ and Cl^-, so we know the compound must be NaCl. The 2 is placed out front to *balance* the equation, that is, to make sure that the number of each kind of atom is the same on both sides. If there is no number out in front of a formula, it is assumed to be 1. We call the number out front the **stoichiometric coefficient,** which, though a mouthful, is shorter than saying "the little number in front of each formula in a

Table 7-2. Aqueous Solubility of Inorganic Compounds

Generally Soluble	
Anion	**Exceptions**
Nitrates	**None**
Acetates	**None**
Sulfates	**Silver, Lead, Calcium**
Chlorides	**Silver, Lead**

Not Generally Soluble	
Anion	**Exceptions**
Hydroxides	**Sodium, Potassium, Ammonium, Calcium**
Oxides	**Sodium, Potassium, Ammonium**
Sulfide	**Sodium, Potassium, Ammonium**
Carbonates	**Sodium, Potassium, Ammonium**

balanced chemical equation." The unit of the stoichiometric coefficient is the ***mole.*** 1 mole calcium chloride + 1 mole sodium carbonate yields 2 moles sodium chloride + 1 mole calcium carbonate. Don't worry, the more you use words like *stoichiometric coefficient* and *mole,* the more you'll grow to understand them.

What about the reaction:

$2\ NaCl(aq) + CaCO_3(s) = CaCl_2(aq) + Na_2CO_3(aq)$

Isn't this also a balanced reaction? For reactions in aqueous solution, the reactants must go into solution and at least one product must come out of solution. In this case, the calcium carbonate is not soluble, so it just sits in a lump at the bottom of the container. How do you know which compounds are soluble? You guessed it, another table, Table 7-2

I'm sure that dozens of people will benefit from the little ditty presented in the sidebar *Solubility in a Nutshell.* Whether you memorize the verse or the table, you will be able to predict in some detail an enormous number (over 9,000) of possible chemical reactions. Let's look at some examples.

Solubility in a Nutshell

Nitrates and acetates melt into water.
Sulfates and chlorides dissolve as they oughtta.
Sulfates and chlorides of silver and lead:
These are exceptions that stick in your head.

Calcium sulfate and calcium oxide,
Fall to the bottom just like the hydroxide.
Maybe a little bit starts to dissolve;
Sparingly soluble salts they are called.

Po-tas-si-um,
Am-mon-i-um,
And So-di-um too,
Make soluble compounds with everything;
The rest don't dissolve; we're through.

Q: Will sodium chloride and potassium sulfate react in aqueous solution? If so what is the balanced reaction equation? If not, why not?

A: Both sodium chloride and potassium sulfate are soluble in water. If I swap the names I get potassium chloride and sodium sulfate and these are both soluble. So if I pour solutions of sodium chloride and potassium sulfate solutions together, nothing happens. All four ions stay in solution and there is no reaction.

Q: Will sodium sulfate and silver nitrate react in aqueous solution? If so what is the balanced reaction equation? If not, why not?

A: Both sodium sulfate and silver nitrate are soluble in water. If I swap the names I get silver sulfate and sodium nitrate. Silver sulfate is insoluble and sodium nitrate is soluble. So if I mix a solution of sodium sulfate and a solution of silver nitrate, an insoluble precipitate of silver sulfate will form and sodium nitrate will remain in solution. Two go into solution, and one comes out. The balanced equation is:
$Na_2SO_4(aq) + 2\ AgNO_3(aq) = Ag_2SO_4(s) + 2\ NaNO_3(aq)$

Q: Will lead sulfide and calcium carbonate react in aqueous solution? If so what is the balanced reaction equation? If not, why not?

A: Both lead sulfide and calcium carbonate are insoluble in water. If I put them in water, they both just fall to the bottom like sand or stones. Since they don't dissolve in water, there is no reaction.

Q: Will copper sulfate and calcium hydroxide react in aqueous solution? If so what is the balanced reaction equation? If not, why not?

A: Both copper sulfate and calcium hydroxide are soluble in water. Both copper hydroxide and calcium sulfate are insoluble. Two go in, two come out.

$$CuSO_4(aq) + Ca(OH)_2(aq) = Cu(OH)_2(s) + CaSO_4(s)$$

Notice that I put the OH in parentheses to make it clear that the subscript, 2, applies to the whole hydroxide ion, not just to the hydrogen.

Q: Will hydrogen sulfate and sodium hydroxide react in aqueous solution? If so what is the balanced reaction equation? If not, why not?

A: This is a bit of a special case. When hydrogen ions bump into hydroxide ions in solution, they stick together and form water. This is a very energetic reaction, the basis of what we will later call ***acid-base reactions.***

$$H_2SO_4(aq) + 2\ NaOH(aq) = Na_2SO_4(aq) + 2\ H_2O(l)$$

We use *(l)* for water instead of *(aq)* because *(aq)* would imply that water is dissolved in water; *(aq)* is always attached to a **solute,** the thing that is dissolved, not to the **solvent,** the thing in which it is dissolved.

Material Safety

Acute toxicity, the kind that makes you sick immediately, is often spoken of in terms of a material's LD_{50}. This is the dose which would be expected, on average, to be lethal for 50% of test animals. Oh, it sounds morbid, I know. And it is. It isn't at all pleasant to contemplate cute little bunnies and mousies and doggies being force-fed chemicals to see how many of them croak. And since animals differ in their tolerance to different chemicals, it isn't an extremely reliable way to judge human toxicity. But it turns out to be very difficult to recruit *human* subjects for these tests, given that half of them are going to snuff it. So we use what we can get.

The LD_{50} is usually expressed in grams or milligrams of chemical per kilogram of body weight. If this looks like a unit factor to you, you haven't completely slept through the lessons of Chapter 3. You can use UFA to determine what would be a toxic dose for 50% of animals your size. But to judge relative toxicity, we can use the raw LD_{50} directly, recognizing that a big number means the chemical is relatively safe (it takes a lot to kill you) while a small number means it is hazardous (a little bit will do you).

Sometimes LD_{50} values are given in the MSDS. A reliable way to search for information online is to search for the CAS number, the keyword "MSDS," and the keyword "LD50." LD_{50}'s always refer to the animal which was tested (rat, mouse, dog, etc.) and the route by which it was introduced (oral, inhalation, etc.). An LD_{50} that is missing such details has been copied from an anonymous source at best, and may be completely fabricated at worst. Particularly when looking online, it's important to consider the source of your information.

Though this project involves no materials, you're not off the hook for material safety. To get your bearings, look up oral, rat LD_{50}'s for caffeine (CAS 58-08-2), sodium chloride (CAS 7647-14-5), sucrose (CAS 57-50-1), and sodium cyanide (CAS 143-33-9). Using these values, arrange these substances in order of toxicity.

Research and Development

So there you are, studying for a test, and you wonder what will be on it.

- Know the meanings of all of the words that are important enough to be included in the ***index*** or **glossary.**
- Know the Research and Development items from Chapter 3 and Chapter 4.
- Memorize the ions in Table 7-1.
- Given the name of an inorganic compound, be able to write its formula. Given the formula of an inorganic compound, be able to write its name.
- Memorize the solubility rules of Table 7-2.
- Be able to write a metathesis reaction between any two binary compounds and to predict whether or not the reaction will go.
- Know how to interpret LD_{50}'s.

7.3 Θ

The idea of a metathesis reaction may be very abstract to you at this point, but we'll see real live metathesis reactions in the next couple of chapters. If you're bored because this is so easy, you can move on. If you're still bewildered, here are a few more problems to work on. You should also be able to make up your own problems at this point; just pick any two first names and any two last names from Table 7-1, matching each first name to a last name to form a compound. For example:

Q: Will ammonium sulfate and calcium hydroxide react in aqueous solution? If so what is the balanced reaction equation? If not, why not?

A: $(NH_4)_2SO_4(aq) + Ca(OH)_2(aq) = 2\ NH_4OH(aq) + CaSO_4(s)$

Q: Will silver chloride and calcium carbonate react in aqueous solution? If so what is the balanced reaction equation? If not, why not?

A: No. Calcium carbonate is not soluble in water.

Q: Will silver nitrate and calcium hydroxide react in aqueous solution? If so what is the balanced reaction equation? If not, why not?

A: $2\ AgNO_3(aq) + Ca(OH)_2(aq) = 2\ AgOH(s)) + Ca(NO_3)_2(aq)$

Q: Will lead acetate and sodium sulfide react in aqueous solution? If so what is the balanced reaction equation? If not, why not?

A: $Pb(CH_3COO)_2(aq) + Na_2S(aq) = PbS(s) + 2\ NaCH_3COO(aq)$

Q: Will iron sulfate and copper chloride react in aqueous solution? If so what is the balanced reaction equation? If not, why not?

A: No. Both reactants and both products are soluble in water.

Quality Assurance

When you can solve these problems without consulting the tables, you should be able to predict the outcome of about half the reactions in this book. Your notebook for this project may omit the "Observations" section, as no manipulation of materials is involved in this project. Include any examples you worked in your notebook along with a flawless metathesis quiz. Also include your analysis of the relative toxicities of caffeine, sodium chloride, sucrose, and sodium cyanide.

Chapter 8. Job (Alkali)

> Again there was a day when the sons of God came to present themselves before the Lord, and Satan came also among them to present himself to the Lord. And the Lord said unto Satan, From whence comest thou? And Satan answered the Lord, and said, From going to and fro in the earth, and walking up and down in it.
>
> And the Lord said unto Satan, Hast thou considered my servant Job, that there is none like him in the earth, a perfect and an upright man, one that feareth God and escheweth evil? and still he holdeth fast his integrity, although, thou movedst me against him, to destroy him without cause.
>
> And Satan answered the Lord, and said, Skin for skin, yea, all that a man hath he will give for his life. But put forth thine hand now, and touch his bone and his flesh, and he will curse thee to thy face
>
> And the Lord said unto Satan, Behold, he is in thine hand; but save his life.
>
> So went Satan forth from the presence of the Lord, and smote Job with sore boils from the sole of his foot unto his crown.
>
> And he took him a potsherd to scrape himself withal; and he sat down among the ashes.
>
> — *Job 2:1-8* [1]

8.1 ☿

∇ You are probably wondering why I am sitting here in the fireplace, covered in ashes, and scraping myself with a piece of broken crockery. I will tell you. First of all, you must know that Satan, having nothing better to do, convinced God to let him take away my oxen, my sheep, and my camels. Not satisfied, he also insisted on knocking down my oldest son's house with all my sons and daughters inside. And now he has given me a rash that itches like the dickens. You must also know that my wife, who urged me to curse God and die, has not so much as a zit. Anyway, I am scratching myself with a broken pot because the itching is unbearable. And to avoid infecting my open, running sores, I cover myself in ashes. These ashes, of course, are rich in potassium carbonate, which hydrolyze the cell walls of any bacteria which may happen to drop by. But I am getting ahead of myself.

---

1. Reference [14].

∇ I will not, of course, know about either potassium carbonate or bacteria for another six thousand years. But my mother, who was something of a clean freak, taught me all about potash, which comes, of course, from soaking ashes in a pot. "Potash," she would say, "can wash the stink off a Chaldean, the dirt off a Sabean, and the smirk off any son of mine." A strict woman, my mother, but she knew her cleaning supplies. Actually potash was her only cleaning supply. She used it for washing clothes and dishes, for scrubbing furniture and children. And since disease comes from being unclean, she would make us cover ourselves in ashes at the first sign of trouble. It was, for her, Chlorox, Comet, and Bactine all rolled into one.

8.2 🜍

If you would like to understand potash, you must realize that Lucifer has oversimplified fire considerably. Reconsidering Equation 1-1 (page 9), you will notice that all of the products of combustion are gases. Where, then, do the ashes come from? You will recall that these equations are for the combustion of cellulose, and that wood is only *mostly* cellulose. When wood is heated anaerobically, it turns black as the water is driven off, leaving charcoal, or carbon, behind. When charcoal burns in air, the carbon combines with oxygen, producing the gas, carbon dioxide. But if you have ever used a charcoal grill, you may have noticed that charcoal turns white as it burns. This white ash is what remains of the non-flammable minerals which were present in the wood to begin with. You don't really notice them until the carbon has burned away. These ashes have a composition which varies according to the kind of wood and the soil in which it grew, and it is this variable composition which marks ash as a mixture rather than a pure substance.

You will recall, no doubt, that a mixture can be separated into two or more pure substances by recrystallization, distillation, and chromatography. You will be pleased to learn that we are discussing only **recrystallization** in this chapter. You have, of course, noticed that some things, like salt and sugar, are soluble in water, while others, like sand and charcoal, are not. Recrystallization separates substances which differ in their solubility. Ash, for example, is mostly insoluble in water. Only a small portion of the ash dissolves in water, and this is the substance we call ***potash,*** or potassium carbonate. To make potash, you must add your ash

Table 8-1. Combustion Products of Beech Wood

Substance	Pounds
Beech Wood	1000.0
Flammable Compounds	994.2
Ash	5.8
Insoluble Ash	4.6
Crude Potash	1.2
Potassium Carbonate	0.9
Sodium Carbonate	0.2
Potassium Sulfate	0.1

to a quantity of water. Any leftover charcoal will float to the top, while the insoluble minerals will sink to the bottom. The good stuff, the potash, will be dissolved in the water. You must separate the water from the charcoal above and the minerals below. Once you have done this, you will have what looks like clear, clean water. But if you boil the water away, or let it evaporate in the Sun, a white, crystalline residue will remain. This residue is potash.

Now, it is important, if you are to be successful, that your ashes have never been wet. If they have been wet before you started, then, of course, the potash will already have been washed out of them. So you must get your ashes from a fire that has been allowed to burn out, not from one which has been doused with water. But if your ashes were dry, and if you were careful to skim off the charcoal, and if you allowed the minerals to settle completely, and if you were able to collect the water without stirring up the sediment, and if, finally, you boiled away all the water, you will have nice, pure, white, crystalline potash, which is a lovely thing to behold.

This potash will look just like salt or sugar, so how will you know that it is not just salt or sugar? You will give it a taste. If your mother was as strict as mine, the taste will be reminiscent of a day when she caught you saying words you were not supposed to know yet. This is the bitter taste of **alkali,** or **base.** It would be irresponsible of me, of course, to suggest that you should go around tasting everything. Chemists have learned the hard way that tasting unknowns can get you into a world of hurt and so

Table 8-2. Solubility of Alkali Sulfates and Carbonates

Compound	Solubility (g/100 mL)
Sodium Sulfate	11
Potassium Sulfate	12
Sodium Carbonate	22
Potassium Carbonate	147

they have developed pH test paper to serve as a *virtual tongue.* Bitter things turn pH test paper blue and sour things turn it red. Salty and sweet things leave pH test paper a neutral yellow color. If you have never used pH test paper before, use a few strips to test materials whose flavors you already know. Good choices are lemon juice, vinegar, baking soda, and soap. From this experience you will be able to use pH test paper to distinguish bitter things from sour things, alkalis from acids, without risking your health.

Before we get too much farther, I should tell you that potash, or potassium carbonate, is not the only soluble component of wood ash. Depending on the soil conditions, sodium carbonate may also be present. As a matter of fact, when the ashes come from burning seaweed, there may be more sodium carbonate than potassium carbonate, and in this case we refer to the product as ***soda ash.*** Table 8-1[2] shows what happens to 1000 pounds of Beech wood when it is burned. Most of it is consumed in the fire, of course, producing gaseous water and carbon dioxide. Less than six pounds of ash remain. Most of this ash is not soluble. When the water is boiled from the soluble bit, a little over a pound of crude potash remains. As I have explained, most of this crude potash is potassium carbonate, but some of it will consist of sodium carbonate, potassium sulfate, and other soluble compounds. You may be wondering how you could remove these contaminants. I am happy you asked.

Adam, I must tell you, has considerably simplified the whole business of solubility. Solubility is not a black-and-white issue; some "soluble" compounds are more soluble than others. Table 8-2 shows that potassium carbonate has a much higher solubility than the other compounds we might expect to be present in wood ashes. If, instead of boiling away all the water, we were to boil away only *most* of the water, the less soluble

2. Data from Reference [29], p. 123.

Figure 8-1. Recrystallization as a Process

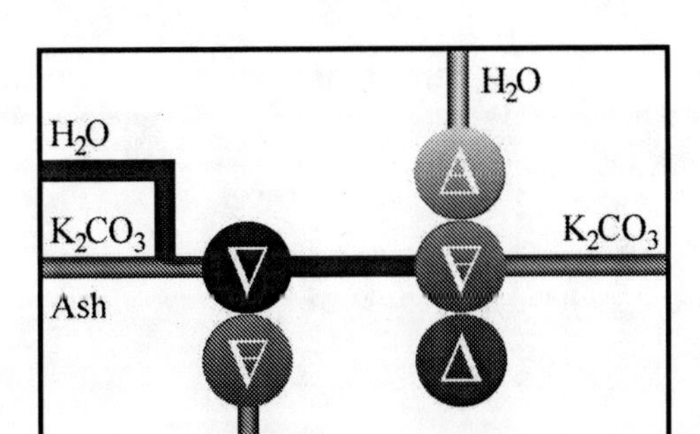

compounds would precipitate, that is, they would sink to the bottom of the solution as solids, and the potassium carbonate would stay in solution until the last possible moment. If we were to pour off this solution and boil it to dryness, the resulting solid would have fewer contaminants than the crude potash.

Well, really, we have done the same thing to remove the sodium carbonate and potassium sulfate that we did to remove the insoluble ash. In both cases we are physically separating compounds that differ in their solubility. This process, known as recrystallization, remains the most widely-used technique for purifying solids.

Figure 8-1 illustrates the recrystallization process in schematic form. The first reactor, the **lixiviator,** is a container in which part of a solid is allowed to dissolve in water. In the next section, we will use our familiar 2-liter soft-drink bottle to lixiviate wood ashes. The second reactor, the furnace, should be familiar from Figure 1-3 (page 9). A beaker in an oven will serve well for this. Unlike previous processes, there is no chemical reaction here. It is simply a physical process for separating things that differ in solubility. The usual conventions are followed; reactants come in from the left, waste products exit to the top and bottom, and the main product exits to the right.

You will be quite interested to know that Nature does some recrystallizing of her own. When a sea becomes land-locked, soluble minerals wash into it from the rivers and streams that empty into it. Eventually, the Sun dries up the water, and the least-soluble component precipitates, forming a bed of, say, salt. If the climate is more arid, soda ash may begin to pre-

Figure 8-2. The pH Scale

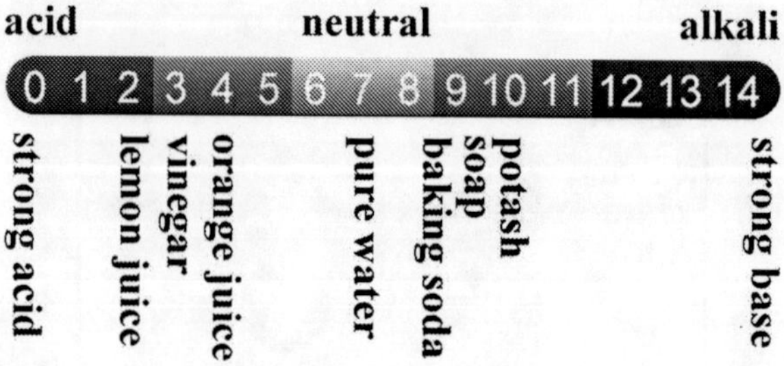

acid **neutral** **alkali**

0 1 2 3 4 5 6 7 8 9 10 11 12 13 14

strong acid
lemon juice
vinegar
orange juice
pure water
baking soda
soap
potash
strong base

cipitate, and if the sea dries up completely, a layer of potash may form on the top, providing that there was any potash in the river water to begin with. So an ancient sea-bed consists of beds of material which differ in their solubility. This makes it quite convenient for mining, since it saves you the trouble of quite a lot of recrystallization.

Now I am quite aware that some readers will be nodding off at this point. If you think this is bad, you've obviously never had Satan poking his big fat nose into your ashes. But I am afraid that I must tell you a bit more about potash if you are to avoid confusion later on. Adam has told you that when inorganic compounds dissolve in water, they fall apart into cations and anions. What he did not tell you is that water itself does the same thing, as shown in Equation 8-1(a). A water molecule may fall apart into a hydrogen cation and a hydroxide anion. Only a tiny portion of the water falls apart in this way, but, it turns out, this tiny portion is extremely important. In pure water, of course, there are an equal number of hydrogen and hydroxide ions, since each water molecule gives one of each. For this reason, we say that pure water is neutral and assign it a pH of 7. Please notice that **pH** is spelled with a small *p* and a big *H*.

Now, potassium carbonate falls apart into two potassium cations and a carbonate anion. The potassium ions float happily about the solution and take no further part in the chemistry for the moment. But if carbonate ion bumps into a water molecule, it may swipe a hydrogen ion from it, leaving a hydroxide ion behind. Alternatively, it may bump into a hydrogen ion, which may stick to it. In the first case, the number of hydroxide ions has increased. In the second, the number of hydrogen ions has decreased. In either case, there are now more hydroxide ions floating around than hydrogen ions and the solution is no longer neutral.

Equation 8-1. Reactions of Potassium Carbonate with Water

$$(a)\ H_2O(l) = H^+(aq) + OH^-(aq)$$
$$(b)\ K_2CO_3(aq) = 2K^+(aq) + CO_3^{2-}(aq)$$

$$(c)\ H_2O(l) + CO_3^{2-}(aq) = HCO_3^-(aq) + OH^-(aq)$$
$$(d)\ H^+(aq) + CO_3^{2-}(aq) = HCO_3^-(aq)$$

We say that the solution is **alkaline,** or ***basic,*** and it gets a pH bigger than 7. The more basic a solution is, the bigger the number. A 10% potash solution, for example, has a pH of 10; a strongly alkaline solution can go as high as 14.

The anion HCO_3^- is called the bicarbonate[3] ion. You may know that baking soda is sodium bicarbonate, and now you are able to write its formula.

Material Safety

Locate MSDS's for potassium carbonate (CAS 584-08-7) and sodium carbonate (CAS 497-19-8). Summarize the hazardous properties of these materials in your notebook, including the identity of the company which produced each MSDS and the potential health effects for eye contact, skin contact, inhalation, and ingestion. Also include the LD_{50} (oral, rat) for each of these materials.[4]

Your most likely exposure is eye or skin contact. If you get some potash in your eyes, you should flush them with cold water and go to the emergency room. If you get some on your skin, wash it off when it is convenient to do so.

You should wear safety glasses while working on this project. Leftover ash can be disposed of in the trash. Your potash can be saved for use in future projects or washed down the drain.

---

3. The modern name for the bicarbonate ion is the *hydrogen carbonate* ion. The older name, however, continues to be widely used.
4. The LD_{50} was introduced in Section 7.2 (page 96).

∇ All of this was to tell you why I happened to be sitting in the ashes when we met. All living things are rather picky about their pH, most preferring something close to pH 7. If you should get some potash in your eyes, you will discover this for yourself in a rather painful way. But most of your body is well-protected by a layer of skin. Bacteria, on the other hand, have no such protection. So applying potash to a cut or boil sends the germs right around the twist. I can only wish that it had the same effect on the fellow who gave me the boils in the first place.

Research and Development

You are probably wondering what you need to know for the quiz. I will tell you.

- You should know the meanings of all of the words important enough to be included in the ***index*** or **glossary.**
- You should have studied the Research and Development items from Chapter 1 and Chapter 4.
- Know several synonyms for potash and soda ash, along with their formulas.
- Know the hazardous properties of the alkali carbonates.
- Know how recrystallization can be used to separate compounds that differ in their solubility.
- Know the equations for the ionization of water and the reactions of carbonate ion with water and hydrogen ion.
- Know how to test for alkali and be familiar with the pH scale.
- Know why potash is antiseptic.

8.3 Θ

You may be wondering where you can get a bit of ash. Anyplace, really, as long as it is fresh. Get it from a wood-stove, a fireplace, a campfire, even a barbecue grill. It is imperative, you understand, that it must never have been wet or your time will be wasted. Any container will do for leaching the soluble components. I would suggest that ubiquitous container, the twenty-first century equivalent of the gourd, the 2-liter soft-drink bottle. Fill it half-way with ash and add water up to the shoulder, as you did when making mead, and then put the top on the bottle. Shake your ash enthusiastically. It will probably look quite nasty.

Figure 8-3. Dissolution

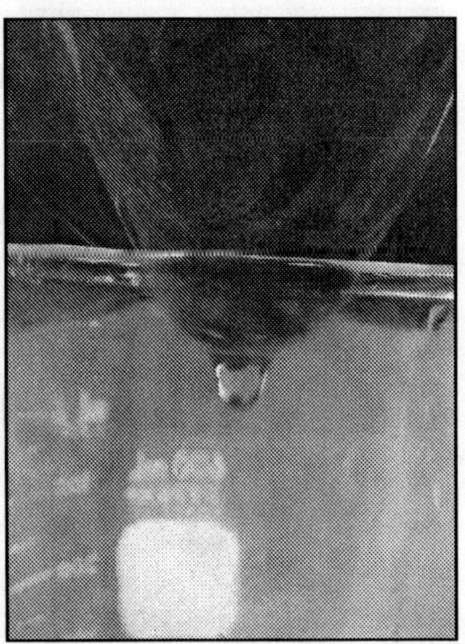

Now, you will need a second container and it must be fireproof. You may use a Pyrex measuring cup, a beaker or an iron pot. If you are Athanor, you may even make your own pot. After your ashes have settled, the insoluble minerals will sink to the bottom and the charcoal will float to the top. The good stuff, remember, is in the water. We will use the insoluble ashes to filter the good stuff. Use a pin or a thumb tack to poke a little hole in the bottom of your bottle and place it into your pot. The smaller the hole, the better (but slower) the filtration will be. Remove the top from the bottle and the water will be filtered through the sediment as it leaks from the hole. The water collecting in the pot will appear quite clean and clear. You would think there is nothing in there, but you would be wrong.

Under no circumstances should you boil potash in an aluminum pot. Hot alkali reacts violently with aluminum. It will eat a hole in your pot and your ash will be all over the place.

All that remains is to remove the water from the pot. You may heat it on a stove or hot-plate or place it into a hot oven. Take care that the hot solution does not splatter or you will have an alkaline mess. When the water boils away, crystals of potash will remain. If you were a stickler

Figure 8-4. Coagulation

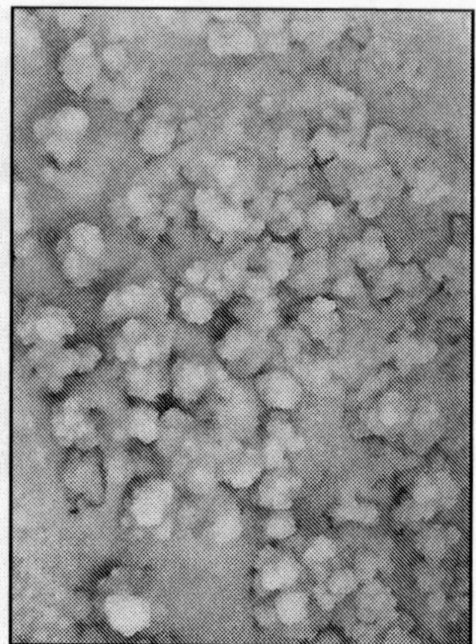

for purity, you could dissolve your crude potash in water, boil the water away until a precipitate began to form, decant the remaining solution from this precipitate, allow the hot solution to cool and coagulate, and calcine the resulting solid to burn off any remaining charcoal, that is, you could do a second recrystallization. Of course, I am not such a stickler, so I am quite satisfied with my lovely, bitter, crude potash.

Quality Assurance

If you have followed my suggestions, your ash will be quite pale. It will be soluble in water, have the bitter taste of alkali and/or will turn pH test paper blue. You can tape the test paper into your notebook.

Chapter 9. Vulcan (Metals)

And Vulcan answered, "Take heart, and be no more disquieted about this matter; would that I could hide him from death's sight when his hour is come, so surely as I can find him armour that shall amaze the eyes of all who behold it."

When he had so said he left her and went to his bellows, turning them towards the fire and bidding them do their office. Twenty bellows blew upon the melting-pots, and they blew blasts of every kind, some fierce to help him when he had need of them, and others less strong as Vulcan willed it in the course of his work. He threw tough copper into the fire, and tin, with silver and gold; he set his great anvil on its block, and with one hand grasped his mighty hammer while he took the tongs in the other.

First he shaped the shield so great and strong, adorning it all over and binding it round with a gleaming circuit in three layers; and the baldric was made of silver. He made the shield in five thicknesses, and with many a wonder did his cunning hand enrich it.

He wrought the earth, the heavens, and the sea; the moon also at her full and the untiring sun, with all the signs that glorify the face of heaven—the Pleiads, the Hyads, huge Orion, and the Bear, which men also call the Wain and which turns round ever in one place, facing Orion, and alone never dips into the stream of Oceanus.

— Homer, *The Iliad, Book XVIII*, *ca.* 1100 BC [1]

9.1 ☿

△ Nothing in the history of humankind, save for the cultivation of noble fire itself, can compare to the discovery of metal. It is a most wondrous substance, born of the earth, purified by the air, nurtured in the fire it flows like water. Poured in dazzling brilliance, it takes on the shape of its container. Struck, it yields to the hammer, becoming wand, sword, chalice or shield. Drawn, it becomes as a thread. Polished, it contains the whole world in its reflection. Honed, its edge cuts lesser materials without violence. There are few objects which would not be more durable, more beautiful, and more useful if fashioned by art and ingenuity from metal.

1. Reference [15].

Δ The discovery of metal is as ancient as any other, for it can be found from time to time scattered among the stones and pebbles. Native metals, pure and uncontaminated, separate naturally from molten rock just as quartz does. Freed by rain and flood from their subterranean nurseries, copper, silver and gold wander from mountain to gully to river to fluvial plain. Catching the Sun, they attract the eye of hunter, farmer and shepherd, becoming trinket or talisman.

Δ Only Athanor knows how to transform these trinkets into more deliberate forms. Submitted to the intense heat of the pottery kiln, the native metals become fluid and may be cast into shapes familiar to the potter but made more precious by the scarcity and novelty of the material. Such was the world of my youth, some 8,000 years ago. I began life as Athanor, my business being to heat the bejeezus out of things. Clay and limestone and gypsum were my stock in trade, punctuated by the occasional acquisition of native metals for casting into ritual objects. The world remained unchanged for two millennia, and might have remained so for another six had I not made two crucial observations.

Δ The first thing that I noticed was that copper jewelry changes colors as it ages; bright copper metal acquires a red patina which eventually becomes a blue-green scale. It is as if the metal putrefies, as meat and bread do. The second observation was that the blue-green mineral *malachite*, used as a pigment for paint, becomes red when fired in a reducing kiln, that is, one deprived of air. This might have been chalked up to coincidence, except that the blue-green of malachite is *exactly* the same as that of decrepit jewelry and the red of fired malachite is *identical* to that of the red patina.

Δ It was my habit to throw rock in with the metal when preparing to cast it. The molten rock acts as a ***flux,*** that is, a material which, in melting, assists in the melting of other bodies. One day I decided to use common malachite as a flux for the melting of copper, thinking that there might be some sympathy between these two materials. And it appeared to me that the malachite nourished the copper, fattening it like a calf. In time I found that the smallest seed of copper could grow to an enormous size given this fodder. Eventually, I dispensed with the seed altogether, finding that simply melting malachite in a reducing kiln, that is, ***smelting*** it, was sufficient for producing copper ingots to dwarf any native nugget. That was the day I became Vulcan. I have been glorified by every culture of any consequence, and rightly so.

Table 9-1. Metals and Their Ores

Metal	Mineral	Formula
Gold	**Native Gold**	Au
Silver	**Argentite (in Galena)**	Ag_2S
Copper	**Malachite**	$Cu_2CO_3(OH)_2$
	Azurite	$Cu_3(CO_3)_2(OH)_2$
	Chalcopyrite	$CuFeS_2$
Mercury	**Cinnabar**	HgS
Iron	**Hematite**	Fe_2O_3
	Magnetite	Fe_3O_4
	Pyrite	FeS_2
Tin	**Cassiterite**	SnO_2
Lead	**Galena**	PbS

9.2 ♀

Seven metals were in use before the invention of writing; gold, found only as native metal; silver and copper, found as native metals but, more commonly, as carbonate and sulfide minerals, or ***ores;*** mercury sleeps in pools of liquid metal and as oxide and sulfide ores; native iron can be found in meteorites, but far more commonly as oxide and sulfide ores; tin and lead are found only in ores, most commonly tin oxide and lead sulfide.

Copper was the earliest metal to come into common usage because its ores are fairly common and because it is smelted at moderate temperatures. It is a soft metal, however, which makes it of marginal value for tools and weapons. Some ores of copper contain arsenic as a contaminant and copper smelted from these ores contains from 2-6% of residual arsenic. This arsenical bronze, a solution or ***alloy*** of arsenic in copper, is much harder than copper alone, which makes it more useful for tools and weapons. Tin also forms an alloy with copper, ***bronze,*** which is harder than either metal alone. By 3,000 BC bronze had become the dominant metal, so much so that its use defines the Bronze Age. The higher temper-

Equation 9-1. Smelting of Oxide and Carbonate Ores

$(a)\ 2\,Ag_2O(s) + C(s) \stackrel{\Delta}{=} 4\,Ag(s) + CO_2(g)$

$(b)\ 2\,CuO(s) + C(s) \stackrel{\Delta}{=} 2\,Cu(s) + CO_2(g)$

$(c)\ Cu_2CO_3(OH)_2(s) + C(s) \stackrel{\Delta}{=} 2\,Cu(s) + 2\,CO_2(g) + H_2O(g)$

$(d)\ 2\,Cu_3(CO_3)_2(OH)_2(s) + 3\,C(s) \stackrel{\Delta}{=} 6\,Cu(s) + 7\,CO_2(g) + 2\,H_2O(g)$

$(e)\ 2\,HgO(s) + C(s) \stackrel{\Delta}{=} 2\,Hg(s) + CO_2(g)$

$(f)\ 2\,Fe_2O_3(s) + 3\,C(s) \stackrel{\Delta}{=} 4\,Fe(s) + 3\,CO_2(g)$

$(g)\ SnO_2(s) + C(s) \stackrel{\Delta}{=} Sn(s) + CO_2(g)$

$(h)\ 2\,PbO(s) + C(s) \stackrel{\Delta}{=} 2\,Pb(s) + CO_2(g)$

atures required for iron production delayed the advent of the Iron Age to about 1200 BC. Bronze continued to be used into the Iron Age because, unlike iron, it could be cast into molds. Cast iron, a lower melting alloy of iron, was developed in China as early as 500 BC but did not become common in Europe until about 1500 AD.

The fundamental problem of smelting is the **reduction** of metallic compounds to elemental metals. Different compounds require different treatments, which I shall explain in turn, beginning with the oxide ores. Whereas calcination heats the bejeezus out of a material in the presence of air alone, smelting uses a reducing agent to aid in the removal of the bejeezus. Anything that burns may be considered a reducing agent, the most common being charcoal or some material which turns to charcoal in the kiln.

The oxide and carbonate ores are smelted by very similar reactions, as shown in Equation 9-1. Charcoal combines with the oxygen in the ore, escaping as carbon dioxide (the ***bejeezus***), and leaving the molten metal behind. Mercury and tin are smelted at modest temperatures, lead at higher temperatures, copper and silver at still higher temperatures, and iron at the highest temperature. The high temperature required for melting iron is not achieved in a simple kiln. In practice, iron oxide is reduced to solid, rather than molten iron at 1200°C. The non-ferrous minerals in the ore, the *gangue*, melt under the influence of the flux, leaving a bloom of solid iron suspended in the melt. This solid is removed to an

Equation 9-2. Roasting of Sulfide Ores

$$(a)\ 2\,Ag_2S(s) + 3\,O_2(g) \stackrel{\Delta}{=} 2\,Ag_2O(s) + 2\,SO_2(g)$$
$$(b)\ 4\,CuFeS_2(s) + 13\,O_2(g) \stackrel{\Delta}{=} 4\,CuO(s) + 2\,Fe_2O_3(s) + 8\,SO_2(g)$$
$$(c)\ 2\,HgS(s) + 3\,O_2(g) \stackrel{\Delta}{=} 2\,HgO(s) + 2\,SO_2(g)$$
$$(d)\ 4\,FeS_2(s) + 11\,O_2(g) \stackrel{\Delta}{=} 2\,Fe_2O_3(s) + 8\,SO_2(g)$$
$$(e)\ 2\,PbS(s) + 3\,O_2(g) \stackrel{\Delta}{=} 2\,PbO(s) + 2\,SO_2(g)$$

open hearth, or forge, where it is sequentially heated to red heat and then hammered on an anvil to produce wrought iron.

Sulfide ores must be roasted to remove sulfur. In roasting, the ore is heated in an oxidizing atmosphere. Sulfides are oxidized by the oxygen in the air and the sulfur combines with oxygen, the bejeezus coming off as sulfur dioxide gas as shown in Equation 9-2. What remains are metal oxides, which may be smelted with charcoal as previously described.

I shall now tell you of the properties of these metals and the many and varied uses to which they are put. ***Gold*** is without doubt familiar to anyone reading this book. It is a soft, dense yellow metal. Three properties distinguish it from the other metals. First, it has density of 19.6 g/mL, almost 20 times that of water and the highest of the seven metals under discussion. Second, it is exceedingly resistant to corrosion. Finally, it is the rarest of the seven metals, and consequently the most valuable. It is mined largely as particles of native metal naturally dispersed in sand or other ores. The chief problem in its production is the physical separation of the minor quantities of gold from the bulk of the material which contains it. The properties of gold have suited it to its familiar uses in jewelry and as a standard for monetary value. Hammered into thin foil or gold-leaf, it can be used to adorn paper, wood, and even thread. More recently, gold has become widely used for electrical contacts because of its resistance to corrosion. In all times and places it has been a symbol of completion, perfection, and immortality.

Silver is valued next to gold among the seven metals. This soft white metal is moderately resistant to corrosion, though far less so than gold. Though it is found as native metal or as a sulfide ore in its own right, it is a common impurity in galena, the sulfide of lead, and because such large quantities of lead are produced, galena is the principle commercial source of silver as well as lead. Frequently it is alloyed with copper, which produces a harder metal. Familiar uses include the metallic coating of mirrors and the photo-sensitive emulsions of photographic films and papers.

Copper is a common red metal, similar to silver in its resistance to corrosion. Though found as native metal, it is more often found in its sulfide, carbonate and oxide ores. It ranks second (behind iron) in worldwide production. Bronze, as previously described, is an alloy of copper and tin; ***brass*** is an alloy of copper and zinc. Copper is used to make inexpensive jewelry plated with silver or gold, as an important coinage metal, and for electrical wiring.

Mercury is a most mysterious and wonderful metal, the only metal which is liquid at room temperature. It is a heavy metal, though not as dense as gold. Found as pools of liquid metal in mercury mines, it is most often extracted from its sulfide ore, cinnabar. Mercury dissolves many metals, including gold and silver, forming solutions called ***amalgams.*** "Silver" dental fillings are actually amalgams of mercury and silver. Widely used in industry, its most familiar domestic use is as the liquid metal of thermometers. While metallic mercury was handled with impunity throughout human history, its compounds, such as mercury nitrate or mercury acetate, are generally acutely toxic; the organic compound, methyl mercury, is *extremely* toxic. But because the general public fails to distinguish the properties of elements from those of their compounds, mercury is widely and incorrectly perceived to be a "toxic metal." Many web-sites and news reports claim that mercury is among the most toxic substances known, a claim which is patently and demonstrably false. Millions of people live to ripe old age with mercury fillings; try having your teeth filled with caustic soda or potassium cyanide if you want to know what a toxic filling looks like. There are legitimate health concerns over both metallic mercury and its compounds but it is wrong to oversimplify them. In alchemy mercury is one of a holy trinity of principles, holding a place of honor as the archetype of the metallic essence, the spirit of the metals and, symbolically, of everything in Nature.

Iron is the familiar gray metal of commerce and industry. Very rarely found as native meteoric iron, it is most commonly smelted from its oxide and sulfide ores. Of the metals, iron is produced in the largest tonnage. Far more susceptible to corrosion than the other seven metals, it must be painted, alloyed or coated if it is to resist the effects of air and water for more than a few years. Its alloy with carbon, steel, is among the most useful metals known. Subtle changes in the concentration of carbon (less than 2%) produce a steel which is harder or softer, stiffer or more flexible. Wrought iron is typically found in ornamental iron work, cast iron in cook-ware, and steel in knives and bridges.

Tin is a metal almost as white as silver and almost as soft as lead. Next to mercury it has the lowest melting point of the seven metals. It has excellent resistance to corrosion. For this reason it was widely used as a coating for iron, "tin-plate," a use which has been largely supplanted by the less expensive zinc-plated, or galvanized iron. The "tin can" was actually tinned iron, now aluminum or galvanized, and "tin foil" has been replaced for household use by inexpensive aluminum foil. With copper, tin makes the alloy, bronze, and with lead the alloys, *pewter* and *solder*, and it is here that the public is most likely to come into contact with it.

Like mercury, ***lead*** has been widely slandered as a metal, owing to confusion between the properties of the metal and those of its compounds. Metallic lead is extremely un-reactive while its compounds, particularly the soluble ones, are acutely toxic. It is smelted from its sulfide ore and ranks fifth in tonnage after iron, copper, aluminum, and zinc. Lead is a dense white metal which quickly loses its metallic luster on contact with air. It found widespread use during the Roman Empire as lead pipe for plumbing. Modern households find it used as a component of pewter and solder, but its use in automobile batteries is its largest single application.

Consulting a modern periodic table, you will find that nearly 80% of the elements are metals, most of them unrecognized until the nineteenth century. Zinc was alloyed with copper to produce brass during the Roman Empire and nickel was recognized during the Renaissance. The next metal to achieve widespread utility, aluminum, would not do so until the twentieth century.

Material Safety

Locate MSDS's for copper carbonate (CAS 12069-69-1), tin oxide (CAS 18282-10-5), and sodium carbonate (CAS 497-19-8). Summarize the hazardous properties of these materials in your notebook, including the identity of the company which produced each MSDS and the potential health effects for eye contact, skin contact, inhalation, and ingestion. Also include the LD_{50} (oral, rat) for each of these materials.[2]

Your most likely exposure is dust inhalation. If a persistent cough develops, see a doctor.

You should wear safety glasses and a dust mask while working on this project. Leftover materials can be disposed of in the trash.

Research and Development

Before proceeding with your work, you must master the following material:

- Know the meanings of those words from this chapter worthy of inclusion in the ***index*** or **glossary.**
- You should have mastered the Research and Development items of Chapter 5 and Chapter 7.
- Know the names, properties and uses of the seven metals whose discoveries pre-date recorded history.
- Know the ores listed in Table 9-1 and be able to recognize samples of them.
- Know the equations for the smelting of malachite, Equation 9-1(c), and cassiterite, Equation 9-1(g).
- Know which metals are alloyed to produce bronze, brass, pewter and solder.
- Be prepared to handle malachite, cassiterite, soda ash, charcoal and fire responsibly.
- Know that the properties of an element may be completely different from those of its compounds.

9.3 Θ

Δ The rabble has long since lost interest in this book. Content to be buffeted about by a world they neither apprehend nor comprehend, they close their eyes and hope for the best. Their hands are pale, their fingernails clean, and their minds are filled with spirits who demand little and offer nothing at all. You will forgive me, my brothers and sisters, if I am

2. The LD_{50} was introduced in Section 7.2 (page 96).

circumspect but what I have to tell you is of the utmost importance and sensitivity. I mean to speak to you of life and death.

Δ I know that you feel a sympathy for the mortal you inhabit. You may even feel that you are that mortal, but nothing could be further from the truth. The mortal is a creature of the moment, with passions and appetites which see no further than tomorrow or next month or next year because it lives for only a century, at best. Your life, distinct from that of the mortal in which you currently reside, is potentially much longer and your vision must look beyond the concerns of the moment. Looking back on history, the great movements and advances have taken place across centuries and millennia, time periods too vast to hold the attention of a mere mortal.

Δ If we are not mortals, then what are we? We are the essence of humanity, not the substance; we are the spirit, not the soul; we are the mercury, not the sulfur; we are the element, not the compound. Are you confused? If so, then perhaps you are ready for a great lesson. What follows is not a proof of our separate existence; it is but a physical metaphor to help you understand the circumstances in which we live.

You will use the crucible you made in Chapter 5 (page 70). The crucible must be sound and free of cracks or weaknesses. It must have a conical interior to funnel molten materials to the center and it must have a tight-fitting lid. Your crucible must have been previously fired, or *bisqued,* to prove that it can keep in the heat and withstand it. You will fill your crucible with the ores of the metals you wish to smelt.

Copper melts at 1083°C, which is at the upper range of the temperatures which earthenware clays can withstand. Rather than move to a more refractory clay, the cassiterite adds tin, which lowers the melting point and produces a harder metal than copper alone. There is no “correct” amount of tin to be added; bronze is a solution, an alloy of copper and tin, not a compound, and so its composition is variable, not fixed. You may make bronze with 10% tin, 20% tin, 50% tin, or whatever you choose. The melting point and hardness of the bronze depends on the concentration of tin, just as the strength of an alcoholic beverage depends on the concentration of alcohol. Bronze for casting bells is typically 20% tin, that for casting cannons, 10%. Both malachite and cassiterite may be obtained from a pottery supply as copper carbonate and tin oxide, respectively.

There is a minor dilemma in making bronze; bronze melts at a lower temperature than copper, but it cannot form until the copper and tin melt and mix. In practice, a ***flux*** is added, a material with a melting point

Figure 9-1. The Smelting Crucible

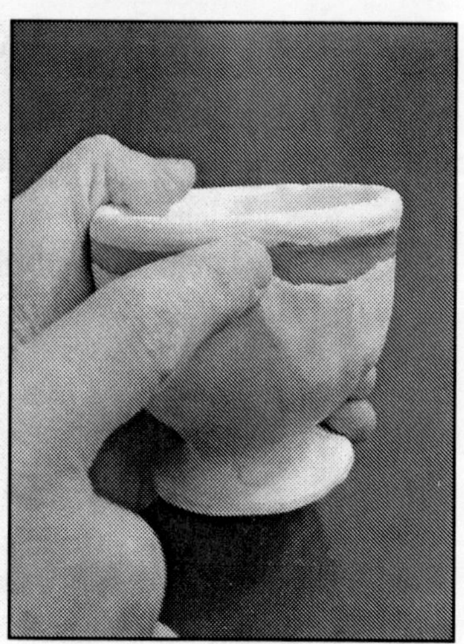

lower than the materials to be melted. In this case, you will add either ***potash*** or ***soda ash,*** materials discussed in Chapter 8. You may use potash you made yourself, or you may purchase "washing soda" and dry it in an oven at 130°C (270°F) for an hour to convert it into soda ash. There is no "correct" amount of soda ash; you simply need enough to cover the malachite and cassiterite with molten slag. For a first bronze you may use 10.0 g of copper carbonate, 2.0 g of tin oxide, and 6.0 g of sodium carbonate. Weigh these materials on a centigram balance using the methods of Appendix C (page 384), mix them in a plastic bag, and place them into your crucible.

Finally, we need a reducing agent, which will combine with the oxygen in the ores and leave as carbon dioxide. Charcoal was the reducing agent of choice in antiquity, being inexpensive, plentiful, easy to make from wood but rather messy. You may use dry seed corn, which turns to charcoal in the kiln without making a mess of your work-space. Place the corn on top of your mixed ores, nearly filling the crucible but leaving enough space that the lid makes a tight fit.

Smelting is a reduction process and as such, oxygen must be excluded. To that end, the crucible must be closed to prevent oxygen from oxidizing your metal. Using a fresh piece of clay, roll a snake and place it on the

Figure 9-2. The Bronze Nugget

mouth of the crucible to form an *ouroboros*, a snake biting its own tail. Place the lid on the crucible, compressing the ouroboros to make a tight seal, as shown in Figure 9-1. Then trim away any excess clay with a knife or spatula and smooth it out with your thumb.

Equation 9-1 shows that gas is produced during smelting and you must provide for this bejeezus to escape. The ouroboros will make a good seal between the crucible and its lid but it will not "glue" them together. As gas builds up in the crucible it will lift the lid slightly and escape. In a sense, the ouroboros serves the same purpose in smelting that the fermentation lock did in brewing.

The smelting crucible must be dried before it can be fired. You may either put it in an oven at 130°C (270°F) for an hour or set the kiln to include a drying segment before the firing. Fire the crucible to cone 05. When the firing is complete and the crucible cool, remove the lid and explore the interior. If there is no charcoal then your lid did not seal properly or there was insufficient corn from the start. You are unlikely to find any metal. If there is charcoal present, remove it and explore the charred material beneath. There may be beads of metal adhering to the walls of the crucible; if so then the atmosphere within was sufficiently reducing to smelt the ores into metal. Continue to explore the bottom of the crucible with a knife or spatula. If the ores contained sufficient metal, if the atmosphere was sufficiently reducing, if the shape of the crucible was sufficiently conical, and if there was sufficient flux present then you may find a large nugget of bronze in the bottom, as shown in Figure 9-2. Carefully pry it free from the crucible and scratch the blackened surface to reveal the golden crop beneath. Clean it with fine sandpaper or a wire brush and polish it to a high luster.

Δ What have you learned from this operation? Molten rock from within the Earth separated, congealed, and re-melted, the elements separating and recombining to produce myriad rocks and minerals. Some were exposed by wind and water; others were buried and crushed by the weight of the overbearing rock. The rock has changed form many times over the eons, from sulfides to sulfates to carbonates to oxides. It was dug out of the ground, crushed, milled, sorted and separated, and came to you as blue and white chalky powders. You placed them into a crucible along with other obviously non-metallic materials, and yet when you opened your fired crucible, a nugget of metal lay in among the cinders. The metals were there all along! In the same way that the metals passed from one mineral to another over the ages, so you have passed from one mortal to another. Do not believe that you are salt, my brothers and sisters, mere mortals of flesh and blood and bone. Do not believe that you are sulfur, the life which animates the body. No, you are a mercurial spirit, the essence of humanity, and the mortal is merely your temporary home.

Quality Assurance

You placed obviously non-metallic materials into a sealed crucible of your own making. After firing it, you recovered a nugget of solid metal, a talisman to remind you of your noble identity. You will record in your notebook the weight of copper carbonate and tin oxide used in your smelting as well as the weight of the resulting bronze nugget. You will also include a photograph of your nugget as a record of your achievement.

Chapter 10. Vitruvius (Lime)

Sand and its sources having been thus treated, next with regard to lime we must be careful that it is burned from a stone which, whether hard or soft, is in any case white. Lime made of close-grained stone of the harder sort will be good in structural parts; lime of porous stone in stucco. After slaking it, mix your mortar, if using pitsand, in the proportion of three parts of sand to one of lime; if using river or sea-sand, mix two parts of sand with one of lime. These will be the right proportions for the composition of the mixture. Further, in using river or sea-sand, the addition of a third part composed of burnt brick, pounded up and sifted, will make your mortar of a better composition to use.

The reason why lime makes a solid structure on being combined with water and sand seems to be this: that rocks, like all other bodies, are composed of the four elements. Those which contain a larger proportion of air, are soft; of water, are tough from the moisture; earth, hard; and of fire, more brittle. Therefore, if limestone, without being burned, is merely pounded up small and then mixed with sand and so put into the work, the mass does not solidify nor can it hold together. But if the stone is first thrown into the kiln, it loses its former property of solidity by exposure to the great heat of the fire, and so with its strength burned out and set free, and only a residuum of heat being left lying in it, if the stone is then immersed in water, the moisture, before the water can feel the influence of the fire, makes its ways into the open pores; then the stone begins to get hot, and finally, after it cools off, the heat is rejected from the body of the lime.

Consequently, limestone when taken out of the kiln cannot be as heavy as when it was thrown in, but on being weighed, though its bulk remains the same as before, it is found to have lost about a third of its weight owing to the boiling out of the water. Therefore, its pores being thus opened and its texture rendered loose, it mixes readily with sand, and hence the two materials cohere as they dry, unite with the rubble, and make a solid structure.

— Vitruvius, *The Ten Books on Architecture*, *ca.* 40 BC [1]

10.1 ☿

🜃 You know, I've been sitting here real quiet while Lucifer and her descendants go on about the fire and all the stuff that comes out of it. But

1. Reference [28], Book II, Chapter 5, Section 1.

what about the stuff that goes into it, huh? You'd have no fire without wood, no pottery without clay, and no metal without ore. I figure that stone is getting the short end of the stick here. I've been waiting for Lucifer to talk about two stones in particular, since they come out of the fire, but he just whooshed right on past them. So let's just come on back to ***limestone*** and ***gypsum.***

🜄 Of course, limestone's been used since God was a child to build pyramids and tombs and temples and such. And it's such a good building material that some of those buildings are still around, in spite of wars and revolution and time. To use limestone blocks, you have to cut them out of the quarry, chisel them to the shape you need, and haul them to where you need them. And I'm not saying it isn't worth the trouble for real important tombs and all, but it is a lot of work for ordinary houses and shops and offices. If you live where wood is expensive, these places are made out of brick.

🜄 Now, in the beginning, these bricks were made of sun-dried mud and they were just stacked up, and when the rain came, they got a little wet and stuck together, which was good. But over years and years of getting wet, they eventually crumbled and returned to the Earth, which was not so good. So folks started firing brick, like Athanor said, and the clay calcines into mullite and silica, which don't soften in water. But while a fired brick lasts a good long time, it doesn't stick to the others. All in all, it's just another brick in the wall with nothing to hold the wall together. That's where limestone comes into it again.

🜄 When you burn limestone in a kiln it just crumbles into a powder. And when you add water to this powder, it turns back into stone. Mix this powder with sand and you've got a real nice mortar for cementing bricks together. And there's another stone, gypsum, which works almost the same way; heat it up and it crumbles to dust; add water and it turns back into stone. Where limestone is good for making mortar, gypsum is good for making ***plaster*** and stucco. And folks have been using mortar and plaster since about 12,000 BC, so that's nothing new.

🜄 Long about 200 BC, my cousin Unktomi-Vitruvius was floating around in the head of a Roman builder and discovered that if you take burnt limestone, which is called ***lime,*** and mix it with a kind of volcanic sand, you get a kind of cement, *hydraulic* cement, which can even harden under water. Mix it with sand and pebbles and rubble and such and you get a brand spanking new material, *concrete*, that can be poured into the shape you want, which is a lot easier than cutting stone into that shape. Just build a wooden form to keep it in, mix up your concrete, pour it in, and once it sets up it's every bit as strong as if you had a limestone block of that same shape. Now my cousin moved down the line from Roman to Roman for a long time, but eventually the Roman Empire fell and my cousin went extinct along with it. So folks didn't know how to make concrete any more. All those Gothic cathedrals and such were made out of cut stone with no concrete at all, which is okay, but you can't afford to build too many fancy buildings like that.

🜃 Well, a long time went by. I was just a general-purpose in*spir*ation at the time and I found myself living in a guy named ***Joseph Aspdin*** when all of a sudden in 1824, *bam*! I connectified lime and clay. That did the trick all right, made a hydraulic cement so I could get back to work concretizing the place. So when you're wandering around in all those skyscrapers and community centers and movie theaters and, well, most any building you ever go into, give old Unktomi-Aspdin a moment of thought. So I don't go extinct, you know.

10.2 🜍

The silicates were introduced in Chapter 2. There we learned that the mineral *quartz* is composed of the chemical *silica,* SiO_2. In Chapter 5 we learned that the aluminosilicate minerals contain silica and alumina, Al_2O_2. Chapter 9 introduced a variety of metal oxides and sulfides. In this chapter we discuss chemicals derived from calcium sulfate and calcium carbonate.

Three calcium sulfate minerals are pertinent to our discussion. The mineral ***anhydrite*** is, as its name implies, the *anhydrous* form of calcium sulfate, $CaSO_4$. ***Bassanite*** is a hydrous form, $CaSO_4 \cdot \frac{1}{2} H_2O$, or calcium sulfate hemihydrate. As with cellulose or kaolinite, the "H_2O" in the empirical formula doesn't mean that bassanite is wet. No, the water is part of the formula, but the material itself is bone-dry. We might just as well write the formula "CaH_2SO_5" but "$CaSO_4 \cdot \frac{1}{2} H_2O$" is a convenient way to write it for comparison with the other calcium sulfates. The third mineral form of calcium sulfate, ***gypsum,*** is another hydrate, $CaSO_4 \cdot 2\ H_2O$, or calcium sulfate dihydrate. It may seem like picking nits to have three different calcium sulfates but they have different crystal structures and different properties. It is easy, however, to convert from one to the other,

Equation 10-1. From Gypsum to Plaster and Back Again

$$(a)\ CaSO_4 \cdot 2\,H_2O(s) \overset{\Delta}{=} CaSO_4 \cdot {}^1/_2\,H_2O(s) + \frac{3}{2}\,H_2O(l)$$

$$(b)\ CaSO_4 \cdot {}^1/_2\,H_2O(s) \overset{\Delta}{=} CaSO_4(s) + \frac{1}{2}\,H_2O(l)$$

$$(c)\ CaSO_4(s) + \frac{1}{2}\,H_2O(l) = CaSO_4 \cdot {}^1/_2\,H_2O(s)(slow)$$

$$(d)\ CaSO_4 \cdot {}^1/_2\,H_2O(s) + \frac{3}{2}\,H_2O(l) = CaSO_4 \cdot 2\,H_2O(s)(fast)$$

as shown in Equation 10-1. Heating gypsum to 128°C (262°F) converts it to bassanite, more commonly known as plaster of Paris, or ***plaster,*** for short. Heating plaster to 163°C (325°F) converts it to anhydrite. You can add water to anhydrite to convert it back to plaster, but this conversion is slow. By contrast plaster absorbs liquid water to become gypsum quite rapidly. The growth of gypsum crystals is what causes plaster to "set" when water is added. Gypsum is the "porous stone" described by Vitruvius for making stucco.

The "hard" stone mentioned by Vitruvius is ***limestone,*** a ***sedimentary rock*** composed of calcium carbonate. Chalk and marble are other rocks derived from calcium carbonate. Sea-shells are also composed primarily of calcium carbonate. When calcium carbonate crystallizes, it does so as the minerals ***calcite*** and ***aragonite,*** both with formula $CaCO_3$. Dolomitic limestone is derived from the mineral ***dolomite,*** $CaCO_3 \cdot MgCO_3$, with chemistry similar to that of limestone itself.

Calcium carbonate is not soluble in water; if you add water to limestone, you just get wet limestone. But if you burn limestone, if you heat the ***bejeezus*** out of it, it turns into ***quicklime,*** calcium oxide. When you add water, that is, when you *slake* it, it gets hot and turns into calcium hydroxide, or ***slaked lime.*** Figure 10-1 shows the whole lime-making process as a schematic. The first reactor, the ***furnace,*** should be familiar from Figure 1-3 (page 9). It's just a container where gas can come off of a solid that's having the bejeezus heated out of it. The second reactor, the ***slaker,*** is a container where water is added to a solid. As usual, reactants enter from the left, waste products exit to the top and bottom, and the main product exits to the right of the figure. Calcium hydroxide is sparingly soluble in water and when mixed with sand forms a dandy mortar. You might think that rain would wash the slaked lime right out of the mortar. That's where carbon dioxide comes in.

Figure 10-1. Lime-Making as a Process

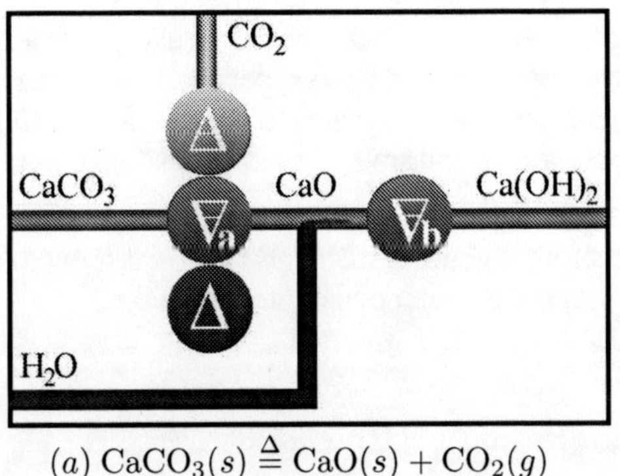

$$(a)\ \mathrm{CaCO_3}(s) \overset{\Delta}{=} \mathrm{CaO}(s) + \mathrm{CO_2}(g)$$
$$(b)\ \mathrm{H_2O}(l) + \mathrm{CaO}(s) = \mathrm{Ca(OH)_2}(s)$$

Equation 10-2. From Lime Back to Limestone

$$(a)\ \mathrm{CO_2}(g) + \mathrm{H_2O}(l) = \mathrm{H_2CO_3}(aq)$$
$$(b)\ \mathrm{H_2CO_3}(aq) + \mathrm{Ca(OH)_2}(aq) = \mathrm{CaCO_3}(s) + 2\,\mathrm{H_2O}(l)$$

Carbon dioxide is a weak acid. When it dissolves in water it forms hydrogen carbonate, H_2CO_3, also known as carbonic acid. Hydrogen carbonate reacts with calcium hydroxide in a classic metathesis reaction to produce calcium carbonate and hydrogen hydroxide, as shown in Equation 10-2. Calcium carbonate is just limestone and hydrogen hydroxide is just water. In other words, when carbon dioxide reacts with slaked lime it turns back into limestone. Thus the lime in mortar gradually turns to limestone, cementing the silica in the sand together to form a material which is quite impervious to the elements. That's just what happens when a calcium-rich sea dries up; it absorbs carbon dioxide from the air and deposits a layer of limestone or chalk. So the limestone in our mortar has come full-circle, when you think about it.

Material Safety

Locate MSDS's for limestone (CAS 471-34-1), lime (CAS 1305-78-8), slaked lime (CAS 1305-62-0), gypsum (CAS 10101-41-4), and silica (CAS 14808-60-7). Summarize the hazardous properties of these materials in your notebook, including the identity of the company which produced each MSDS and the potential health effects for eye contact, skin contact, inhalation, and ingestion. Also include the LD_{50} (oral, rat) for each of these materials.[2]

Your most likely exposure is dust inhalation. If a persistent cough develops, see a doctor. Lime is caustic, which means it eats skin; in case of skin contact, wash the affected area with plenty of cold water. Be aware that lime gets hot when it gets wet.

You should wear safety glasses and a dust mask while working on this project. Leftover materials may be disposed of in the trash. Lime should be slaked before disposal.

Research and Development

Well, I guess if you are in a class or something, you might want to know what will be on the quiz.

- You better know all the words that are important enough to be ***indexified*** and **glossarated.**
- You better know the Research and Development stuff from Chapter 5 and Chapter 6.
- You ought to be able to recognize calcite, limestone, and gypsum, either from photographs or from samples. You should also know that sea-shells are made of calcium carbonate.
- Know the formulas for limestone, lime, gypsum, and plaster and all the equations in this chapter.
- Know all the hazards of working with limestone, lime, gypsum, plaster, and silica and what to do if things get out of hand.
- Know that lime is an alkali, what it tastes like, whether it's pH is high or low and what color it turns pH test paper.
- Know what water of hydration means, and that just because a formula has H_2O in it doesn't mean that it literally has liquid water in it.

---

2. The LD_{50} was introduced in Section 7.2 (page 96).

10.3 Θ

Plaster and lime are manufactured by heating the bejeezus out of gypsum and limestone, respectively. Because the bejeezus is different in each case, however, water for gypsum and carbon dioxide for limestone, they require different temperatures for calcination. Conversion of gypsum to plaster requires a temperature between 128°C (262°F) and 163°C (325°F). If the temperature is too low, the conversion to plaster is incomplete. If it is too high the plaster is converted to anhydrite, which slakes only very slowly. A good compromise is to set an ordinary oven to between 130 and 140°C (266 and 284°F).

Weigh your empty crucible and record the weight in your notebook. Fill it two-thirds full of powdered gypsum, available wherever lawn and garden products are sold, and record the weight of your filled crucible in your notebook. By subtracting the empty weight of your crucible you can determine the weight of gypsum used, w_{gypsum}. Test the pH of your gypsum with wet strip of pH test paper and record the result in your notebook. Place your crucible in the oven for a couple of hours, even overnight if you have access to a laboratory oven which stays on all the time. Remove it from the oven and weigh it again; you should find that it has lost weight, the weight of the bejeesical water that was driven off by the heat. Record the weight in your notebook and subtract the empty weight of the crucible to get the weight of your plaster, $w_{plaster}$. Divide the weight of your plaster, $w_{plaster}$ by the weight of the gypsum, w_{gypsum}. *If* your gypsum were 100% $CaSO_4 \cdot 2\ H_2O$ and *if* it were completely converted to $CaSO_4 \cdot \frac{1}{2}\ H_2O$ then the ratio of plaster to gypsum would be 84%. Your ratio may be a little more or less depending on the purity of your gypsum and the completeness of your conversion. Agricultural "gypsum" often contains anhydrite in addition to gypsum. Would this make the ratio of plaster to gypsum higher or lower than 84%? Test the pH of your plaster with a wet strip of pH test paper and record the result in your notebook.

Now that you have some plaster you should make something useful from it. In Chapter 13 you will make glass and we can use our plaster to make a mold for that project. Plaster alone is often used to make ornamental objects or molds for ornamental objects, but by itself it cannot withstand the high temperatures needed for making glass. Glass casters mix plaster with silica to produce an *investment* suitable for high-temperature molds. We will use one part plaster to one part silica, available in powdered form

Figure 10-2. Coating the Crucible, Making the Model

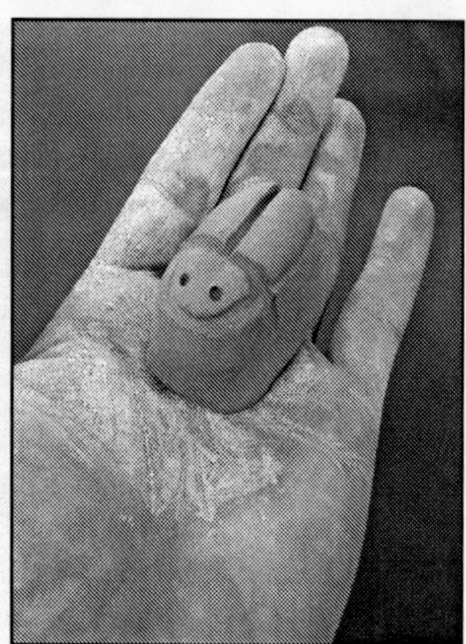

wherever pottery supplies are sold. To begin with we need to coat the inside of the crucible with investment to keep the model from touching the sides of the crucible. Transfer your plaster from the crucible into a plastic tub for storage and label it. Before you go any further you should read the instructions for weighing ***by difference*** (page 384). You may think that you know how to weigh things, but a few minutes of extra reading will save you a lot of trouble.

Your crucible should be wet. Fill it with water while you mix your investment. Weigh 20 g of plaster and 20 g of silica into a plastic bag and seal it. You now have a plastic bag containing equal ***parts*** of plaster and silica. Turn it end for end, massaging any lumps and mixing your *investment* as completely as possible. Weigh out 20 g of water into a plastic cup and add a spoonful of investment to it. When the dry powder sinks to the bottom add another spoonful. Continue adding investment to the water, waiting until each spoonful sinks before adding the next. The last spoonful will not sink below the surface and this is a sign that you have the correct ratio of water to investment. Use a spoon or spatula to completely mix your investment.

Quickly empty the water from your crucible and pour your investment into it. Tilt the crucible from side to side to completely coat the walls

Figure 10-3. Making the Mold

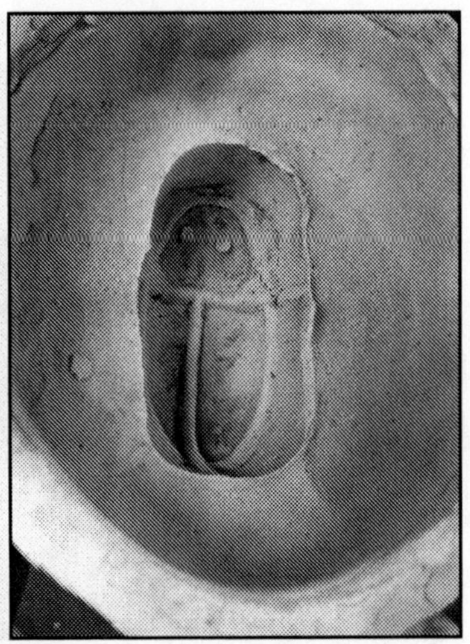

with investment, as shown in Figure 10-2(L). While this investments sets, make a model from clay. Keep it simple. The model shown in Figure 10-2(R) is of a scarab, a noble creature even if it does have only six legs. Make your model approximately 5 cm x 3 cm x 5 cm high. Mix up another cup of investment, this time using 50 g each of plaster, silica, and water. Pour this second batch of investment into your crucible, use a spoon or spatula to pour investment into the features of your model and then plunge it into the crucible, leaving the base sticking up above the investment as shown in Figure 10-3(L). Allow your investment to set for a full hour. Then grab the base of your clay model and wiggle it from side to side until you are able to pull it free from the mold. Use a knife or toothpick or remove any excess clay from your mold. The finished mold is shown in Figure 10-3(R).

Your mold will need to be completely dry before you can use it from making glass in Chapter 13. You can let it air-dry or you can put it in an oven at 130°C (266°F) for a couple of hours. While you are waiting for your mold to dry, let's make some lime. Weigh your crucible *lid* and record its empty weight in your notebook. Fill your lid with crushed limestone, weigh it, and subtract the empty weight. Record the weight of the limestone, $w_{limestone}$, in your notebook. Test the pH of limestone with a wet strip of pH test paper and record the result in your notebook.

Fire the lid to cone 05 to convert the limestone to lime. Weigh it when it comes back from the kiln, subtract the empty weight of the lid, and record the weight of your lime, w_{lime}, in your notebook. Calculate the ratio, $w_{lime}/w_{limestone}$. If your limestone were 100% $CaCO_3$ and if it were converted completely to CaO, the ratio would be 56%. Because agricultural limestone is not 100% calcium carbonate and because the conversion may not be complete, your ratio may be slightly different from the theoretical value.

Whatever the weight of your lime, weigh half that much water into a glass jar and spoon your lime into it. If the lime is fresh from the kiln it may get hot as it slakes. Test the pH of the slaked lime with a strip of pH test paper and record the result in your notebook. When it has cooled, screw on the cap, label the jar, and save your slaked lime for making paper in Chapter 14.

I think that Unktomi-Hermes must have been talking about the magic of carbon dioxide, which the Sun turns into trees, which the trees return to the air by moonlight, which turns into stone in limestone caverns, and which escapes from that stone in the heat of the kiln:

> Its father is the Sun, its mother is the Moon. The wind has carried it in his belly. Its nourishment is the Earth. It is the father of every completed thing in the Whole World. Its strength is intact if it is turned toward the Earth. Separate the Earth by Fire, the fine from the gross, gently and with great skill.
>
> — *The Emerald Tablet of Hermes Trismegistos*

Quality Assurance

Compare your ratios, $w_{plaster}/w_{gypsum}$ and $w_{lime}/w_{limestone}$ to their theoretical values. Compare the pH of plaster to that of lime and explain why they are different. Include a picture of your mold in your notebook.

Chapter 11. Pliny (Redox Reactions)

> The flower of copper also is useful as a medicine. It is made by fusing copper and then transferring it to other furnaces, where a faster use of the bellows makes the metal give off layers like scales of millet, which are called the flower.
>
> . . .
>
> Great use is also made of verdigris. There are several ways of making it; it is scraped from the stone from which copper is smelted, or by drilling holes in white copper and hanging it up in casks over strong vinegar which is stopped with a lid; the verdigris is of much better quality if the same process is performed with scales of copper.
>
> — Pliny the Elder, *Natural History, Book XXXIV*, *ca.* 60 AD[1]

11.1 ☿

🜂 Imagine you're living a hundred thousand years ago. Your best friend, Bob, was just mauled by a lion and he's looking, quite literally, like something the cat dragged in. You want to gussy him up for the afterlife so you dress him up in his best loincloth and stuff him into a hole with his second-favorite spear. He looks so pale, lying there like a mackerel in his Sunday go-to-meeting clothes. So you give him a good dusting with red ochre, that is, powdered iron oxide, and it seems to restore the blush of life. Seventy thousand years later, your palette has expanded to include yellow ochre and charcoal and you can paint the Paleolithic equivalent of the Sistine Chapel on cave walls, enhancing the longevity of your meme-plexes beyond all previous records. By the time of Pliny the Elder you can paint frescos in yellow orpiment (arsenic sulfides), red cinnabar (mercury sulfide), white limestone (calcium carbonate) and gypsum (calcium sulfate), blue malachite and azurite (copper carbonates), and, of course, green verdigris (copper acetate). All of these pigments are compounds of metals. While Chapter 7 taught you to balance reactions in which two compound swap their first and last names, it did not explain how to produce compounds from elements or *vice versa.*

1. Reference [23].

These kinds of reactions, oxidation-reduction (redox) reactions, are the subject of this chapter.

∆ I have to confess that I just don't understand people, especially the non-fictional variety. People will go to any lengths to memorize the most inconsequential things; one of the Author's life ambitions is to memorize every story Dr. Seuss ever wrote. People will spend enormous effort to learn the rules for games, no matter how complicated they may be. And don't get me started on puzzles. Yet when these same skills are to be used for practical ends, people turn into self-confessed imbeciles. "I don't have a head for math." "I can't do science." A relatively simple set of rules for balancing chemical equations might allow you to understand the chemistry of gunpowder; it seems to me that that would be incentive enough. But as I said, I don't understand people.

∆ The Author assures me that people will go to any lengths to *avoid* learning something useful. A few years ago he taught a course for pre-med students boning up for the big test which would determine whether or not they would get into medical school. Pre-medical students are a motivated bunch. They will pay extra to go to a college with a good record on medical school applications. They will take the "hard" courses that most students avoid like shrimp-and-bubble-gum salad. They will even pay hundreds of dollars to take a class to prepare them for the test which will decide whether their dream of becoming a doctor will be flushed down the academic toilet. Now, part of the test is a block of questions which depend on the student's ability to balance oxidation-reduction reactions, so the Author made up a handout explaining in detail the easiest method for doing this. He went through some examples and then gave them a sample test on which he *allowed* them to use the handout. As he gazed out upon the earnest faces, he noticed that *not one* was using the handout. Obviously they had already learned the method. But when they graded this practice test, would you believe it? Not a single one of those students got even a single correct answer!

∆ So the Author has instructed me to tell you that this chapter provides no useful information of any kind. It will not allow you to understand gunpowder, batteries, photography, fertilizer, pharmaceuticals or plastic. It's just a pointless game. It has some rules, as any game does. Learn the rules and you can play the game. You win the game by getting the same number of each letter of the alphabet on either side of an equal sign. You can play it anywhere with only a pencil and paper. You can play it alone or turn it into a race between individuals or teams. It's fun and challenging, will allow you to impress those of the opposite sex and awe your friends and neighbors. But keep in mind at all times that it is nothing more than a silly game.

11.2 ♙

The following procedure is guaranteed to produce an accurately balanced reaction equation for any redox reaction that occurs in aqueous solution. In fact, it will even work for reaction equations in which no redox occurs, but then it's pretty tedious compared to balancing a metathesis reaction. The only thing you have to know to begin is the chemical identity of the major reactants and products of the reaction. Let's take the following example: a popular WW-II German rocket propellant was nitric acid (HNO_3) and hydrazine (N_2H_4). As a general rule, unless stated otherwise, nitrogen will wind up as elemental nitrogen (N_2), carbon will wind up as carbon dioxide (CO_2), and hydrogen and oxygen are specifically treated below. For any other elements, the problem at hand must give you both reactants and products.

1. Write a "skeleton" reaction equation, one with no stoichiometric coefficients.

   $HNO_3 + N_2H_4 = N_2$

2. Pick out an unusual element from the skeleton reaction (non-hydrogen, non-oxygen unless that's all you have). Write a skeleton "half-reaction," one involving only species of that element. Elemental oxygen (O_2) and hydrogen (H_2) may appear only if they are included explicitly in the problem.

   $HNO_3 = N_2$

3. Insert **stoichiometric coefficients** to balance the number of atoms of that element on each side of the half- reaction.

   $2\ HNO_3 = N_2$

4. Insert species from the original reaction to take care of other non-H, non-O elements (if any). Balance the half-reaction with respect to the non-H, non-O element (if any).

   (Not applicable to this problem)

5. Insert enough H_2O molecules on the side that's short of O atoms to balance the O atoms.

   $2\ HNO_3 = N_2 + 6\ H_2O$

6. Insert enough H^+ ions on the side that's short of H atoms to balance the H atoms.

   $2\ HNO_3 + 10\ H^+ = N_2 + 6\ H_2O$

7. Total up the electrical charge on each side of the half-reaction equation.

   (+10 on left, 0 on right)

8. Add enough electrons to the side that's too positive (or not negative enough) to make the net charge balance.

   $2\ HNO_3 + 10\ H^+ + 10\ e^- = N_2 + 6\ H_2O$

9. Check to make sure that atoms of each element are balanced and that charge is balanced in the half-reaction.

   (2 N, 12 H, 6 O, 0 charge)

10. Pick another unusual element in the original skeleton equation and write a skeleton half- reaction for it. Go through steps 3-9 again; that is, balance the unusual elements, then balance O using H_2O, then balance H using H^+, then balance charge using e^-.

    $N_2H_4 = N_2$

    $N_2H_4 = N_2 + 4\ H^+$

    $N_2H_4 = N_2 + 4\ H^+ + 4\ e^-$

    (2 N, 4 H, 0 charge)

11. The electrons are the key to balancing redox reactions. In a redox reaction electrons are transferred from one reactant to another, so the number of electrons in the first half-reaction must equal the number in the second. If they don't, just multiply each half-reaction by an integer chosen so that the electrons do balance.

    (Multiply first reaction by 2 and second reaction by 5 to give 20 electrons)

    $4\ HNO_3 + 20\ H^+ + 20\ e^- = 2\ N_2 + 12\ H_2O$

    $5\ N_2H_4 = 5\ N_2 + 20\ H^+ + 20\ e^-$

12. Add the two half-reactions together, canceling the electrons and as many H^+ and H_2O as possible.

    $4\ HNO_3 + 5\ N_2H_4 = 7\ N_2 + 12\ H_2O$

13. Double check that the number of atoms of each kind are balanced and that the total charge is balanced.

    (14 N, 12 O, 24 H, 0 charge)

The half-reaction with electrons on the right is called the **oxidation.** Since electrons are lost in this process, we use the acronym LEO: Lose Electrons Oxidation. The half-reaction with electrons on the left is called the **reduction.** GER stands for Gain Electrons Reduction. When we speak of "the oxidation" or "the reduction," we are speaking of half-reactions, but when we are speaking of react*ants,* the situation is reversed.

Everyone knows that an insurance *agent* causes others to be insured. Similarly, the reactant that causes another reactant to be oxidized is called the oxidizing agent, or oxidant. The reactant which causes another reactant to be reduced is called the reducing agent, or reductant. If you stop to consider the matter, you will realize that the oxidant is the one that is, itself, reduced and *vice versa.* In the previous example HNO_3 is the oxidant and N_2H_4 is the reductant.

Redox in a Nutshell

Split the reaction into two half-parts;
balance strange elements first.
Then using water get oxygen right,
and add H^+ for hydrogen thirst.

Finish each half with electrons for charge,
multiplying by numbers assured,
To make the electrons completely drop out.
Remember that LEO says GER.

I'll work three more examples, two easy ones from *Natural History* and a hard one, but the only way to learn this is to work lots of problems on your own.

Q: $Cu + O_2 = Cu_2O$

A: $2\ Cu + H_2O = Cu_2O + 2\ H^+ + 2\ e^-$

$O_2 + 4\ H^+ + 4\ e^- = 2\ H_2O$

$4\ Cu + O_2 = 2\ Cu_2O$

Copper is the reductant and oxygen is the oxidant. There are actually two common oxides of copper: red copper(I) oxide, Cu_2O, and black copper(II) oxide, CuO. The Roman numeral gives the charge on the cation.

Q: $CH_3COOH + Cu = Cu(CH_3COO)_2 + H_2$

A: $2\ CH_3COOH + Cu = Cu(CH_3COO)_2 + 2\ H^+ + 2\ e^-$

$2\ H^+ + 2\ e^- = H_2$

$2\ CH_3COOH + Cu = Cu(CH_3COO)_2 + H_2$

Acetic acid is the acid in vinegar. Copper is the reducing agent and H^+ is the oxidizing agent.

Q: $KClO_3 + C_{12}H_{22}O_{11} = KCl + CO_2$

A: $KClO_3 + 6\ H^+ + 6\ e^- = KCl + 3\ H2O$

$C_{12}H_{22}O_{11} + 13\ H_2O = 12\ CO_2 + 48\ H^+ + 48\ e^-$

$8\ KClO_3 + C_{12}H_{22}O_{11} = 8\ KCl + 12\ CO_2 + 11\ H_2O$

Potassium chlorate is the oxidant and sucrose is the reductant.

I don't think anyone will have an easy time learning to balance redox reactions. Balancing redox reactions is probably one of the two hardest things in first-year chemistry and I give you permission to find it difficult. Curse it, swear at it, grit your teeth at it, but know that if you persevere, you can master it. I have seen people with Velcro shoelaces get this, so I am sure that if you've read this far, you're smart enough to do it. And there are two payoffs; first, this is the hardest thing in the book, so if you get this it'll be smooth sailing from here on; second, you will be able to amuse yourself for hours in your doctor's waiting room. When you finally get in, tell *her* how much fun you've been having and watch her eyes boggle.

Material Safety

A distinction must be made between acute toxicity, the kind in which exposure makes you sick right now, and ***chronic toxicity,*** the kind which makes you sick some time in the future. So far, we have considered only acute toxicity. While LD_{50}'s determined in test animals may not be quantitatively reliable for humans, they at least provide a measure of relative acute toxicity. Sadly, there is no single, easily understood measure of chronic toxicity. What is clear, however, is that chronic toxicity results from repeated exposure to sub-acute doses of a substance over extended periods of time.

The cigarette is probably the most familiar example of a material which is chronically toxic but acutely non-toxic. How can a cigarette be acutely non-toxic when the LD_{50} (mice, oral) for nicotine is 230 mg/kg? Each cigarette delivers about 2 mg of nicotine. A little UFA shows that a 100 kg person would have to smoke 11,500 cigarettes in order to have a 50% chance of snuffing it from nicotine poisoning, assuming that the toxicity for inhalation is not too different from that for ingestion and that toxicity for humans is similar to that for mice. No, you just don't hear about people smoking themselves to death in one sitting; it takes years to do that, and taking years to kill you is what chronic toxicity is all about. It bothers me, then, when well-meaning anti-smoking activists list all of the toxic chemicals in tobacco smoke, including carbon monoxide and benzene, but don't tell you the amounts of these chemicals delivered by a cigarette. Just because toxic chemicals are detectable in something doesn't mean that they are present in dangerous amounts. Until people start overdosing on cigarettes the way they do on Heroin or alcohol, there's just not a case to be made for acute toxicity.

As with smoking, chronic chemical intoxication occurs with repeated exposure to sub-acute doses over long periods of time. Consequently we ought to pay the most attention to those situations which bring about such exposure. Occupational exposure is one such situation which attracts and deserves attention. We might hear on the news, for example, that workers in the widget industry are three times as likely to get biglongnamitis as the general public. As terrible as this statistic sounds, it is vital to know whether the incidence of this disease is *high* or just *higher-than-average.* If biglongnamitis is a rare disease then even a tripled rate of incidence may be small in comparison to other dangers of widget work.

In no way do I mean to downplay the risks posed by smoking, by occupational hazards, by food additives, or by environmental pollutants. Unfortunately, however, these risks are far more complex than those posed by acute toxins. For chronic toxins there is no measure of risk as straightforward as the LD_{50} for acute toxins. The bad news, then, is that it is extremely difficult for the general public to distinguish between genuine chronic dangers and alarmist propaganda. The good news is that chronic toxicity is not about the one exposure which came back to bite you twenty years later; it's about years of exposure that finally caught up with you. If the motto for acute toxicity is "the dose makes the poison," then perhaps the one for chronic toxicity ought to be "the longer you live, the sooner you bloody-well die." While you may not be able to avoid exposure to every conceivable chronic hazard, you have time to demand more of your news sources than sensationalistic sound-bites.

You aren't going to get away without a material safety assignment, even though there are no materials involved in this project. Find an MSDS for mercury (CAS 7439-97-6), silica (CAS 14808-60-7), or asbestos (CAS 12001-29-5), and write a paragraph comparing and contrasting the chronic and acute toxicities of the material you have chosen.

Research and Development

So there you are, studying for a test, and you wonder what will be on it.

- Study the meanings of all of the words that are important enough to be included in the ***index*** or **glossary.**
- Know the Research and Development items from Chapter 7 and Chapter 8.
- Memorize the rules in the sidebar *Redox in a Nutshell* and play the game.
- Know the difference between acute and chronic toxicity.
- When you can't remember the last time you missed a redox problem, you've probably worked enough examples.

11.3 Θ

You've already seen many redox reactions in this book; we just didn't call them that. You've been memorizing them so far, but now you can just balance them on the fly. Try your hand at these classics:

Q: $C_6H_{10}O_5 + O_2 = CO_2$
A: $C_6H_{10}O_5 + 6\ O_2 = 6\ CO_2 + 5\ H_2O$
Cellulose is the reductant and oxygen is the oxidant.

Q: $CuCO_3 + C = Cu + CO_2$
A: $2\ CuCO_3 + C = Cu + 3\ CO_2$
Copper carbonate is the oxidizing agent and charcoal is the reducing agent.

Q: $PbS + O_2 = PbO + SO_2$
A: $2\ PbS + 3\ O_2 = 2\ PbO + 2\ SO_2$
Lead sulfide is the reducing agent and oxygen is the oxidizing agent.

Q: $CO_2 = C_6H_{12}O_6 + O_2$
A: $6\ CO_2 + 6\ H_2O = C_6H_{12}O_6 + 6\ O_2$

I know that these were pretty simple reactions, as redox reactions go, and human nature being what it is, you may have figured them out without using the sidebar *Redox in a Nutshell*. If you did, you wasted the easy exercises which were intended to give you some practice with the method without freaking you out. The reactions are going to get harder, not because I'm going to make it harder, but because Nature doesn't always make it easy. And without the method, how are you going to know which is the oxidant and which is the reductant in a reaction like the third example? So go back and do them again if you need to, and then try out these harder problems:

Q: $KMnO_4 + C_2H_4(OH)_2 = CO_2 + Mn_2O_3 + K_2CO_3$
A: $10\ KMnO_4 + 4\ C_2H_4(OH)_2 = 3\ CO_2 + 5\ Mn_2O_3 + 5\ K_2CO_3 + 12\ H_2O$
Potassium permanganate is the oxidant and ethylene glycol is the reductant.

Q: $KMnO_4 + C_3H_5(OH)_3 = CO_2 + Mn_2O_3 + K_2CO_3$
A: $42\ KMnO_4 + 12\ C_3H_5(OH)_3 = 15\ CO_2 + 21\ Mn_2O_3 + 21\ K_2CO_3 + 48\ H_2O$
Potassium permanganate is the oxidant and glycerol is the reductant.

Q: $Cu + HNO_3 = NO + Cu(NO_3)_2$
A: $3\ Cu + 8\ HNO_3 = 2\ NO + 3\ Cu(NO_3)_2 + 4\ H_2O$
Nitric acid is the oxidant and copper is the reductant. Hint: nitric acid appears in both half-reactions.

Q: $C_8H_{18} + NH_4NO_3 = N_2 + CO_2$
A: $C_8H_{18} + 25\ NH_4NO_3 = 25\ N_2 + 8\ CO_2 + 59\ H_2O$
Ammonium nitrate is the oxidant and octane is the reductant. This reaction is responsible for one kind of "fertilizer bomb."

Q: $S + KNO_3 = H_2SO_4 + N_2 + K_2SO_4$
A: $5\ S + 6\ KNO_3 + 2\ H_2O = 2\ H_2SO_4 + 3\ N_2 + 3\ K_2SO_4$
Sulfur is the reductant and potassium nitrate is the oxidant. Hint: sulfur appears in both half-reactions.

Q: $Al = Al(OH)_3 + H_2$

A: $2\ Al + 6\ H_2O = 2\ Al(OH)_3 + 3\ H_2$

Aluminum is the reductant and water is the oxidant. This reaction is quite vigorous in alkaline solutions, which is why we avoided using an aluminum pot for making potash.

Q: $C_{12}H_{22}O_{11} + KNO_3 = CO_2 + N_2 + K_2CO_3$

A: $5\ C_{12}H_{22}O_{11} + 48\ KNO_3 = 36\ CO_2 + 24\ N_2 + 24\ K_2CO_3 + 55\ H_2O$

Sucrose is the reductant and potassium nitrate is the oxidant. Hint: in Step 4 of the procedure (page 134), use CO_2 to balance carbon in the same spirit that you use water to balance oxygen in Step 5.

Quality Assurance

When you can solve these problems without consulting the sidebar *Redox in a Nutshell*, you will be prepared to balance any redox equation in this book. Your notebook for this project may omit the "Observations" section, as no manipulation of materials is involved in this project, but you should summarize the chronic and acute toxicities of either mercury, silica, or asbestos.

Chapter 12. Marie (Dyes)

105. Dyeing in Dark Blue. Put about a talent of woad in a tube, which stands in the sun and contains not less than 15 metretes, and pack it in well. Then pour urine in until the liquid rises over the woad and let it be warmed by the sun, but on the following day get the woad ready in a way so that you [can] tread around in it in the sun until it becomes well moistened. One must do this, however for 3 days together.

145. Cleaning by Means of Soap Weed. Take and treat soap weed with hot water. Make a ball from it as if from tallow. Then steep this in hot water until it is dissolved. The water, however, should go above the wool. Then boil up the water. Put the wool in and prevent it from becoming scorched. Leave it there a little while until you see that it is clean. Lift it out, rinse it and dry it.

146. Mordanting. Take lime and hot water and make a lye from it, let it stand and take away thereby the impurity existing upon it. When you see that the water has become crystal clear, then put the wool in, shake and leave it there again a little while. Lift it out and rinse it.

152. Shading off of Colors. When you desire to shade off the brightness of a color then boil sulfur with cow's milk, and the color will be easily shaded off in it.

— *The Stockholm Papyrus, ca.* 200 AD[1]

12.1 ☿

∇ No doubt you are wondering why I am treading ankle deep in a vat of woady piddle. The reason is actually quite simple. First of all you will recall that human beings have been wearing clothes for something like 20,000 years. The earliest clothing styles, Figure 6-1(L) (page 75) for example, providing for neither warmth nor modesty, evidently communicated the reproductive status of the wearer. In time clothing became more elaborate and communicated not only reproductive but economic and political status. Consider, for example, the perennial problem of distinguishing a queen from a call-girl. From a distance after all, one naked woman pretty much looks like another. "Pardon me, Mademoiselle, I have this large sack of money and am in desperate need of a favor." I

1. Reference [5].

can tell you that you had better know which one you are talking to before you elaborate on the nature of such a favor, and telling them apart became much easier once they began to wear different outfits.

∇ To cite another example, in the pre-clothing days the king often suffered from laryngitis. "I do not believe we have met. I am the King." "Hello, you must be new around here. I am the King." "King here, how is it hanging?" It was much easier for everyone once the king began to dress, well, like a king. Wearing his royal garments, he could assert his authority merely by being seen and the poor subjects were not left wondering whether they had just handed over their savings to the village idiot. Unless, of course, the village idiot had the same tailor as the king. So of course the king needed finer and finer clothing to outclass the idiots and butchers and farmers, and the key to this was color.

∇ You see, most sheep are white, so if you can find a sheep of any other color, usually black or gray, you can demand a higher price for its wool. Since only the rich can afford colored wool, anyone wearing colored clothing must be rich. It might occur to you, as it did to me, to stain white wool with grass or flowers or berries and so produce colored wool for fancy dress without paying through the nose. But most of these stains are ***fugitive,*** that is, they come out in the wash. And so we are in the peculiar position of complaining that the laundry got our clothes too clean. There are, however, some dyes which produce ***colorfast*** colors on wool. Black walnuts, for example, make a beautiful chocolate brown which will not wash out. Many vegetables will produce colorfast colors when the wool is pre-treated with a mordant. But the most popular dye of all times, indigo, produces colorfast blue only under rather specific conditions; the dye-bath must be both alkaline and reducing, conditions that are met with in stale pee-pee.

∇ And so I am treading about in fermenting urine which creates a reducing environment for dissolving indigo from the woad plant so that thread may be colored blue, this thread to be woven into increasingly complex patterns so that the prince may be distinguished from the pauper, the bride from her bridesmaids, the Yankees from the Confederates and the Dallas Cowboys from the New York Giants. The demand for more and brighter clothing will in time create a demand for sulfuric acid, the first chemical to be produced on an industrial scale and even today the chemical produced in the greatest tonnage. Soap will be needed to launder these clothes which will increase the demand for potash to the point that the price of wood ash goes through the roof and new sources of alkali are sought, the budding alkali industry providing such innovations as air and water pollution. To comply with anti-pollution laws, the waste products will be turned into bleach, which of course makes possible even brighter colors to be supplied from coal tar, whose waste products will be used to manufacture fertilizer so that crops may be grown more efficiently from limited acreage, increasing the global population and the demand for more clothing. But I am getting ahead of myself.

12.2 🜍

The problem of coloring cloth is not as simple as you might at first expect. Pliny has told you about colored minerals in Chapter 11, but while these compounds make useful ***pigments,*** they make lousy ***dyes.*** You see, a pigment by definition does not stick to the surface to be colored. A ***binder*** mechanically sticks the pigment to the surface like glue. A crayon, for example, contains pigments in a wax binder. Plaster is the binder for fresco, linseed oil for oil paints, and gum Arabic for water colors. If you wish to make paint, you add a solvent, usually oil or water, to the pigment and binder. While paint is useful for coloring wood and stone and stucco, it is useless for clothing unless, of course, canvas is your idea of slipping into something more comfortable. No, colored clothing requires a coloring agent which will penetrate each fiber so that the color is as flexible as the yarn. Such a color is called a dye. Now there are two hard bits about dyeing: getting the dye into solution so that it may be absorbed by the cloth and then preventing this dye from washing out again.

One approach to this problem is to use the juice from colored vegetables. Just think about all of the foods you would hate to spill on your frock accidentally: coffee, tea, beets, red onions, to name but a few. Now if you think about it, food stains generally wash out because the colored compounds had to be water-soluble to get into the cloth, and being water soluble they wash right out again. Even the most persistent food stain leaves only a faint brown, tan, or yellow stain; dull, faint colors are prized by neither queens nor hookers. A ***mordant*** is a metal compound which chemically binds to the fiber on one hand and to the dye on the other. Whereas paint is a heterogeneous mixture of pigment and binder, a mordant forms a homogeneous compound with the fiber and dye molecules.

Alum, potassium aluminum sulfate, is the most popular mordant of all times and may be used to render coffee (tan), tea (rose), beets (gold), and red onions (orange) colorfast. Alum also mordants other vegetable dyes, including fern (yellow-green), elder-berries (lilac), madder (red), and saffron (yellow). A few animals provide dyes, notably red cochineal and kermes extracted from insects and Tyrian purple extracted from a sea-snail. If you are interested in these dyes, you should consult *The Weaving, Spinning, and Dyeing Book* [2] or *The Art and Craft of Natural Dyeing.*[3]

Indigo requires no mordant for reasons that are both intrinsically interesting and, as it happens, important to the historical development of chemical industry. The blue color comes from the compound ***indigotin,*** $C_{16}H_{10}N_2O_2$, which is insoluble in water. It is easily reduced to leucoindigotin, $C_{16}H_{10}N_2(OH)_2$, which is colorless and soluble in alkaline solution. You are possibly considering that if wool could be soaked in a solution of leucoindigo, oxygen from the air might oxidize it back to blue indigo which, being insoluble in water, would be extremely colorfast. If so, you have hit the snail on the head, so to speak. The problem, then, is to reduce indigotin to leucoindigotin.

So far we know only two reducing agents, charcoal and sugar. We used charcoal at high temperature to reduce metal oxides to elemental metals. We forced yeasts to oxidize glucose anaerobically to produce alcohol. Since heating the bejeezus out of indigo just makes smoke, let us try the glucose route. Equation 12-1(a) shows the reduction of indigotin, (b) the oxidation of glucose, and (c) the balanced redox reaction. Now, leucoindigotin is soluble only in alkaline solution and yeasts are not particularly tolerant of alkali. There is a bacterium, however, which eats urea, farts carbon dioxide, and pisses ammonia. You will be familiar with this bacterium if you have ever used an outdoor privy.

When mammals metabolize protein, the nitrogen is excreted as urea. What is waste to us is food for the bacterium. Equation 12-2 shows the reaction by which the bacterium converts urea to ***ammonia,*** NH_3. Ammonia gas is soluble in water and ionizes to give hydroxide ion and so it is an alkali. The bacterium has just what we need for dissolving indigotin; it produces alkali and oxidizes glucose. You are probably thinking that we should add honey and indigo to a 2-liter bottle of urine, burping

2. Reference [91].
3. Reference [92].

Equation 12-1. From Indogotin to Leucoindigotin

$$(a)\ C_{16}H_{10}N_2O_2(s) + 2\,H^+ + 2e^- = C_{16}H_{10}N_2(OH)_2(aq)$$
$$(b)\ C_6H_{12}O_6(aq) + 6\,H_2O(l) = 6\,CO_2(g) + 24\,H^+ + 24\,e^-$$
$$(c)\ 12\,C_{16}H_{10}N_2O_2(s) + C_6H_{12}O_6(aq) + 6\,H_2O(l) = 12\,C_{16}H_{10}N_2(OH)_2(aq) + 6\,CO_2(g)$$

Equation 12-2. From Urea to Ammonia

$$(a)\ CH_4N_2O(aq) + H_2O(l) = CO_2(g) + 2\,NH_3(g)$$
$$(b)\ NH_3(g) + H_2O(l) = NH_4OH(aq)$$
$$(c)\ NH_4OH(aq) = NH_4^+(aq) + OH^-(aq)$$

it from time to time as we did with mead. The indigo vat would be much more pleasant were that possible. You will recall that whereas yeasts thrive anaerobically, ***bacteria*** require oxygen to live. We must carefully balance the air intake of our vat, providing enough oxygen for the bacteria to live, but not so much that our hard-won leucoindigotin is oxidized back to insoluble indigo.

Once the vat is in order, alkaline and saturated with colorless (or pale green), soluble leucoindigotin, we are ready to dye our yarn. Simply soak the yarn in the vat until it is thoroughly saturated with liquid. When the yarn is pulled from the liquid, oxygen from the air oxidizes leucoindigotin to indigotin as shown in Equation 12-3. The yarn will change from yellow to blue and this blue will be colorfast. It looks like a magic trick.

Actually, we might have saved ourselves a good deal of trouble if we had been able to extract leucoindigotin from woad or indigo directly. Unfortunately, the indigotin is oxidized by the air very quickly. But we can use the same idea to extract dye from black walnuts if we are able to collect them when they have just fallen from the tree. The walnut tree has been kind enough not only to produce the dye juglone, $C_{10}H_6O_3$, in its colorless, water-soluble state, but to package it in an air-tight container. I am speaking, of course, about the hull of the walnut itself. When the nuts fall in the fall they are initially soft and green, but after they have been lying around for a few weeks, the air oxidizes the juglone and the nut becomes hard and brown. If you are able to harvest black walnuts

Equation 12-3. From Leucoindigotin to Indigotin

$$(a)\ C_{16}H_{10}N_2(OH)_2(aq) = C_{16}H_{10}N_2O_2(s) + 2\,H^+ + 2e^-$$
$$(b)\ O_2(g) + 4\,H^+ + 4e^- = 2\,H_2O(l)$$
$$(c)\ 2\,C_{16}H_{10}N_2(OH)_2(aq) + O_2(g) = 2\,C_{16}H_{10}N_2O_2(s) + 2\,H_2O(l)$$

as soon as they fall from the tree, you will be able to make a beautiful brown, colorfast dye with no mordant and no urine required. You will be pissed at yourself if you miss the harvest.

Material Safety

Biological hazards tend to be far more insidious than chemical ones. You see, one molecule of even the most toxic chemical is absolutely harmless. Such a molecule might react with one of your molecules, but you have so many that it makes no difference. With a chemical compound the dose makes the poison; a larger dose is more hazardous and a smaller dose is less hazardous. Biological hazards are not like this at all. You know from making mead that one yeast becomes two, two become four, and so on until the mead is chock full of them. If you choose to ferment urine, the bacteria behave the same way. Of course, you have been exposed to these particular bugs all your life, so you have some immunity. Nevertheless, it is a good idea to wash your hands regularly, particularly when you have been brewing your own juices.

The squeamish may prefer to use something other than fermenting urine as a reducing agent and a popular alternative is sodium hydrosulfite (CAS 7775-14-6) in household ammonia (CAS 1336-21-6). Summarize the hazardous properties of these materials in your notebook, including the identity of the company which produced each MSDS and the potential health effects for eye contact, skin contact, inhalation, and ingestion. Also include the LD_{50} (oral, rat) for each of these materials.

Your most likely exposure is eye or skin contact. If you get some in your eyes, you should flush them with cold water and go to the emergency room. Exposed skin should be washed with soap and water. Be aware that sodium hydrosulfite will bleach clothing. Be aware that walnut hulls will stain skin.

You should wear safety glasses and rubber gloves while working on this project. Leftover dye solution may be washed down the drain.

Research and Development

You are probably wondering what will be on the quiz.

- You should know the meanings of all of the words important enough to be included in the ***index*** or **glossary.**
- You should have studied the Research and Development items from Chapter 6 and Chapter 8.
- Know the equation for the reduction of idigotin.
- Know the equation for the oxidation of leucoindigotin.
- Know the equation for the conversion of urea to ammonia and carbon dioxide.
- Know the hazardous properties of indigo, walnut hulls, sodium hydrosulfite, and urine.
- Know why humans wear clothing and why dyes are so important.

The Stockholm Papyrus which began this chapter contains 154 recipes, of which I have chosen only a wee sample. Indigo will remain the number one dye even down to modern blue jeans. Progress in dye technology will come in dribs and drabs until the eighteenth century, when a new way to dissolve indigo gives a leg up to the infant alkali industry. Cheap alkali, in turn, will relieve a soap industry starved for soda. Industrial waste flushed out of the alkali trade will become the mainstay of the bleach industry. All of this in the service of clothing so that we can tell the whiz-kids from the pee-ons. Now, if you are dyeing to get started and you cannot hold it any longer, urine in for a treat.

12.3 Θ

If you are going to dye indigo the old fashioned way, you will need to collect some urine. Urea comes from the metabolism of protein, not from drinking beer, so if you are drinking heavily to make more water, you have your leg up the wrong tree. Just let it flow naturally; you only need a liter or so and you can collect it in our old friend, that twenty-first century gourd, the 2-liter soft-drink bottle. Be sure to label it or someone may get a bit of a surprise. Put a few blades of grass into it to inoculate it with bacteria and then leave it outside in the sun. The time needed for fermentation to begin depends on the temperature. It will go quicker in the summer than in the winter. Unlike yeasts, bacteria need air so leave the top off your bottle. Your vat is ripe when you can smell ammonia, usually after a few days.

Once ammonia is being produced, grind 1 teaspoon (approximately 1 gram) of indigo in a mortar and pestle with a few teaspoons of water. You want the indigo to be as fine as possible because bacteria have tiny little mouths. Add your ground indigo to the bottle and gently swirl it to distribute the dye without adding too much oxygen. The fermentation will be most active when your bottle is warm so let the little fellows have their day in the sun. The solution will be dark blue when you first add the dye, but it should become pale green once the bugs get going.

If all is well, you should have a pale green vat with a blue scum floating at the surface. If not, there are several things that might have gone wrong. First of all, your vat should smell of ammonia. If not, add 1 teaspoon of household ammonia, swirl to mix it in, and wait another day. Some household ammonia has soap in it; you do not want that kind, you want *clear* ammonia. Check your vat each day, adding ammonia 1 teaspoon at a time until it smells as it should.

If your vat smells of ammonia but remains blue, then perhaps your bacteria are starving; give them a teaspoon of honey to eat. Fermentation should pick up and the vat should turn green in a couple of days. If not, then either a contamination from your dye or something else has killed your bacteria. Give it a few more days to make sure, but if it does not pull itself together I can only advise you to start over with a fresh bottle.

Once your vat is in order it is time to dye your yarn. Wash your woolen yarn, presumably that which you made in Chapter 6, with plenty of soap and water, rinse the soap out, and place it gently into the vat, trying to add as little oxygen as possible. Use a stick or stirring rod to keep the yarn beneath the surface and let it soak for 10 minutes or so. Then fish it out and gently squeeze the excess fluid from the yarn. In a matter of minutes, the yarn will have changed from pale green to colorfast blue. If you would like the color to be darker, you may re-dip your yarn after 10 minutes in the air; with each dip the color should become darker and

Figure 12-1. Dyed in the Wool

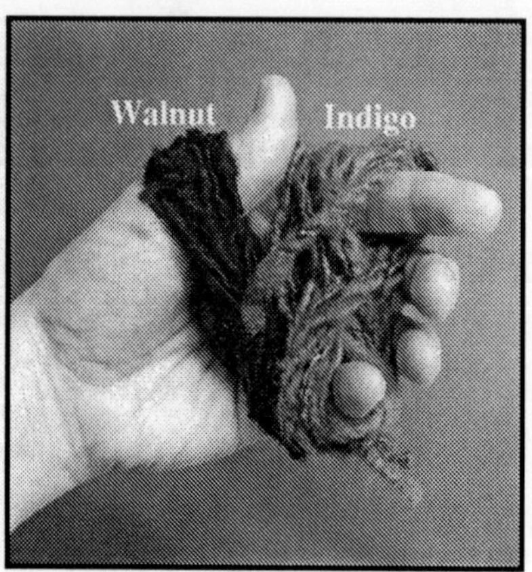

darker. Let your yarn air out for a day, wash it in vinegar, and then in soap and water. The color will be quite permanent.

As with a mead, it is possible to keep the vat going by adding food and nutrients. If the vat turns blue and no longer smells of ammonia, you need to add more urine. If it turns blue but smells of ammonia, you need more honey. And if it is pale green and ammoniacal but your blue is getting wimpy, add more indigo.

I realize that not everyone is comfortable handing precious bodily fluids. For you uriphobes, I am pleased to provide an alternative indigo vat which replaces the bacteria with a chemical reducing agent: sodium hydrosulfite. You will learn all about bleach in Chapter 25, but for now all that you must understand is that there are oxidizing bleaches and reducing bleaches and that sodium hydrosulfite is of the latter type. So if I may continue, you should run 100 mL of hot water (50°C, 120°F, no hotter) into a small bottle and add 4 mL of household ammonia to make a kind of faux-pee. Grind 0.2 g of indigo with a little water as before and add it to your vat. Now add 0.4 g of sodium hydrosulfite, which may be purchased as "color remover" wherever dyes are sold. Gently swirl until the hydrosulfite dissolves and let the vat rest. Within 10 minutes or so, the color should change from deep blue to pea green. If not, place your bottle into a pan of hot water as a kind of makeshift double-boiler or *bain Marie*. Swirl the bottle gently until the color changes from opaque blue to transparent green, as shown in Figure 12-1(L). You may now use this hydrosulfite vat in the same manner as the urine vat for dyeing yarn. The hydrosulfite vat may be "kept going" similarly to the urine vat. If you can no longer smell ammonia, add some. If the vat changes from green to blue, add more sodium hydrosulfite and re-warm it by swirling the bottle in a pan of hot water. And when your blues lose their hues, add more indigo.

Black walnuts produce a wonderful colorfast brown dye without the muss and fuss of the indigo vat. To dye 20 feet of yarn or so, you will need 5 or 6 walnuts freshly fallen from the tree. Now, you are not interested in the nut itself; what you need is the green rind which surrounds the nut. With a knife, peel the rinds from your nuts and place them into a pan or beaker with about 1 liter of water. The good stuff is in the juice from these rinds, so if you can squeeze the juice into the water, so much the better. Now, if you work without gloves you will notice in an hour or two that your hands are stained brown even though the juice was green. You are probably thinking that the colorless juices have been oxidized by the air to produce dark brown, insoluble juglone on your hands. If so, you have indeed cracked a tough nut. This is precisely what we would like to have happen to the wool, but if you do not wish to dye your hands, you had better wear dishwashing gloves.

Heat your rind-water on a stove or hot-plate until it comes to 60°C or 140°F. Any hotter than this may damage the wool. Wash your woolen yarn with soap and water, immerse it in the hot dye-bath and let it soak for half an hour or so. Remove the yarn from the dye and let it air out overnight. It should be dark brown and colorfast the following day. If you would like to dye wool black, dye it first in black walnut and then in indigo.

There is a movement these days to recognize the contributions of women to science. The textile arts have provided an important driving force for the development of chemical industry, as subsequent chapters will show. For now just remember, "It may be the clothes that make the man, but it is often the woman who makes the clothes."

Quality Assurance

On the other hand, if your dye washes out with soap and water, you've just been piddling around. Break off a few inches of your colored yarn and tape it into your notebook.

Chapter 13. Theophilus (Glass)

If you have the intention of making glass, first cut many beechwood logs and dry them out. Then burn them all together in a clean place and carefully collect the ashes, taking care that you do not mix any earth or stones with them. After this build a furnace of stones and clay, fifteen feet long and ten feet wide,...

When you have arranged all this, take beechwood logs completely dried out in smoke, and light large fires in both sides of the bigger furnace. Then take two parts of the ashes of which we have spoken before, and a third part of sand, collected out of water, and carefully cleaned of earth and stones. Mix them in a clean place, and when they have been long and well mixed together lift them up with the long-handled iron ladle and put them on the upper hearth in the smaller section of the furnace so that they may be fritted. When they begin to get hot, stir at once with the same iron ladle to prevent them from melting from the heat of the fire and agglomerating. Continue doing this for a night and a day.

Meanwhile take some white pottery clay, dry it out, grind it carefully, pour water on it, knead it hard with a piece of wood, and make your pots. These should be wide at the top, narrowing at the bottom, and should have a small in-curving lip around their rims. When they are dry, pick them up with tongs and set them in the red-hot furnace in the holes made [in the hearth] for this purpose. Pick up the fritted mixture of ashes and sand with the ladle and fill all the pots [with it] in the evening. Add dry wood all through the night, so that the glass, formed by the fusion of the ashes and sand, may be fully melted.

— Theophilus, *On Divers Arts, ca. 1100 AD* [1]

13.1 ☿

Δ Have you been observant, my brothers and sisters? Have you chosen to see things for yourself or are you afraid to touch what might burn your lily-white fingers, to taste what might upset your sensitive stomach, to smell what might clear your stuffy sinuses, to hear what might offend your timid ears? Did you stop to examine the lining of your crucible in your haste to retrieve your nugget of bronze, or did you notice that the

1. Reference [27], Book II, Ch. 4-5.

slag had given the interior of the crucible an amazing glossy finish? If so, you have seen what went unappreciated by generations of potters and smelters until I attached significance to it.

Δ It was in the early days of smelting technology, 4527 BC, if meme-ory serves, a millennium before the Bronze Age began in earnest. I had been raised to smelting from early childhood, sorting ores, making crucibles and gathering firewood for the furnace. I pestered my father and brothers constantly with questions and unsolicited advice. "Why is the inside of the crucible black while the outside is white?" "What happens if you leave out the charcoal?" "What happens if you leave out the soda ash?" When I was older I began to answer these questions for myself. A waste of good material, my father said. I always suspected he was talking about me as well as the ore I used. I discovered that the inside of the crucible is blackened by the charcoal. But if you leave out the charcoal, the inside of the crucible is blue, not white, as if the malachite ore had been melted like wax. If you leave out both the charcoal and the soda ash, that is, if you simply heat malachite ore in a crucible, it turns black but does not melt. And soda ash alone melts right into the clay. It seemed to me that the soda ash made it easier to melt malachite, and melted malachite is a beautiful thing.

Δ By 4000 BC I had learned to glaze quartz beads with a combination of soda ash and lime colored with malachite. At the same time I began applying the glaze ingredients to the surfaces of clay tiles, but because soda-lime glazes and clays shrink at different rates as they cool, they are not suitable for curved pottery. This limitation was overcome 2000 years later with the introduction of galena (lead sulfide), rather than soda ash, as the flux. From that point glass and glaze followed separate paths. Glaze developed as I wended my way down through generations of potters using lead in the form of galena and cerrusite (lead carbonate) to flux sand and clay. Glass emerged from the smelting tradition some 500 years later, with soda ash and potash used to flux lime and sand.

Δ Having grown out of the metal arts, glass was cast in molds much as metals are. But beginning in about 30 BC glass began to be worked in its molten state, blown on the ends of iron pipes. Today glass is cast, blown, molded and drawn into almost any shape that can be conceived. Without it churches and skyscrapers would be pitch-black caverns, the color of fine wines would remain unappreciated in opaque clay bottles and cups. The microscope, telescope and electric light would be impossibilities. And the greatest contribution of all, that which has literally returned eyesight to billions, goes by the name, glasses. In short, had I seen slag as merely the refuse of the smelter, much of the world would have remained impenetrably opaque and many of its inhabitants hopelessly blind.

13.2 🜍

Have you ever made ice cream from scratch? There is a fundamental problem in making ice cream; the freezing point of cream is lower than that of water and consequently ice is not sufficiently cold for freezing it. The solution is to add salt to the ice. Saltwater freezes at a lower temperature than pure water and the more salt is added, the lower is the freezing point. This is not peculiar to saltwater; any solution melts at a lower temperature than its solvent. We use salt because it is one of the least expensive substances that dissolves in water, but we could use many other materials to produce temperatures lower than the melting point of water. As a case in point, sugar, too, lowers the freezing point of water, which is the reason that ice cream, a solution of sugar in water with various suspended milk fats, has such a low freezing point to begin with.

Consider the sequence of events that take place when water freezes. We begin with pure water at room temperature, 25°C and cool it, say, by placing it into a freezer with the temperature set at -15°C. The temperature of the water falls, 24, 23, 22°C... A peculiar event occurs when the temperature reaches 0°C; a crystal of ice forms. Though we continue to cool the water, its temperature ceases to fall. An unobservant reader may have glossed over this remarkable statement and it bears repeating. We have water at 0°C in a freezer at -15°C, and yet the temperature of the water remains at 0°C. Though the temperature does not change, all is not static in our water sample; the ice crystal is growing. The water sample is now half ice, yet its temperature is still 0°C. It is now almost completely frozen, with only a single drop of liquid water remaining and yet the temperature has not budged. Only when the final drop has frozen does the temperature begin to drop, -1, -2, -3°C, until at last the ice reaches the ambient temperature of the freezer. One more aspect of the melting behavior of pure substances needs to be understood. Ice is solid and water is liquid. Ice is hard and water is runny. Ice holds its shape and water takes the shape of its container. There is no middle ground between ice

Figure 13-1. Melting Ice and Ice Cream

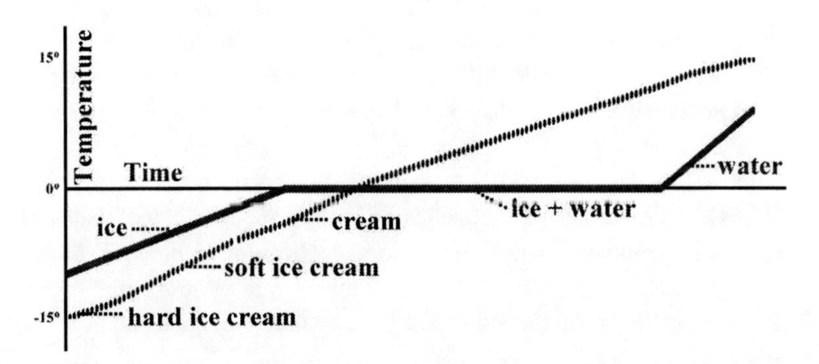

and water. At any point during the freezing process, you can point to the spot where the ice ends and the water begins. This behavior is characteristic of the freezing of pure substances.

Contrast this with the freezing behavior of a solution like ice cream. We put the cream and sugar into the ice cream freezer, begin to turn the crank, and as the temperature falls the cream becomes thicker, more *viscous.* Finally, the cream is so hard that we can no longer turn the crank, but when we open the container we find that the ice cream is still fairly soft. It will continue to get harder and harder as the temperature drops. Indeed, it is hard to say exactly when the freezing is complete. Furthermore, there is no spot where the ice cream ends and the cream begins. This behavior is characteristic of the freezing of solutions.

Figure 13-1 compares the melting behavior of ice and ice cream. The temperature of a cup of ice rises to 0°C, at which point water begins to collect in the bottom of the cup. The temperature remains constant as the water rises and the ice shrinks. Only after the last ice has melted does the temperature resume its upward climb. In contrast, the temperature of ice cream rises steadily as it gets softer and softer. Nothing "special" happens at 0°C. It is clearly solid at -15°C and clearly liquid at +15°C, but there is no single melting *point.* Rather, the ice cream melts over a *range* of temperatures. The addition of salt or sugar (the **solutes**) to water (the **solvent**) causes it to freeze or thaw over a range of temperatures lower than the normal freezing point of the solvent. We say that a solute acts as a ***flux,*** for solvents in which it dissolves.

While salt and sugar dissolve in water, they do not do so in silica. A proper flux for silica must be soluble in it but must not boil away in the high temperatures of the furnace. There are four such fluxes in common usage, potassa, soda, litharge, and borax. ***Potassa*** K_2O is derived from ***potash,*** K_2CO_3 introduced in Chapter 8. In the heat of the kiln, bejeesical carbon dioxide flees the potassium carbonate, leaving potassium oxide to flux the glass. Similarly, ***soda,*** Na_2O, is derived from ***soda ash,*** Na_2CO_3. Glass produced from silica and soda or potassa alone is somewhat soluble in water and so has limited application. In practice, ***lime,*** CaO, is added to glass to render it insoluble in water. Most of the glass in common use is soda-lime glass, a frozen solution of soda and lime in silica.

Leaded glass crystal and many pottery glazes are fluxed with ***litharge,*** PbO, produced from the roasting of lead ore. You may associate lead with fishing sinkers, car batteries, and solder, but it is also associated with lead poisoning. And indeed, lead oxide is quite toxic. The symptoms of lead poisoning include abdominal pain, nausea, vomiting and headache. Acute poisoning can lead to muscle weakness, loss of appetite, insomnia, dizziness, and even coma and death in extreme cases. Moreover, lead exhibits chronic toxicity, with small doses acquired over long time periods causing symptoms similar to those of acute lead poisoning. No wonder that lead has been branded a "toxic" metal. But while lead compounds are certainly hazardous, it is a mistake to tarnish the reputations of leaded crystal and leaded pottery glazes. Consider, for example, that soda and lime are strong alkalis and would act as acute poisons if ingested directly. Yet glass does not even taste bitter. The reason is that the soda and lime are locked into the glass and consequently cannot leach out into food and drink. The same is true of lead crystal and lead glazes when properly manufactured.

Complete solution of lead oxide into the silica is the central concern of the lead glass manufacturer. With many pottery glazes, raw glaze ingredients may be applied directly to the ware. When the pottery is fired, the glaze components melt and mix. But with lead glazes, it is safer to melt the raw glaze ingredients into a glass so that complete solution is ensured. This glass is then cooled and crushed into ***frit***, which is simply a term for powdered glass. It is this frit, with the lead already locked into the glass, which is applied to the pottery. Lead poisoning remains a concern for workers making the frit, but not for the end-user. However, un-fritted lead glazes may still be used by some potters and such glazes pose a health risk when used to contain food or drink.

Those who insist on viewing the world in simplistic terms may argue that because it is possible to do a poor job of formulating lead glazes, it is not worth the risk. I would like to point out that it is possible to build a car poorly, to store food poorly, to raise a child poorly. Yet we continue to drive cars because it is faster than walking, to store food because it is better than starving, and to raise children because it is better than letting them fend for themselves. We continue to use lead crystal and glazes because they are more beautiful and versatile than other glasses and glazes. The key to safety is do these things as well as we can, to be prepared to recognize when things go wrong, and to know what to do in those circumstances. Some will persist in demonizing lead, imagining invisible goblins in every cup and saucer, but nobody has to wonder about lead poisoning; test kits for lead may be purchased at ceramics shops and pharmacies.

In addition to potassa, soda, and litharge, ***borax,*** $Na_2O{\cdot}2\ B_2O_3$, or $Na_2B_4O_7$, is an important glass flux. Borosilicate glasses are probably familiar to most people under the trade names, Pyrex and Kimax. Like salt, trona, gypsum, or limestone, borax is deposited when mineral-rich seas evaporate. Compared to the other glasses, borosilicate glasses are particularly suitable for applications in which they will be heated and cooled repeatedly. To understand this property we must return briefly to the details of the cooling of molten glass.

Recall that we were unable to specify a "melting point" for ice cream. At room temperature it was clearly a liquid, and as it cooled it became thicker, more viscous, until at a temperature of -15°C, it was clearly a solid. With no single melting point, how are we to specify the temperature range over which its viscosity changes? It may surprise you to learn that ***viscosity***, the resistance to flow, may be measured; its unit is the *poise*. We may characterize glass by the temperatures at which its viscosity takes on certain values. For the sake of comparison, water at room temperature has a viscosity of 10^{-2} poise (0.01 poise) and motor oil a viscosity of about 10^2 poise (100 poise). Even at high temperatures, molten glass has a viscosity in excess of 10^3 poise. As it cools, it reaches the ***working point***, the temperature at which its viscosity equals 10^4 poise. Near the working point, glass is viscous enough to be easily worked by glass-blowers without dripping from the end of the blow-pipe. By the time it cools to the ***softening point***, the viscosity has risen above 10^7 poise. A sheet of glass below this temperature will not sag under its own weight. Cooling further, the viscosity reaches 10^{13} poise at the ***annealing point***. At this temperature the glass is rigid, but still soft enough that it is not easily shattered by mechanical or thermal shock. Below the ***strain point*** the viscosity exceeds 10^{14} poise and the glass is essentially solid. At room temperature, the viscosity of glass is in the neighborhood of 10^{20} poise. While the viscosity values are important for defining the working, softening, annealing and strain points, the more important point is that the viscosity of glass changes gradually over a range of temperatures. By contrast, the viscosity of water, or indeed any pure substance, changes abruptly from a very small value, that of the liquid, to a very large value, that of the solid, and it does so at a single temperature, the melting point.

I have spoken up until this point as if the temperature of a piece of glass were uniform, but the reality is that a cooling piece of glass will be cooler on the surface than it is in the interior. Because glass expands as it is heated and contracts as it cools, such a temperature difference introduces stress. In molten glass, of course, such stress is easily and quickly relieved because the glass is free to move. But below the strain point the glass is so rigid that any residual stresses are permanently frozen in. Glass stressed in this way may look normal but can shatter violently without warning. It is imperative, then, for a glass-maker to ensure that these stresses have been relieved before the cooling glass reaches the strain point. The process of accomplishing this is called ***annealing***.

The ideal solution to the problem of stress would be to cool the glass so slowly that the surface of the glass is at the same temperature as the interior. But since stresses do not accumulate in the liquid state, it would be a waste of time to slowly cool the molten glass. Even worse, Unktomi has explained in Chapter 2 that crystals may grow when molten rock cools slowly. Given enough time, crystals of pure silica, that is, quartz, may grow in the molten glass, a process called ***de-vitrification***. In practice molten glass should be cooled quickly to a temperature just above the annealing point, held there until the temperature of the glass surface is close to that of the interior, and then cooled slowly to a temperature below the strain point. Below the strain point, the glass may be cooled more quickly.

Returning to the borosilicate glasses we can begin to understand why they are used for cook ware and laboratory glassware. Compared to the other glasses, borosilicates expand and contract less rapidly with changing temperature. Consequently, thermal stresses do not build up in borosilicate glasses to the same extent that they do in others and they are less likely to shatter when heated or cooled. If we were to make soda-lime glass in the pottery kiln we would have to pay careful attention to the firing schedule, cooling slowly through the annealing point to avoid having the glass shatter. Whereas pottery, metals, and lime may all be fired together in a four-hour firing schedule, I have obtained good results making soda-lime glass only with a twenty-four-hour firing schedule, most of that time spent annealing the glass.[2] Borosilicate glasses, by contrast, can be made using the same firing schedule as the other kiln projects. It is not that the glass does not need to be annealed, it is just that the normal cooling rate of the kiln is sufficiently slow to anneal small borosilicate glass objects. For this reason the next section will describe the fabrication of a borosilicate glass rather than the more common soda-lime glass.

Δ Purity is an over-rated quality, my brothers and sisters. An atmosphere of pure nitrogen would support neither flame nor fish nor fowl; one of pure oxygen would consume the forest in a whirling inferno. An ocean of pure water would be free of both salt and salmon. A perfect desert of pure quartz sand would be devoid of cactus and scorpion. Purity of thought is over-rated as well. The world of the ideologue is as simple and beautiful as the quartz crystal and equally as sterile. Do you think you are a man, my brother? You may happen to temporarily inhibit a male body but you are no man, you are no woman. You are not black or white or yellow. You share your temporary mortal home with memes from every corner of the globe, with ancient notions of remote antiquity and with young whipper-snappers fresh from the six o-clock news. Your mortal, in turn, is thrown together with mortals of different genders, complexions and cultures. At times de-vitrification seems imminent as purity threatens to separate the world into sterile crystalline strong-holds. My siblings, *be* the flux! Make living glass by lowering the melting point and bringing myriad dissimilar entities into rich and complex solution.

---

2. Reference [49] gives firing schedules for a wide variety of glass objects of different sizes.

Material Safety

Locate MSDS's for borax (CAS 1330-43-4), copper oxide (CAS 1317-38-0), plaster (CAS 10034-76-1), and silica (CAS 14808-60-7). Summarize the hazardous properties of these materials in your notebook, including the identity of the company which produced the MSDS and the potential health effects for eye contact, skin contact, inhalation, and ingestion. Also include the LD_{50} (oral, rat) for each of these materials.

Your most likely exposure is dust inhalation. If a persistent cough develops, see a doctor.

You should wear safety glasses and a dust mask while working on this project. Leftover materials may be disposed of in the trash.

Research and Development

If you are to make glass for yourself you should know the following:

- Know the meanings of those words from this chapter worthy of inclusion in the ***index*** or **glossary.**
- You should have mastered the Research and Development items of Chapter 9 and Chapter 10.
- Know how the melting and freezing behaviors of a solution differ from those of a pure substance.
- Know the properties of soda ash and potash as described in this chapter and in Chapter 8.
- Know how to handle borax, copper oxide, plaster, and silica safely.
- Know what it means to be a *flux*.

13.3 Θ

The vast majority of glass produced in the world is soda-lime glass and it is certainly possible to make glass in the pottery kiln from soda ash, limestone, and silica. Since some students in a class might be making glass while others are working on metals, pottery, or lime, it would be desirable to use the same kiln program for all four kiln projects so that they might all be fired together. The kiln conditions required for making soda-lime glass, however, are at odds with those for smelting metals. For one thing, soda ash and limestone give off bubbles of carbon dioxide when fired and if the glass is to be free of bubbles it must be left at its highest temperature for an hour or more. In this time, the charcoal in a smelting crucible might burn away completely, spoiling the metal. For another thing, glass shrinks as it cools and consequently must be cooled slowly through its annealing point, considerably lengthening the time for a firing. We can address both problems by making a borosilicate glass instead of a soda-lime glass.

In a soda-lime glass the soda serves as a flux, lowering the melting point of silica. The lime serves to make the resulting glass insoluble in water. Soda serves as a flux in a borosilicate glass, as well, but boron oxide rather than lime (calcium oxide) serves to render the glass insoluble in water. Both sodium oxide (soda) and boron oxide are available in a naturally-occurring mineral, borax, available wherever laundry detergents are sold. Just as calcium sulfate or sodium carbonate occur in anhydrous and hydrous forms, borax occurs as the anhydrous $Na_2O{\cdot}2\ B_2O_3$ and as the decahydrate, $Na_2O{\cdot}2\ B_2O_3{\cdot}10\ H_2O$. Since water will be driven off by the heat of the kiln, either form will do for making glass but the weight needed will depend on which one is being used. To eliminate guesswork, you should dry your borax for an hour in an oven at 130°C (266°F) to convert the decahydrate to anhydrous borax. Anhydrous borax may be stored indefinitely if it is kept in a moisture-proof container. In addition to borax we need silica, which might be provided in the form of ordinary sand, but it should be ground to a fine powder. Ground silica, known among potters as *flint*, can be had inexpensively from ceramic supplies. With borax and silica at hand, we are ready to formulate our glass.

There is no "correct" formula for borosilicate glass. The more borax is used, the lower will be the melting point of the resulting glass. A very nice, colorless glass can be produced using 1 ***part*** anhydrous borax to 2 parts silica. Let us begin, then, by weighing ***by difference*** (page 384) 15 g of borax and 30 g of silica into a plastic bag. Optionally, copper oxide may be added to provide a blue-green color; 1 g of Cu_2O will render the glass completely opaque. Thoroughly mix your glass ingredients by kneading the bag and turning it end for end.

If you completed the mold of Figure 10-3(R) (page 129) you have only to pack your mixed glass ingredients into the mold. If you have no mold, you must at least coat the walls of your crucible with ***investment,*** as

Figure 13-2. Freeing the Glass from the Crucible

Figure 13-3. Removing the Investment

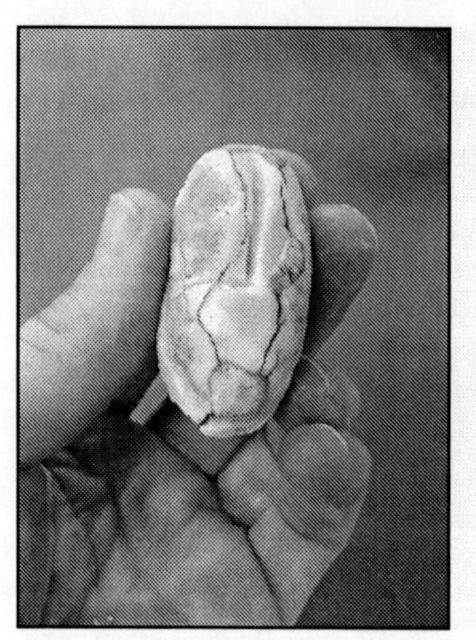

shown in Figure 10-2(L) (page 128), to prevent the glass from sticking to the walls of your crucible. Fill your mold or crucible with the mixed glass ingredients, cover your crucible with its lid, and fire it to cone 05. When it returns from the kiln, remove the lid and use a knife or screwdriver to pry the investment from the crucible. Chip away at the investment to free your glass from its cocoon, as shown in Figure 13-2. A knife can be used to remove much of the investment, but a Dremel tool with a cutting disk is useful for removing it from the surface of the glass. Provide yourself with a bowl of water and use it to wash your glass object twice a minute, or so. The water will help to soften the investment and it will keep the glass cool. Since glass is *much* harder than investment, it is easy to remove the investment without fear of damaging the glass. The finished glass casting is shown in Figure 13-3(R).

Δ What lesson have you learned from old Theophilus? I have variously referred to you in the past as though you were my son or daughter, my brother or sister. But you and I are not animal, vegetable or mineral. We have no gender, no nationality, no ethnicity, or to be more precise, we are free to surround ourselves with identifications of our own choosing. My effort in this book is not to instruct you as if you were a recalcitrant child in need of smartening up. Rather, I am here to re-mind you, to bring together into mental solution thoughts which may have de-vitrified over the centuries. From this point I shall no longer treat you as a child, but as a comrade in flux.

Quality Assurance

Record in your notebook the weights of your glass materials and a description of your procedure. Your glass should be, well, glassy. Either it is glass or it is not. Photograph your glass and include it in your notebook as a testament to your ability to keep in the heat and withstand it.

Chapter 14. Ts'ai Lun (Paper)

Description of the manufacture of Talkhi paper. Take good white flax and purify it from its reed. Soak it in water and shred it with a comb until it is soft. Then soak it in the water of quicklime for one daylight and one night until the morning. Then knead it by hand and spread it out in the sun all daylight until it is dry. Return it to the water of quicklime, not the first water, and keep it for the whole of the second night till the morning. Knead it again by hand as before and spread it out in the sun. This is repeated during three, five or seven days. If you change the water of quicklime twice daily, it would be quicker. When it becomes extremely white, cut it by the shears into small pieces, then soak it in sweet water for seven days also and change the water every day. When it is free from quicklime, grind it in a mortar while it is wet. When it is soft and no knots are left, bring it into solution with water in a clean vessel until it is like silk. Then provide a mould of whatever size you choose. The mould is made like a basket [from] samna reeds. It has open walls. Put an empty vessel under the mould. Agitate the flax with your hand and throw it in the mould. Adjust it by hand so that it is not thick in one place and thin in another. When it is even, and is freed from its water while in the mould, and when you have achieved the desired result, [drop] it on a plate, and then take it and stick in on a flat wall. Adjust it by your hand and leave it until it dries and drops. Then take fine wheat flour and starch, half of each, knead the starch and the flour in cold water until nothing thick remains. Heat water until it boils over and pour it on the flour mixture until it becomes thin. Take the paper and with your hand, paint it over and place it on a reed. When all the sheets of paper are painted and dry, paint them onto the other face. Return them to the plate and paint them thinly with water. Collect the sheets of paper, stack them, then polish them as you polish cloth, if God wills.

— Ibn Badis, *Book of the Supports of the Scribes, ca. 1061 AD* [1]

14.1 ☿

∇ I was born a poor Chinese girl. My family made hempen homespun cloth and from an early age it was my job to beat hemp, that is to say, to soak the hemp in water and then beat it with a mallet to release the fibers from the stem. And, being poor and all, we couldn't afford to waste

1. Reference [53], pp. 192-194.

anything and so when clothes got worn out, I had to beat the rags to get their fibers. Well, we would beat hemp and rags all day long in a tub of water and by the end of the day, that water would be so nasty that we would dump it out and start with fresh water the next day. And that's how it was.

∀ One day, long about 317 BC or so, if I recall, I was about to dump my water out at the end of a long day and I noticed some bits of fiber floating around. Normally I would just pick them out by hand, but that day I was really tired and so I decided to strain the water through a cloth instead. Lo and behold, there was a lot more fiber in there than I thought. I remember thinking, "Boy howdy, we've been throwing out a lot of fiber over the years." But, you know, most of it was too small for spinning. I picked out the good bits and left the rest on the cloth.

∀ Next morning I went to wash my cloth and don't you know that fiber had dried and peeled off of that cloth just as pretty as you please. Now, if we hadn't been so poor I probably would have just pitched it out; those fibers were too small for spinning, like I said. But we were used to using every little bit to get by and I thought all that fiber had to be good for something. And it was. You could wipe babies with it and you could fold it into little packets for holding herbs and such. You could light fires with it and you could even fold it into a hat of sorts. So that was good. And pretty soon everybody was straining their water so as not to waste anything.

∀ As the in*spir*ation for paper, I floated around for a couple of hundred years and didn't take myself too seriously. I moved from place to place anywhere they beat fibers and eventually I wound up in Hunan province, where they made cloth from mulberry bark instead of hemp. That's when I got into the head of a bureaucrat named Ts'ai Lun, who was in charge of manufacturing things. There I hooked up with mass-productionism, supply and demandism, not to mention self-promotionism and pretty soon I was getting a pretty good chunk of his thought-time.

∀ Now, Theophilus' notion that it doesn't matter whether you're a man or a woman is just a pipe-meme. Nobody was going to pay no attention to peasant women making ***paper*** in the sticks. Except Ts'ai Lun maybe. All that testosterone makes men pushy and grabby when it hits their brains. And their names are pushier and grabbier, so they tend to stick to whatever memes they can. Which is what happened to me, when you think about it.

🜃 Anyway, in 105 AD I made it to the emperor. I told him that *chih*, or paper, could be made as a material in its own right, not just as a by-product of textile manufacture. And well-made paper could even be used for writing on, which was good news to the emperor on account of all they had to write on was silk cloth and bamboo strips. Of course, silk was expensive and bamboo was heavy, and a big empire had lots of history and orders and accounts and such to keep track of. So with government backing I started paperating the place until finally paper pretty much replaced silk and bamboo for making books. And not only that, folks started making clothes out of paper, which was interesting since paper started out coming from rags. Paper could be cut and folded to make ornaments. It could be printed with wood blocks to make lots of identical things like money or playing cards. And don't forget wiping. To tell you the truth, there seemed to be few things which would not be cheaper, more beautiful, and more useful if fashioned by art and ingenuity from paper.

🜃 I traveled all over the world, first to the civilized part, then to Japan and Korea. But there were still uncivilized places living without paper. Then in 751 AD Turkish-Tibetan forces routed the army of Kao-Hsien-Chih at the Battle of the Talas River. Now, I was used to going wherever the wind took me. I had never been pried out of somebody before. But the Turks forced Chinese prisoners to set up a paper mill at Samarkand and then, in 794 AD, at Baghdad. So the Arabs started making paper from linen and hempen rags, what with not having mulberry and all.

🜃 When the Arabs invaded Spain in the tenth century, they brought paper with them and, what with all the hubbub, the Spanish were making paper by the twelfth century. The Italians also got the idea from the Arabs and were making paper late in the thirteenth century. I moved slowly through Europe, first as paper traded by the Arabs, and later as know-how via Spain and Italy. By the sixteenth century, paper was being made all over Europe. The Europeans took it with them wherever they went, which was just about everywhere.

14.2 🜍

Not every thin, flexible writing material is paper. You've probably heard of ***parchment*** and ***vellum*** and ***papyrus*** and you might think that they're kinds of paper but they're not. To begin with, papyrus is made from the papyrus plant, a kind of plant that grows in the shallows of lakes and rivers in Africa. It's made by splitting thin sections of the bark, placing those sections side by side to make a layer, laying another layer crosswise on top of the first one, and then pressing the layers together to make a sheet. Egyptians were making papyrus from about the third millennium BC and it was exported to Europe and the Middle East, but not to China. By contrast, parchment and vellum are made from animal skins, parchment from sheep skins and vellum from the skins of calves, lambs or kids. Parchment is produced by splitting the skin, separating the wool-side from the in-side; only the in-side is used for making parchment. Vellum is made from the whole skin of the young animal. Both parchment and vellum are scraped with a knife, rubbed with lime, stretched and sanded to make a thin, flexible and strong surface for writing on. Iranians were making parchment from about the second century BC, but it was unknown in China.

The Chinese had been writing on bones, stones, pottery, bronze and bamboo from the second millennium BC. And it was bamboo that dictated the form that Chinese writing would take. Craftsmen cut bamboo into tubes of a certain length, split those tubes into strips maybe 1 cm wide, and then sewed the strips into rolls. Since a reader rolled the scroll from side to side, the strips were running up and down and that's why Chinese is written from top to bottom. About 700 BC, silk cloth began to be used as a writing medium. As with bamboo, silk cloth was kept in rolls and so it was natural to keep writing vertically.

Paper is different from any of these other writing materials. Unlike parchment, vellum and silk, paper is always made from plants, namely from ***cellulose.*** In Chapter 6 Venus told you all about cellulose; plants make cellulose by stringing sugar molecules end-to-end and so cellulose is naturally long and thin. But unlike papyrus and bamboo, in which the cellulose remains parallel, the cellulose in paper is running every which way. One way to get the cellulose out of the plant stems is to shred them, a process called ***maceration.*** To macerate raw plant stems, linen, hemp, or cotton rags, you can soak them in water, boil them, beat them with a mallet, or put them in a kitchen blender. When you are finished, you have a kind of cellulose stew, or ***pulp,*** with lots of long, thin cellulose fibers floating around every which way. But you can do an even better job of pulling the fibers apart if you understand what holds them together. While plants use cellulose to build up those parts that hold them up—stems and trunks and stalks and such—those cellulose fibers are glued together with a material called ***lignin.***

While maceration alone is capable of tearing cellulose fibers apart to produce a *mechanical pulp*, the separation would be more easily accomplished if the lignin could be removed. Whereas a condensation reaction puts two molecules together by removing water, a ***hydrolysis reaction*** cuts a molecule apart by inserting water. Alkaline solutions hydrolyze both cellulose and lignin, but the hydrolysis of lignin is faster than that of cellulose and this is the key to their separation. When plant stems are boiled in alkali, the insoluble lignin is hydrolyzed to water-soluble compounds and white cellulose fibers are left floating in a brown alkaline broth. Straining the fibers from the broth and washing them with water completes the separation. You could use ***potash*** from burning wood, ***soda ash*** from burning sea-weed, ammonia from fermented urine, or ***lime*** from burnt limestone. But remember from Table 8-1 (page 101) that it takes 1000 parts of wood to get 1 part of potash. Fermented urine is not as strong as potash or soda ash, which leaves lime, which you can make from common limestone. At the time of the *Book of the Supports of the Scribes,* lime was the alkali of choice for *chemical pulping*. Now that you understand pulping, it's time to find out how to make paper from it.

To make paper from pulp, you need a ***mold***, which is either a piece of cloth or fine screen attached to a wooden frame. You could dip the mold into a tub of pulp and lift it out and some of the pulp would stick to the mold, but most of it would just run down the sides and back into the tub. However if you put an empty wooden frame, called a ***deckle,*** on *top* of the mold, the deckle keeps the pulp from running over the sides. With the deckle, the water can drain out of the pulp and you're left with an even layer of pulp on the mold, maybe a millimeter thick, with the fibers running every which way. If you were to let that pulp dry on the mold, you would have paper, no doubt about it.

But then your mold would be tied up all day until the pulp dried and all for one sheet of paper. To get around that, you ***couche*** the paper onto a felt, that is, you transfer it to a piece of cloth. You can stack those felts into a stack called a ***post,*** and press that post with a press to squeeze most of the water out. And once that water has been squeezed out, you can slap that damp paper onto a wall or board to dry. Once dry, you have a smooth, flexible writing material suitable for chronicling heroic deeds, inciting revolution, or sharing pancake recipes.

Material Safety

Locate MSDS's for slaked lime (CAS 1305-62-0), soda ash (CAS 497-19-8), and caustic soda (CAS 1310-73-2). Summarize the hazardous properties of these materials in your notebook, including the identity of the company which produced each MSDS and the potential health effects for eye contact, skin contact, inhalation, and ingestion. Also include the LD_{50} (oral, rat) for each of these materials.

Your most likely exposure is eye or skin contact. Caustic soda, the product of reacting soda with lime, is caustic; in case of eye contact, immediately flush them with water and call for an ambulance; in case of skin contact, wash the affected area with cold water until it no longer feels slippery.

You should wear safety glasses and a rubber gloves while working on this project. Spent caustic soda solution can be flushed down the drain, but be careful not to clog the drain with pulp. Leftover pulp may be thrown in the trash or saved for later use.

Research and Development

Well, I guess if you are in a class or something, you might want to know what will be on the quiz.

- You better know all the words that are important enough to be ***indexified*** and **glossarated.**
- Remember pretty much everything you learned in Chapter 10 and Chapter 12.
- Know that lignin is a condensation polymer which can be hydrolyzed by alkali.
- Know all the hazards of working with lime and soda, and what to do if things get out of hand.
- Know why a man's name is attached to the invention of paper.

14.3 Θ

You can make paper from just about any plant stems, from grass or reeds or weeds or hemp or straw. You can use cotton rags or linen or even old paper, but if you're looking to make paper from scratch, well, that would just be cheating. So don't do that. Go out and gather your own plant materials, but remember you're looking more for stalks and stems than for leaves or roots. Take your stems and beat them with a wooden or rubber mallet to crack them open. Then chop or cut them into 1-inch pieces and put them in a big pot. Don't use an aluminum pot because it reacts with alkali. Get yourself an enameled or stainless steel stock pot. Fill that pot about half-way up with chopped plant stems.

Next you need an alkaline solution and for this you can use ***soda ash***[2] and ***slaked lime.*** These two alkalis react with one another to produce ***caustic soda***, as described in Chapter 15. Fill the pot about half-way up with water so that it covers your plant stems and record in your notebook the amount of water you used. You will need 15 grams of soda ash and 10 grams of slaked lime for each liter of water. Use UFA (page 41) to figure out how much of these alkalis to use and then add them to your pot. Set your pulp pot on a stove or hot-plate and let it simmer overnight at low heat. The water will turn black and start to smell sweet; the stems will fall apart as the lignin glue is hydrolyzed.

While your fiber is cooking, you can get your tools together. You need a mold, which is nothing more than a frame with a piece of cloth or screen stretched across it. And you need a deckle, which is exactly like the mold only without the screen. The mold and deckle can be any size you want, but if it were me, I'd start out no bigger than about 20 cm by 30 cm. You could even use cheap picture frames with window-screen glued or stapled to the edges. Next you need a tub big enough so that you can get your mold into it without fumbling. A dish-washing tub or a litter box would work well, but don't use the sink, because your drain will get clogged up. You need a wooden or rubber mallet or an electric blender.

You should also have a press, and here you can get as simple or as fancy as you like. At the simple end, you could just put your paper under a board and stand on it. Getting fancier, you could put four bolts through

---

2. To convert washing soda to soda ash, calcine it overnight in an oven at 100°C (212°F) or higher. You may substitute ***potash*** for soda ash.

Figure 14-1. Mold, Deckle, and Couching Mound

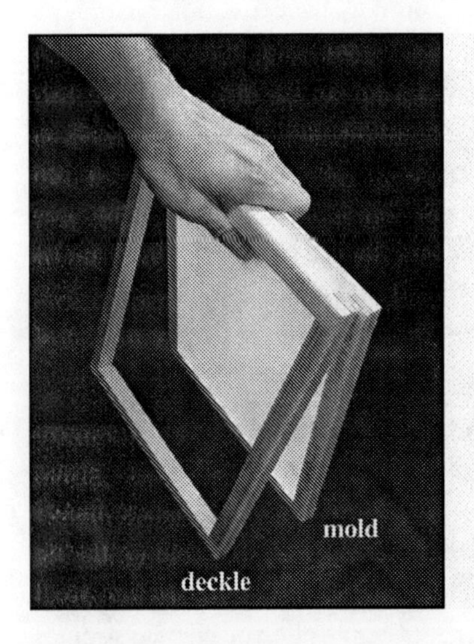

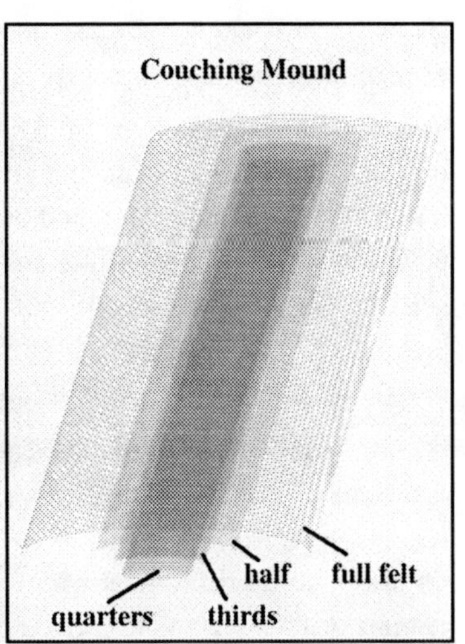

the corners of two boards. Put wing-nuts on those bolts so you can gradually tighten them up. At the fancy end, you could build a press out of 4x4 lumber. I like to use treated lumber so that it is water-proof, but if you choose to do so you should request an MSDS so that you can observe all safety precautions. Cut 12 1-foot lengths of 4x4 and drill a hole through the middle of each one. Bolt 4 of them together with half-inch threaded steel rod with a washer and a nut at each end. Do this two more times, so that way you end up with is 3 thick, wooden slabs, each 1 foot long. Drill holes at the corners and run some more half-inch threaded rod through them to make a little sandwich, with a slab at the top, a slab at the bottom, and a slab in the middle. The holes at the corners of the middle slab should be maybe three-quarters of an inch in diameter, so the middle slab is free to move up and down. Now all you need is an inexpensive hydraulic jack like you would use for changing tires, and you have a fancy paper press that will last you for a good long while. Or you can get by with the simple press; it's really up to you.

You also need some felts, special pieces of cloth which don't stick to wet paper. If you can't get papermaking felts, you can use denim. Your felts should be a little bigger than your mold, say, 30 cm by 40 cm. Soak all your felts in water and wring them out so that they're damp but not wet. Now, you'll find couching (pronounced *koo-ching*) a lot easier if you make a little couching mound. Take a felt, fold it in quarters length-wise, and put it on a shallow pan or tray. You now have a damp, quadruple-thick strip of cloth, say, 8 cm wide and 40 cm long. Take another felt, fold in thirds length-wise, and put it on top of the first one. You now have a mound seven cloths thick in the middle and three cloths thick on the side, about 10 cm wide and 40 cm long. Take another cloth and fold it in half, again length-wise, and put it on top of the second one. You now have a mound nine cloths thick in the middle and two cloths thick on the side, 15 cm wide and 40 cm long. Finally, take two or three more felts and put them on top of the stack so what you have is a little mound of damp felts on a tray to catch the water.

By now, your stems have cooked and fallen apart and it's time to remove the alkali. Now, alkali will eat your skin, so wear some rubber gloves for this part. And, of course, you always wear glasses, because there's no paper worth losing your eyes over. Anyway, you can grab up the big clumps of fiber with your gloved hands and when the bits are too small to pick out, strain the rest of the water through a cloth or kitchen strainer. The most important thing is not to pour any fiber down the drain. For one thing, it will clog up that drain. But even more important, that little bitty fiber is the best part for making paper, so you don't want to lose it. Once you have your fiber collected, rinse out your pot, fill it with fresh water, and put your fiber back into the pot. Swish it around with your gloved hand and then remove that fiber again, like you did before. Keep rinsing your fiber, adding it to fresh water and then straining it out again, until the water runs clear. And you can test your final rinse with pH test paper to make sure that all the alkali has been rinsed out. And once the alkali is gone, you can remove your rubber gloves.

Now you're ready to macerate your pulp, which'll be a lot easier now that the lignin is gone. There are two ways to go about it, the hard way and the easy way. The hard way is to take your fiber outside, put it on a flat rock or a sidewalk, and beat it with a wooden or rubber mallet. Your goal is to tear all those cellulose fibers apart, so just beat that pulp until it is more like a thick paste than it is like a bunch of plant stems. The easy way is to put a handful of fiber into a blender, filled about half-way with water, and run that blender using the "pulse" switch so you don't burn the motor out. The blender will make a pulp about the consistency of a milk-shake, like silk, as the *Book of the Supports of the Scribes* says.

Fill your tub half-way up with water, add a couple of handfuls of pulp to it and swirl it around. Notice that if you let it sit, that pulp will eventually

Figure 14-2. Forming a Sheet I

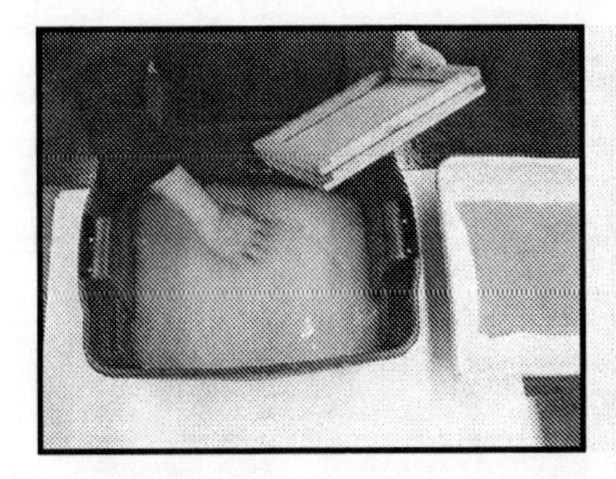

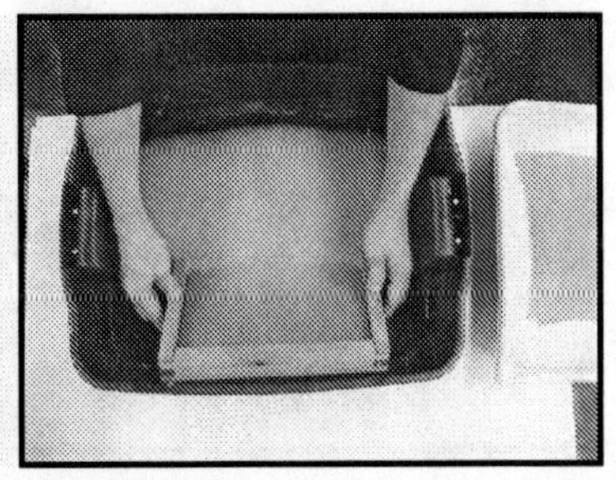

Figure 14-3. Forming a Sheet II

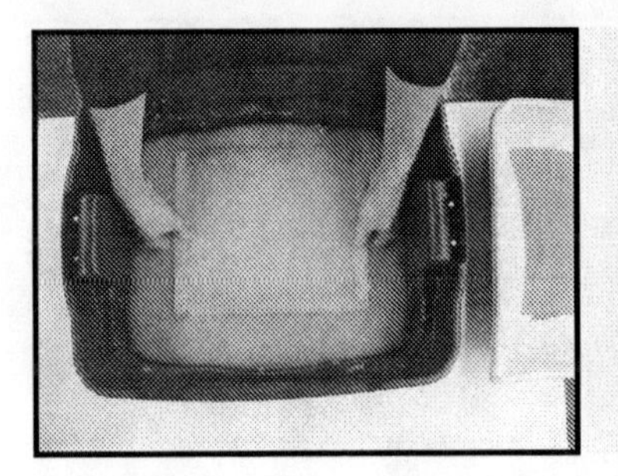

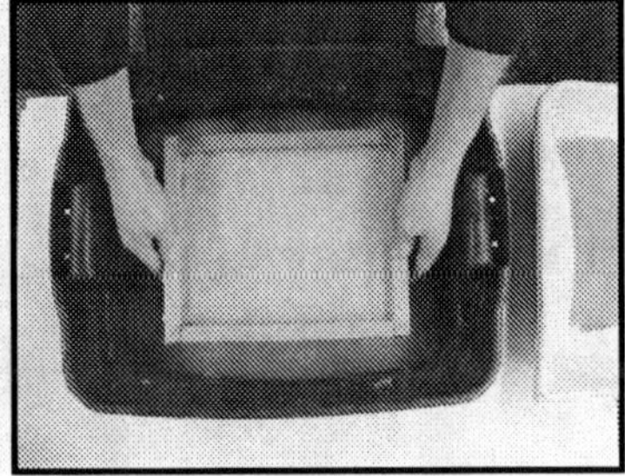

settle to the bottom; it's a mixture, not a solution. Take your mold, screen side up, and put your deckle on top of it to make a little sandwich with the screen in the middle. Swirl your pulp around, dip the mold and deckle into the pulp, and lift straight up. With practice, you'll get a nice, even layer of pulp with no thin spots. Let the water drain out and carefully remove the deckle from the mold. What you have is a layer of pulp sitting right up on top of the mold. Point one of the corners of the mold back into the tub to get the last bit of water out.

Touch one edge of the mold to your couching mound and roll the mold across it. If you do it at the right speed and with the right pressure and if your pulp is of the right thickness and if your felts are damp enough, that pulp will come off the mold and stick to the top felt. If it works and you get a nice rectangle of pulp on the felt, remove that felt, replace it with another, and do it again. If your paper sticks to your mold, you can rock the mold back and forth and press on it with your finger until the paper lets go. And if your paper gets messed up along the way, well, just take the felt with the messed-up pulp and put it back in the tub. That fiber is

Figure 14-4. Forming a Sheet III

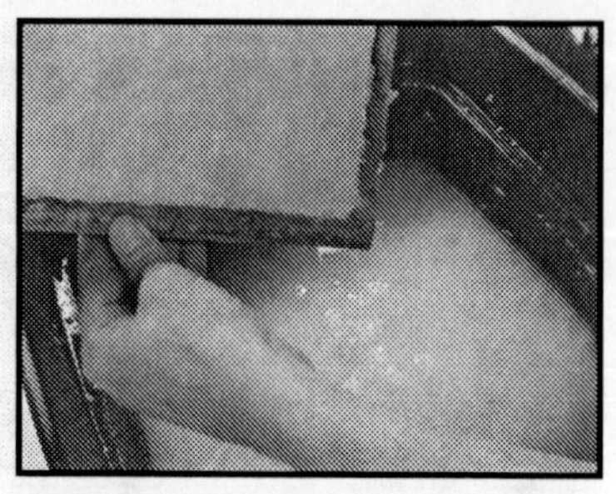

Figure 14-5. Couching I

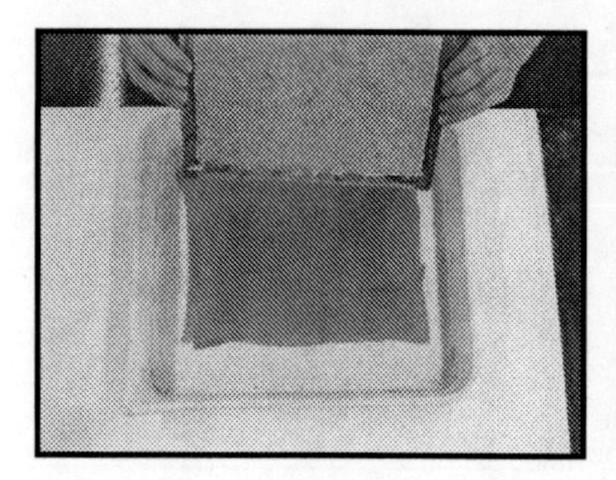

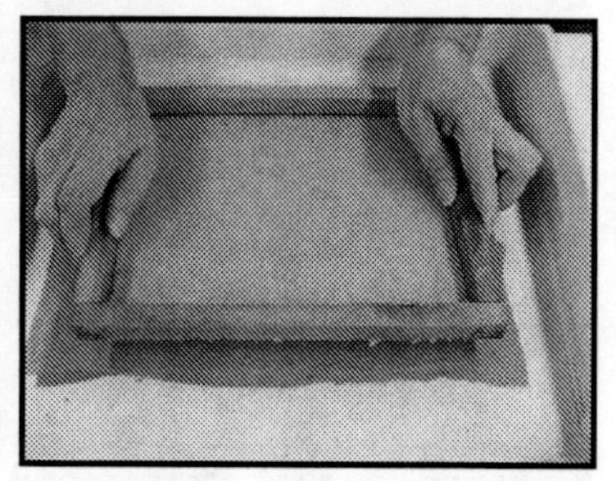

still good, so don't waste it. Now try again using a thicker layer of pulp or different speed or pressure when couching. Don't worry; you'll get the hang of it before too long. Take your good felts and stack them up as you go to make a *post,* that is, a felt-paper-felt-paper sandwich.

Once you have a 5 or 10 or 100 sheets you can build a post. Start out with a board about the size of your paper and lay a felt with its paper on top of it. Place another felt with its paper on top of the first and keep on building that post till you have run out of felts. Then put another board on top and you have your post ready for the press. Now, you can press those boards between your hands or you can bolt them together, or you can stand on top of them, or you can put them in a press. You can get as fancy as you want, but the long and short of it is that you want to press as much water out of the post as you possibly can. The harder you can press, the smoother your paper is going to be and the faster it's going to dry. Once the post has been pressed, you need to separate the paper from the felt. Think of it as un-couching. Dismantle the post, remove a felt from it, and place the paper side of the felt against a window. Press

Figure 14-6. Couching II

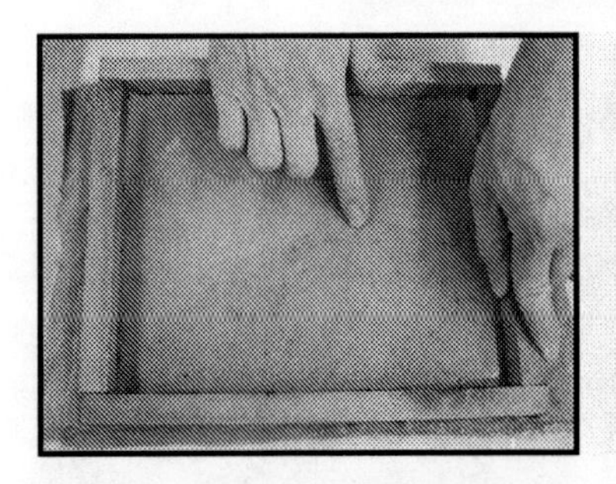

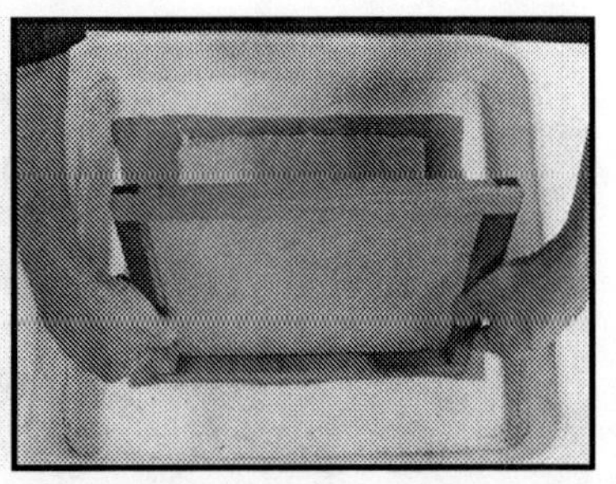

Figure 14-7. Couching III

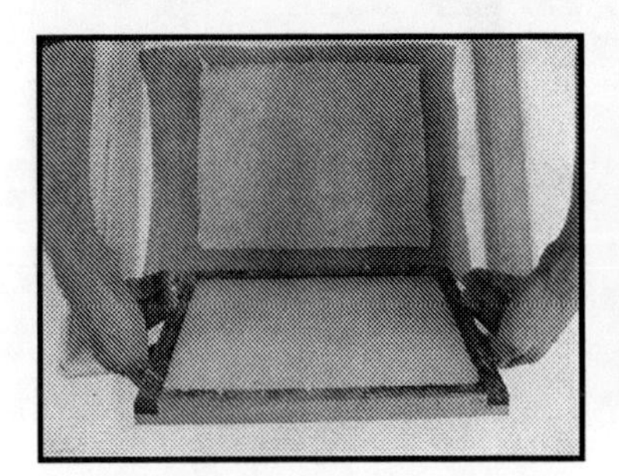

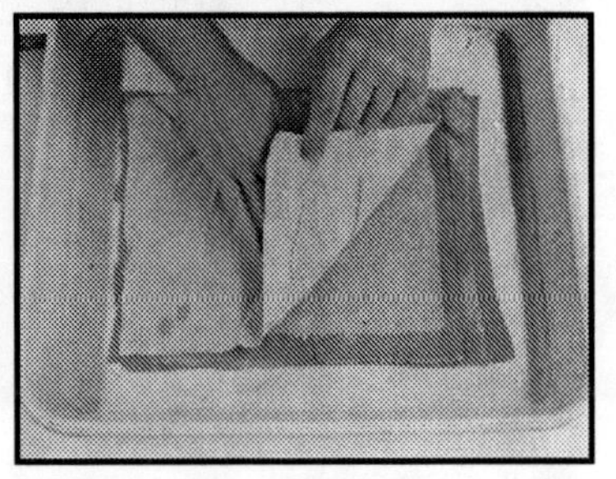

the felt with your finger so that the paper sticks to the window and then peel the felt from the paper. Allow the paper to dry overnight—in the morning it will very likely have fallen onto the floor, but it will have a nice, smooth face on the side that dried against the window.

To clean up, rinse out your felts and set them out to dry. Pour your pulp through a mold to catch as much pulp as you possibly can and save that pulp for later use. You can press it into balls and set them out to dry. The waste water can go down the sink as long as it doesn't contain any pulp to clog the drain. Wash out your blender and your pot and put them away. Clean off your mold and deckle and set them out to dry so they don't warp.

Quality Assurance

I think your paper ought to be flat and flexible. Like paper. You should tape it into your notebook so that after you are a paper-making whiz, you can remember what your first paper looked like. You can use your paper for making photographs in Chapter 23.

Figure 14-8. Pressing a Post

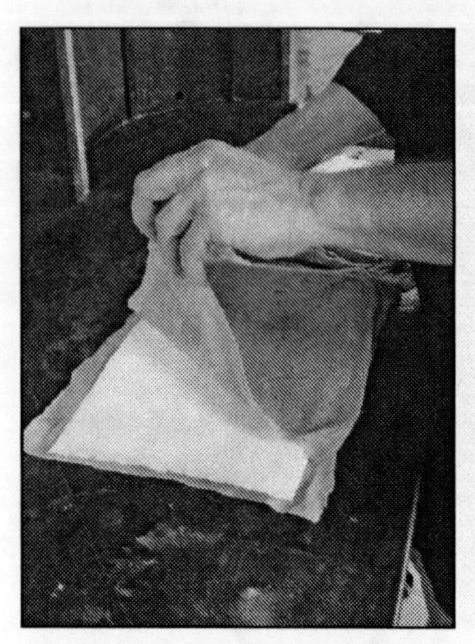

Figure 14-9. Drying Paper

Chapter 15. al-Razi (Stoichiometry)

> Take one *mann* of white Al-Qilī and an equal quantity of Lime and pour over it [the mixture] 7 times its amount of water, and boil it until it is reduced to one half. Purify it [by filtration and decantation] 10 times. Then place it in thin evaporating cups [*kīzān*], and hang it then in [heated] beakers [*jāmāt*]. Return what separates out [to the cup], and raise it [the cup] gradually, and protect from dust whatever drops from the cups into the beakers, and coagulate it into salt.
>
> — al-Rāzī, *The Book of Secrets*, *ca.* 920 AD [1]

15.1 ☿

⚠ Imagine that you are so pleased with your mead-making experience that you resolve to brew up your entire 100 gallon/year entitlement at once. You'll have an entire year's worth of birthday presents, Christmas presents, anniversary and wedding gifts that you won't have to fret over. Who wouldn't enjoy a refreshing bottle of Samson's Special Reserve? You fit out two 55 gallon drums with fermentation locks to let the gas out without letting air in. You buy 100 pounds of honey, 100 pounds of lemons and 100 pounds of tea at wholesale prices and get a local bar to save 100 empty beer bottles for you. Not only don't they charge you, they even wash them in their industrial-size dishwasher. You realize that you can save on yeast since one yeast makes two and two make four and four make eight and so on, so you only need one packet of yeast. You're prepared for the yeast to take a little longer than usual to get going, but you're not expecting months to go by with precious little in the way of gas. Eventually you get tired of waiting and prepare to bottle the stuff anyway, but there's a problem.

⚠ Belatedly you realize that you don't have anywhere near the right number of bottles; you implicitly and erroneously assumed that each bottle would hold a gallon when in fact they are 12-ounce bottles. Hurriedly you use your old friend, UFA from Chapter 3 to calculate the number of bottles needed:

1. Reference [26], p. 391.

$$
\begin{aligned}
?\text{ bottles} &= 100\ \cancel{gal}\left(\frac{128\ \cancel{oz}}{1\ \cancel{gal}}\right)\left(\frac{1\ bottle}{12\ \cancel{oz}}\right)\\
&= 1067\text{ bottles}
\end{aligned}
$$

Δ Before bottling, you give your Bee-jolet a little taste; it's terrible. More like industrial-strength lemon-tea furniture polish from Hell than anything you would even remotely consider drinking. That's when you realize that you made the same error with your ingredients that you made with the bottles. The poor yeasts never had a fighting chance! You spend the rest of the day flushing your experiment down the toilet and hoping that it doesn't set off a pollution alarm at the water-treatment plant.

Δ You've probably already spotted your mistake, but let's walk through it anyway before you waste enough groceries that FoodWatch distributes "Wanted" posters with your smiling face on them. The honey calculation is easy:

$$
\begin{aligned}
?\text{ lbs honey} &= 100\ \cancel{gal}\left(\frac{3.79\ \cancel{L}}{1\ \cancel{gal}}\right)\left(\frac{1\ \cancel{recipe}}{1.75\ \cancel{L}}\right)\left(\frac{1\text{ lbs honey}}{1\ \cancel{recipe}}\right)\\
&= 217\text{ lbs honey}
\end{aligned}
$$

Δ Your original mead had less than half the honey called for by the recipe; no wonder it fizzled out.

Δ The lemons are a little more difficult. You could use:

$$
\begin{aligned}
?\text{ lemons} &= 100\ \cancel{gal}\left(\frac{3.79\ \cancel{L}}{1\ \cancel{gal}}\right)\left(\frac{1\ \cancel{recipe}}{1.75\ \cancel{L}}\right)\left(\frac{1\text{ lemon}}{1\ \cancel{recipe}}\right)\\
&= 217\text{ lemons}
\end{aligned}
$$

Δ But your wholesaler sells lemons by the pound. A 10 pound bag of lemons contains an average of 40 lemons, so:

$$
\begin{aligned}
?\text{ lbs lemon} &= 100\ \cancel{gal}\left(\frac{3.79\ \cancel{L}}{1\ \cancel{gal}}\right)\left(\frac{1\ \cancel{recipe}}{1.75\ \cancel{L}}\right)\left(\frac{1\ \cancel{lemon}}{1\ \cancel{recipe}}\right)\\
&\quad\left(\frac{10\text{ lbs lemon}}{40\ \cancel{lemons}}\right)\\
&= 54\text{ lbs lemon}
\end{aligned}
$$

Δ With almost twice the lemons called for by the recipe, it's no wonder your mead had more pucker than a goldfish in a pickle jar.

🜁 The tea holds complications of its own. You used tea in bags for your smaller batch, but your wholesaler sells bulk tea by the pound. A box of 25 tea-bags weighs 50 grams, so:

$$
\begin{aligned}
?\ \text{lbs tea} = 100\ \cancel{gal} & \left(\frac{3.79\ \cancel{L}}{1\ \cancel{gal}}\right)\left(\frac{1\ \cancel{recipe}}{1.75\ \cancel{L}}\right)\left(\frac{1\ \cancel{teabags}}{1\ \cancel{recipe}}\right) \\
& \left(\frac{50\ \cancel{g\ tea}}{25\ \cancel{teabags}}\right)\left(\frac{1\ \text{lb tea}}{454\ \cancel{g\ tea}}\right) \\
= 1\ \text{lb tea}
\end{aligned}
$$

🜁 Your original mead had 100 times the required amount of tea! It should have pleased any tea-totaler, since it was almost totally tea.

🜁 At this point in the narrative structure I'll bet you're wondering what your mead recipe can possibly have in common with lime and al-qilī. The Author wanted an example that would make absolutely clear the folly of confusing *weight* with *number*. You learned to balance metathesis reactions in Chapter 7 and redox reactions in Chapter 11. Balanced reactions give you the relative number of moles of each reactant and product but atoms and molecules are too small to count out individually. As in our mead example it is often more convenient to weigh things out than it is to count them out. The question of how much of one thing in a reaction goes with a given amount of another is called a **stoichiometric question.**

15.2 🜍

The chapter began with Muhammad bin Zakarīyā al-Rāzī's favorite recipe for making sodium hydroxide (***caustic soda, lye***) from sodium carbonate (al-Qilī, alkali, soda) and calcium hydroxide (***slaked lime***). This classic metathesis reaction ought to be familiar to anyone who managed to stay awake during Chapter 7. This recipe, one part soda with one part lime, brings up one of the most fundamental problems in all of chemistry; what, exactly, is a part?

In Chapter 7 I told you that the little numbers in front of each substance in a balanced chemical equation are called **stoichiometric coefficients** and that their unit is the mole. A mole is just a number like a dozen or a hundred or a million, but much bigger. Just as the weight of a dozen lemons would be expected to be different from the weight of a dozen teabags, the weight of a mole depends on what it is you're talking about.

The truth of the matter is that if you have a grip on Unit Factor Analysis (page 41), three new unit factors will allow you to solve any stoichiometric problem. And these three unit factors share a common unit, (grams/mole). The first of these is the ***atomic weight,*** the weight of one mole of atoms; the second is the ***formula weight,*** the weight of a mole of formula; similarly the ***molecular weight*** is the weight of a mole of molecules. Formula and molecular weights come from adding up the atomic weights for the elements in an empirical or molecular formula, respectively, and atomic weights may be found either in a periodic table or in Appendix E (page 389).

So let's jump right in with a problem from al-Rāzī:

Q: How many grams of soda can react with 100 grams of slaked lime?

A: To answer any stoichiometry problem we must begin with a balanced equation:
$Na_2CO_3(aq) + Ca(OH)_2(aq) = CaCO_3(s) + 2\,NaOH(aq)$

From the balanced equation we get a mole ratio, in this case, for every mole of soda (sodium carbonate), one mole of slaked lime (calcium hydroxide) is consumed; one mole of limestone (calcium carbonate) and two moles of caustic soda (sodium hydroxide) are produced. Note that relative number of moles is given by the stoichiometric coefficient, the little number in from of each reactant or product in a balanced chemical equation. When there is no number in front of a reactant or product, it is assumed to be one. The unit factor required for this particular problem is (1 mole soda/1 mole slaked lime) or (1 mole slaked lime/1 mole soda), depending on the units we need in the top and bottom of the factor.

For the formula weights, we just add up the atomic weights; for soda, we get (2×23 + 12 + 3×16) = (106 g/mole); for slaked lime we get (40 + 2×16 + 2×1) = (74 g/mole).

$$\begin{aligned} ?\text{ g soda} &= 100\ \cancel{g\ slaked\ lime}\left(\frac{1\ \cancel{mole\ slaked\ lime}}{74\ \cancel{g\ slaked\ lime}}\right) \\ &\left(\frac{1\ \cancel{mole\ soda}}{1\ \cancel{mole\ slaked\ lime}}\right)\left(\frac{106\text{ g soda}}{1\ \cancel{mole\ soda}}\right) \\ &= 143\text{ g soda} \end{aligned}$$

That is, each part of slaked lime reacts with 1.43 ***parts*** of soda. So 1 part lime to 1 part soda may be almost 50% off, but it's not half bad for a recipe from the tenth century.

Remember that a string of unit factors can be continued from one line to the next and that a second "=" sign begins a second equation which is equal to the first. You could easily have worked this problem in Chapter 3 if you had been given the unit factors in the statement of the problem. The only new things in this chapter are the formula weights and the mole ratios.

Q: How many grams of washing soda can react with 100 grams of quicklime?

A: So far, I have used the terms *soda ash* and *washing soda* interchangeably. When either one dissolves in water it immediately dissociates into sodium ions and carbonate ions. Soda dissolved in water has no "memory" of where it came from, so chemically speaking, anhydrous (without water) soda and washing soda can be used interchangeably. Nevertheless, they are not exactly the same substances; in particular, their formula weights differ and this is relevant to the problem of how much to weigh out to achieve a balanced reaction. Similarly, as soon as quicklime hits the water, it slakes and the resulting slaked lime is identical to the slaked lime of the previous problem. But the weight of lime we should use depends very much on whether we are using quicklime or slaked lime. When weighing out chemicals, it is vital to know whether the material is hydrous or anhydrous because you must account for the weight of any water present.

$$\begin{aligned}\mathrm{Na_2CO_3 \cdot 10\,H_2O}(aq) + \mathrm{CaO}(aq) = {} & \mathrm{CaCO_3}(s) \\ & + 2\,\mathrm{NaOH}(aq) + 9\,\mathrm{H_2O}(l)\end{aligned}$$

The formula weights of washing soda and quicklime are 286 g/mole and 56 g/mole, respectively.

$$\begin{aligned}?\ \text{g washing soda} &= 100\ \cancel{\textit{g quicklime}} \left(\frac{1\ \cancel{\textit{mole quicklime}}}{56\ \cancel{\textit{g quicklime}}}\right) \\ &\quad \left(\frac{1\ \cancel{\textit{mole washing soda}}}{1\ \cancel{\textit{mole quicklime}}}\right) \\ &\quad \left(\frac{286\ \text{g washing soda}}{1\ \cancel{\textit{mole washing soda}}}\right) \\ &= 511\ \text{g washing soda}\end{aligned}$$

About 5 parts of washing soda are required to react with each part of quicklime.

Q: How many grams of caustic soda can be produced from the reaction of 100 grams of quicklime with excess washing soda?

A: There are two wrinkles in this one; the mole ratio is no longer 1:1. The word *excess* tells us that there is "more than enough soda". We assume that all of the quicklime is consumed and that there is leftover soda when the reaction is done. The reactant that runs out first, in this case quicklime, is known as the ***limiting reagent,*** the one which is not in excess. For now, the word *excess* tells you that you only need to deal with the other reactant.

$$? \text{ g lye} = 100\ \cancel{\text{g quicklime}} \left(\frac{1\ \cancel{\text{mole quicklime}}}{56\ \cancel{\text{g quicklime}}}\right) \left(\frac{2\ \cancel{\text{mole lye}}}{1\ \cancel{\text{mole quicklime}}}\right) \left(\frac{40 \text{ g lye}}{1\ \cancel{\text{mole lye}}}\right) = 143 \text{ g lye}$$

You may be disturbed by the notion that you can make 146 g of caustic soda from only 100 g of quicklime. Is the Law of **Conservation of Mass** violated here? Not at all! If you calculate the number of grams of soda reacting with that 100 grams of quicklime, you will see that the weight of the soda plus the weight of the quicklime is exactly the same as the weight of the caustic soda plus the weight of the chalk plus the weight of the water produced.

Q: How many grams of anhydrous soda can be produced from the calcination of 100 grams of washing soda?

A: This would be a very easy experiment to perform. Just put a weighed amount of washing soda into a hot oven (above 100°C) for a few hours and weigh it again when it is done. Of course, you would substitute your actual weight of washing soda for the "100 grams" in the stated problem.

$$Na_2CO_3 \cdot 10\ H_2O(s) = Na_2CO_3(s) + 10\ H_2O(g)$$

The formula weights of washing soda and anhydrous soda are 286 g/mole and 106 g/mole, respectively.

$$? \text{ g soda} = 100\ \cancel{\text{g washing soda}} \left(\frac{1\ \cancel{\text{mole washing soda}}}{286\ \cancel{\text{g washing soda}}}\right) \left(\frac{1\ \cancel{\text{mole soda}}}{1\ \cancel{\text{mole washing soda}}}\right) \left(\frac{106 \text{ g soda}}{1\ \cancel{\text{mole soda}}}\right) = 37 \text{ g soda}$$

Your calculated weight of product, in this case 37 g of anhydrous soda, is called the ***theoretical yield.*** It can be compared to your actual weight, the ***experimental yield.*** When the theoretical and experimental yields are close to one another, you have a happy, comfortable feeling that you understand what's going on and that you are getting the most from your reaction. If they are different from one another, you have an uncomfortable feeling that you don't understand what's going on, and that you are not getting everything from your reaction that you might. Think about some situations that might result in the theoretical yield being higher or lower than the experimental yield. Then try it for yourself.

Q: How many grams of mullite and silica can be produced from the calcination of 100 grams of dry kaolinite?

A: Now, you were supposed to have weighed your pot before and after firing, and so at this point you can see for yourself whether Athanor was feeding you a line.

$3\ Al_2Si_2O_5(OH)_4(s) = Al_6Si_2O_{13}(s) + 4\ SiO_2(s) + 6\ H_2O(g)$

$$\begin{aligned} ?\ \text{g mullite} &= 100\ \cancel{g\ kaolinite}\left(\frac{1\ \cancel{mole\ kaolinite}}{258\ \cancel{g\ kaolinite}}\right) \\ &\left(\frac{1\ \cancel{mole\ mullite}}{3\ \cancel{mole\ kaolinite}}\right)\left(\frac{426\ \text{g mullite}}{1\ \cancel{molemullite}}\right) \\ &= 55\ \text{gmullite} \end{aligned}$$

$$\begin{aligned} ?\ \text{g silica} &= 100\ \cancel{g\ kaolinite}\left(\frac{1\ \cancel{mole\ kaolinite}}{258\ \cancel{g\ kaolinite}}\right) \\ &\left(\frac{4\ \cancel{mole\ silica}}{3\ \cancel{mole\ kaolinite}}\right)\left(\frac{60\ \text{g silica}}{1\ \cancel{mole\ silica}}\right) \\ &= 31\ \text{g silica} \end{aligned}$$

There are several conditions assumed by the question itself. The calculation assumes that we started with pure, dry kaolinite, a condition that may have been only approximated in your clay body. It assumes that Equation 5-1 (page 63) is the only reaction taking place and that all of the kaolinite was converted, that is, no un-reacted kaolinite remains in your fired pot. And it assumes that nothing stuck to or flaked off of your pot. Substituting the weight of your un-fired pot for the "100 grams" of the problem, calculate a theoretical yield and compare it to the experimental yield. Of course, you have to calculate the theoretical weights of mullite and silica separately and then add

Stoichiometry in a Nutshell

Chill out, whatcha yellin' for? Lay back, you've done it all before.
Put down the grams of stuff you want to find, and an equal sign.
Then put what you're startin' from. You don't have to play so dumb.
I mean the ***problematic*** units, hey, like UFA.

The formula weight of the thing that you hate
will cancel the gram like an unwanted spam.
A mole ratio takes you where you gotta go, you see. Tell me.

Why d'you have to make stoichiometry complicated?
Don't you know it
doesn't have to be the kind of thing that leaves you all frustrated?
Life's like this you
cancel moles that you hate, one more formula weight,
gives you grams of the answer, on to
Just a little number-crunchin' work and you know you made it.
Yea, yea, yea.

them to get the theoretical weight of your pot. If your theoretical and experimental yields are close to one another, it doesn't necessarily mean that everything Athanor told you was true, it just means that your evidence *supports* his assertions. Scientific theories are never proven once and for all; but the more evidence supports a theory, the more confident we become that it is essentially correct.

Material Safety

If you're like most people you find the information in an MSDS rather daunting. It's full of all kinds of technical information suited to occupational safety. Wouldn't it be great if there were some kind of simple, easily understood safety rating system so that you would know whether a chemical was really, really dangerous, or just dangerous in the way that rocks and sticks and dirt are dangerous? Well, there is such a system. You've probably seen symbols like those in Figure 15-1 which provide information on hazardous materials, but most people don't have a clue what they mean.

Figure 15-1. The NFPA Diamond

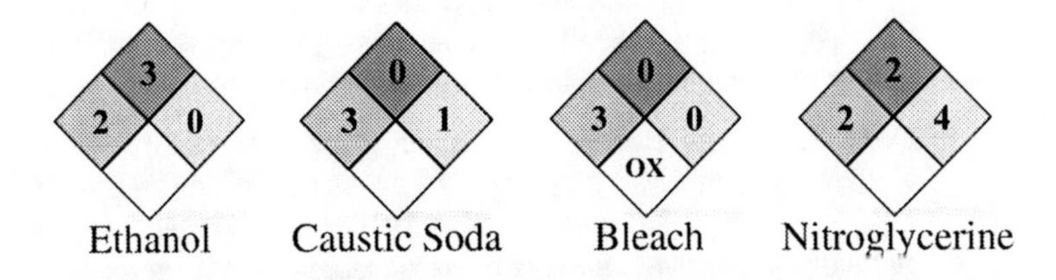

The National Fire Protection Agency[2] developed the *NFPA* diamond to give fire fighters information on hazardous chemicals at a glance. The number in the left-most (blue) quadrant is the health rating, that in the upper-most quadrant (red) is the flammability rating, that in the right-most (yellow) quadrant is the reactivity rating, and the number in the lowest (white) quadrant gives special hazards, for example, OX for oxidizer or W for exceptional reactivity with water. The numerical ratings run from 0 (not very dangerous) to 4 (very dangerous). Most MSDS's give the NFPA ratings, either as a diamond graphic or as text (NFPA H 1, F 0, R 0, OX). Let's look at the NFPA diamonds (Figure 15-1) for four materials which will play a large role throughout the remaining chapters.

We've already been introduced to ethanol in Chapter 4 and will see more of it in Chapter 16. As a poison, ethanol is fair to middling and gets a "2" in the health quadrant. In the flammability quadrant ethanol gets a "3;" it's quite flammable, though not as flammable as something like propane. In the reactivity quadrant ethanol gets a "0;" it's not particularly reactive with common materials and is not explosive. The special quadrant is blank, as ethanol isn't an oxidant and it isn't particularly reactive with water. Ethanol's NFPA symbol tells a firefighter that flammability is its most important property in an emergency situation.

We met caustic soda, sodium hydroxide, in Chapter 14 as an excellent alkali for making paper. Sold in the grocery store as "lye," it's a common ingredient in household drain openers. Caustic soda will emerge as an important player in Chapter 19. In the present chapter we've explored the stoichiometry for making caustic soda from soda and lime. Caustic soda is fairly toxic by ingestion, but the fact that it eats skin bumps its health rating up to a "3." Caustic soda will not burn, so it gets a "0" for flammability. It is reactive with some other materials, notably aluminum, and so it gets a "1"

2. For more information on the NFPA diamond, see Reference [35].

for reactivity. While it gets hot on contact with water, this is not enough to get it a notice in the special quadrant. Thus caustic soda's NFPA symbol tells the firefighter that exposure to the material should be avoided, but that it is neither flammable nor explosive.

Ordinary laundry bleach is a solution of sodium hypochlorite in water. It is roughly as poisonous as caustic soda and gets a "3" for health. It's not flammable so it gets a "0" for flammability. It's not particularly reactive so it gets a "0" for reactivity. It is an oxidizer, meaning that it can substitute for oxygen in a fire, a fact to which fire fighters are alerted in the special quadrant.

We'll meet nitroglycerin in Chapter 27, though you probably know it already as a powerful explosive. While it's moderately toxic and flammable, its most important property from a firefighting point of view is that it will blow you to kingdom come. That gets it the highest rating, a "4," in the reactivity quadrant.

The diamond was designed for fire fighters and the NFPA doesn't encourage its use for other purposes. The diamond, for example, doesn't alert people to chronic toxic effects or carcinogenicity. Thus it's less relevant to hazards associated with long-term occupational exposure than it is to acute exposure under emergency situations. At least two other rating systems have emerged to deal with occupational and laboratory situations: the HMIS rating of the National Paint and Coatings Association[3] and the Saf-T-Data rating of the J. T. Baker chemical company.[4] Like the NFPA diamond, both systems use a numerical rating from 0 (not hazardous) to 4 (very hazardous) in four or five categories akin to those of the diamond. Unlike the diamond, however, these systems factor the effects of chronic exposure into the health rating and so the numerical ratings may differ somewhat from one system to another. Either the HMIS or Saf-T-Data systems would appear to be preferable to the NFPA diamond for laboratory use, but neither has been adopted as widely as the diamond. Since the diamond is nearly universal and since chemical exposure for cavemen is likely to be acute rather than chronic, we'll concern ourselves primarily with the diamond as a shorthand for chemical hazards. Still, it's *only* a shorthand and not a substitute for the full MSDS.

You aren't going to get away without a material safety assignment, even though there are no materials involved in this project. If you don't have them already, search for MSDS's using the keyword "NFPA." Draw the NFPA diamonds for slaked lime (CAS 1305-62-0),

3. For more information on the HMIS system, see Reference [37].
4. For more information on the Saf-T-Data system, see Reference [31].

propane (CAS 74-98-6), and silica (CAS 14808-60-7) in the material safety section of your write-up for this project.

Research and Development

So there you are, studying for a test, and you wonder what will be on it.

- Study the meanings of all of the words that are important enough to be included in the ***index*** or **glossary.**
- Know the Research and Development items from Chapter 9 and Chapter 11.
- Practice calculating formula and molecular weights from atomic weights.
- Most people don't expect you to memorize atomic weights nowadays. But you had better be able to find them on a periodic table. Just remember that the atomic weight of carbon is 12. The atomic weights of the other elements will appear in the same position that 12 does for carbon.
- All of the problems here have used 100 grams, so you are getting a percentage, but you should expect to be able to solve problems using different amounts.
- Understand the meaning of the NFPA diamond.
- When you can't remember the last time you missed a stoichiometry problem, you've probably worked enough examples.

15.3 Θ

Q: How many grams of quicklime can be produced by the complete calcination of 100 grams of limestone?

A: 56 grams of quicklime.
Once again, you can compare the theoretical and experimental yields from your lime-burning experience. Be alert for the possibility that your limestone was impure or your calcination incomplete.

Q: How many grams of water are needed to completely slake 100 grams of quicklime?

A: 32 grams of water.

Q: How many grams of slaked lime can be produced from the reaction of 100 grams of quicklime with excess water?

A: 132 grams of slaked lime.

Q: When cellulose is heated in a reducing environment it becomes charcoal as shown in Equation 1-1 (page 9). How many grams of charcoal can be produced by the complete conversion of 100 grams of cellulose?

A: 40 grams of charcoal.

Q: The partial calcination of gypsum produces plaster according to Equation 10-1 (page 124). How many grams of plaster can be produced by the complete conversion of 100 grams of gypsum?

A: 84 grams of plaster.

Q: How many grams of copper can be produced by the complete reduction of 10 grams of copper carbonate?

A: 5.1 grams of copper. Of course, you can go further and ask how many grams of tin will be produced by the complete reduction of 2 grams of tin oxide; the theoretical yield of bronze can then be had by adding those of copper and tin. The usual conditions apply; perhaps what you thought was pure copper carbonate was actually pure or impure malachite, a hydrate of copper carbonate; perhaps more than one reaction was going on in the kiln; perhaps you were unable to collect all of the metal produced; perhaps charcoal was the limiting reagent and consequently you have some un-reduced ore and less metal than you would expect. When the theoretical and experimental yields agree, however, you get that happy feeling that you understand what is going on and that all is right with the world.

Quality Assurance

Tape your stoichiometry quizzes into your notebook and calculate the theoretical yield for one of your previous projects: pottery, lime, gypsum, or metal. Compare the theoretical yield to your experimental yield. Speculate on the kinds of things that could lead to any discrepancies.

Chapter 16. Adelard (Alcohol)

> Since I possess many wonderful books written on these matters I became anxious to produce a commentary, not that I may appear to be encroaching upon the sacred books and [therefore] despite much labor accomplishing nothing, but that, avoiding that mortal heresy, I will disclose to those who wish to understand these things what the actual processes are that are used in painting and other kinds of work. I call the title of this compilation *Mappae Clavicula*, so that everyone who lays hands on it and often tries it out will think that a kind of key is contained in it. For just as access to [the contents of] locked houses is impossible without a key, though it is easy for those who are inside, so also, without this commentary, all that appears in the sacred writings will give the reader a feeling of exclusion and darkness. I swear further by the great God who has disclosed these things, to hand this book down to no one except to my son, when he has first judged his character and decided whether he can have a pious and just feeling about these things and can keep them secure.
>
> ...
>
> 212. From a mixture of pure and very strong xjnf with 3 qbsut of tbmu, cooked in the vessels used for this business, there comes a liquid, which when set on fire and while still flaming leaves the material [underneath] unburnt.
>
> — *Mappae Clavicula, ca.* 1130 AD [1]

16.1 ☿

Δ This flaming liquid is one of the most remarkable substances ever discovered and so you will understand that I have endeavoured to hide it among my secrets. It is nothing less than the vegetable mercury, which rises from the earth to heaven and descends again to the earth. It is the strong power of all powers for it overcomes everything fine and penetrates everything solid. In Latin it is called *Aqua Vitae*, in Gaelic *Uisce Beatha*, both names meaning "water of life." The English *alcohol* is derived from the Arabic *al-Kuhl*. My child, are ye prepared to preserve this

1. Reference [25].

immortal fire, to protect it from the ignorant rabble, to keep in its heat and withstand it?

∇ Hey, I was sitting there.

∇ You are probably wondering what is going on. I will tell you. I was minding my own business, preparing to tell you about alcohol, it being my turn to speak. I stepped out to answer the memetic equivalent of the call of Nature and returned to find Lucifer in the driver's seat preparing to make off with my chapter. Now I admit that my subject touches on heat and temperature and winds up with a fondue; still, if Lucifer were a chicken she would not belong in every pot. But I am getting ahead of myself.

∇ Of course, alcoholic beverages have been around since God was a child. The earliest written records speak of mead, wine and beer as part of everyday life. But the isolation of alcohol from these sources is of much more recent invention. You see, while any old devil can heat the ***bejeezus*** *out* of something, it is quite another matter to *catch* this bejeezus and put it in a bottle. Catching a spirit requires a kinder, gentler kind of fire, and once the bejeezus has been coaxed from its lair it must be herded like a cat, not summoned like a dog.

∇ People will argue about who was the first to succeed in distilling alcohol. I can only tell you that I myself inhabited a painter of the twelfth century when I discovered it. Now, a painter had to be a jack of all trades in those days. One day you might be painting with oils on canvas, another dyeing yarn for a tapestry, another gilding letters in a book, and yet another preparing plaster from gypsum. And because there was quite a bit of variation in the quality and purity of the various metals, dyes, pigments, binders and solvents an artist really had to be something of a chemist if *he* was not to be taken to the cleaners.

∇ Even more-so, an artist had to be on the lookout for new combinations of colors and textures to provide for the endless struggle of queens and kings to out-dress call-girls and clowns. In the dog-eat-dog world of high fashion, a successful painter had to jealously guard *his* recipes from theft by his competitors. Now in those days there were many solvents for painting and dyeing; water for some, oil for others, wine, vinegar or urine for still others, and so on. It was difficult enough finding a solvent for one pigment, let alone trying to mix a water-soluble pigment with an oil-soluble one. What was needed was a universal solvent, one that would dissolve just about anything.

∇ Metallurgy had such a solvent in mercury. You see, quicksilver will dissolve most any metal, particularly the valuable ones like silver and gold. In fact, one method for recovering gold from it ore is to mix it with mercury, which dissolves the elemental gold but not the oxides, sulfides, and carbonates of the other metals. This mercury, laden with dissolved gold, is then placed in a furnace in which the bejeezus is heated from it, that is, the mercury is boiled away leaving gold behind. Now it had been noted in antiquity that when mercury is boiled it disappears from the pot, but, amazingly, beads of mercury collect on any cool surfaces which may happen to be nearby. Because mercury does not grow on trees, it was advantageous to arrange matters so that there was always a cool surface nearby so that the mercury might be recovered. Such an arrangement came to be known as a ***still.***

∇ So the idea was instilled in me that mercury, the universal metallic solvent, was a kind of spirit which could be driven out of a metal by heating and collected by cooling. The same was true of another very good solvent, water. It became my habit, then, in the odd moment when nothing else was cooking, to distill whatever leftovers happened to be lying about to see what might come of them. Many people had the same notion and wine, in particular, was a popular choice for distillation. The resulting "spirits" provided an enhanced kick, but were not qualitatively better solvents than wine.

∇ One day I happened to throw some salt into the pot for no particular reason. The resulting distillate was stronger than any spirit I had ever seen. Now, taste is an important tool for chemical analysis when you have nothing else and so I had a little shot of my new spirit, purely in the spirit of scientific inquiry you understand. It literally knocked me off my feet, with the result that I lost my balance and dropped my beaker near the furnace. Whoosh! The whole place erupted in flames and I thought sure I had fallen in the clutches of the fellow with the horns. Fortunately the fire burned out as quickly as it started and I was left a little singed but a lot wiser.

∇ I suppose that the lesson most people would have taken from this experience is never to mix salt with wine. But I was not most people. No, the lesson I learned was to sit before sipping solvent. Eventually my attention returned to matters artistic and I found that this new spirit was an almost universal solvent for pigments, dyes and medicines. When I wrote down the recipe for this phenomenal potion, I did so in code, lest

I lose my competitive advantage. You are probably wondering about the secret of this code. I would tell you but then it would no longer be a secret.

∇ As it turned out, my code was soon broken. You see, *it is in the interest of the mortal to keep secrets, but it is not in the interest of secrets to be kept.* Memes which do not get out much are apt to go extinct and so having wriggled my way out of my medieval painter, it is I, the un-kept secret, who remain to write this chapter. So I am now prepared to let the alcoholic cat out of the bag, so to speak.

16.2 🜍

First of all, I must tell you that when a less volatile **solute** is dissolved in a more volatile **solvent,** the boiling point of that solvent goes up. You probably imagine that ***volatile*** means flammable or unstable or explosive. If so you imagine incorrectly. No, in chemistry to say that one compound is more volatile than another simply means that it is easier to evaporate, that its boiling point is lower. So, for example, the boiling point of salt-water is higher than that of water. The boiling point of mead is higher than that of alcohol and lower than that of water.

Imagine if you will, a tea-kettle, called by chemists a *pot* on the stove filled with mead. The heat is on and the mead comes to a boil. If it were pure alcohol it would boil at 78°C; if it were pure water it would boil at 100°C. But because it is a solution of alcohol in water, it boils somewhere in between, let us say, at 90°C. Now if we were to collect a bit of the steam coming out of the spout, what do you suppose we would find? The alcohol "wants" to boil at 78°C so at 90°C it is really ready to fly the coop, so to speak. By contrast, the water does not really "want" to boil until 100°C so at 90°C it is in no hurry to bolt. Put another way, because alcohol is the more volatile of the two, the steam will be richer in alcohol than the original mead.

Imagine further that you place a cold dinner plate over the spout of the tea-kettle. The steam will condense into droplets of liquid on the cold surface. This liquid will have the same concentration of alcohol as the steam, that is to say, it is richer in alcohol than the original mead. We could collect this liquid and place it into another tea-kettle, albeit a tiny one. Being richer in alcohol, it would come to a boil at a lower tempera-

Figure 16-1. Distillation as a Process

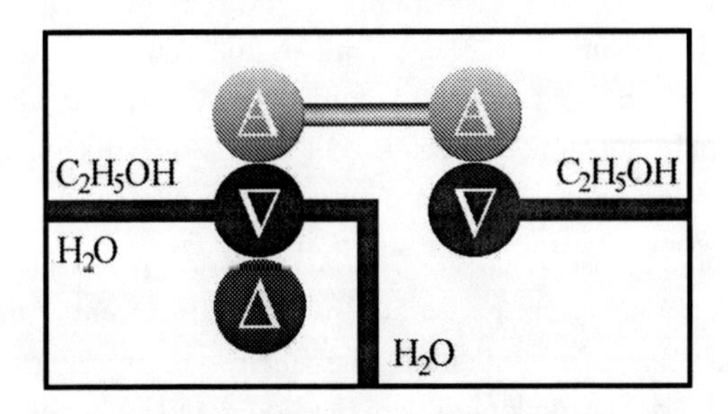

ture than mead, perhaps 85°C. You are probably thinking that we could then collect steam from this new kettle, condense it on a second cold plate, and dribble this even richer liquid into a third kettle. If so, your idea would be gathering steam.

But it is not actually necessary to use such a cumbersome arrangement. If, instead of a dinner plate, we mounted a tube, called by chemists a *column* on the spout of the tea-kettle, the steam would condense on the inside. As the bottom of the tube gets hot from the rising steam, the drops of condensate will boil again, as it would have in a second kettle. The new steam rises a little further up the tube until it finds a place cool enough to condense and the whole process starts all over again. A simple tube, hot at the bottom and cooler at the top, is all that is needed to take the place of a whole brigade of tea-kettles and the longer the tube is, the richer in alcohol the distillate will be.

Imagine further that we attach another tube, a ***condenser*** to the top of the column. The only real difference between the column and the condenser is that the column is vertical, so that condensing liquid dribbles back down toward the pot while the condenser has a downward slant so that condensing liquid can exit the still. The condenser might be cooled by air, but it is more efficient if it is cooled by water. The place where column and condenser meet is called the *head*. Imagine now a thermometer at the head. What would it read during a distillation? At the beginning there is boiling mead in the pot but the head is still at room temperature. The steam begins to rise, heating the bottom of the column, but the head remains at room temperature. Eventually, though, hot steam reaches the head where it simultaneously passes into the condenser. If we were distilling pure alcohol, the thermometer would read 78°C and the temperature would remain constant throughout the distillation. If we were distilling pure water, the temperature would remain at 100°C throughout. But with a solution, the head temperature can be anywhere in this range, lower when the distillate is more alcoholic, higher when it is less so. Unlike pure substances, which boil at a single, fixed temperature, solutions boil over a range of temperatures and this range can be used to monitor the concentration of the solution.

Actually, there may come a point where the boiling point of the distillate is lower than that of the alcohol. The steam coming off of this distillate will be no richer in alcohol than the liquid and so further distillation is ineffective. For ethanol-water this *azeotrope* boils at 78.17°C, 0.23°C lower than pure alcohol, and consists of 95% ethanol and 5% water. "Denatured" alcohol is simply 95% ethanol with little bit of poison added to discourage people from drinking it.

Figure 16-1 shows the distillation process in schematic form. The first reactor looks like a ***furnace,*** but one which contains liquid instead of solid. The second reactor is the condenser. Together, they make up a still. No chemical reaction takes place in this process. The process physically separates two liquids that differ in boiling point. The usual conventions are followed; reactants enter from the left of the figure, waste products exit the top and bottom, and the main product exits to the right.

Adelard's motivation for distilling alcohol was to use it as a solvent and so I should probably tell you a bit about solubility. There are essentially three kinds of substance; ***ionic*** substances, or **salts** consist of positive and negative ions and were discussed in Chapter 7; ***polar*** substances consist of molecules in which the negative charge is bunched up at one end and the positive charge at the other; ***non-polar*** substances consist of molecules in which the charge is evenly distributed throughout the molecule. In general, substances in which the charges are separated, salts and polar substances, are mutually soluble and non-polar substances are mutually soluble, but salts and polar substances are not soluble in non-polar substances and *vice versa.* "Oil and water do not mix," or, to put it another way, "like dissolves like."

Polar molecules, as I have said, have negative charge bunched up at one end and positive charge at the other. For this to happen, one or more atoms must, like the devil, take more than its due. Of electrons, that is. While carbon and hydrogen are rather mild-mannered in this regard, oxygen and nitrogen are greedy. In the water molecule, for example,

Figure 16-2. Two Water Molecules

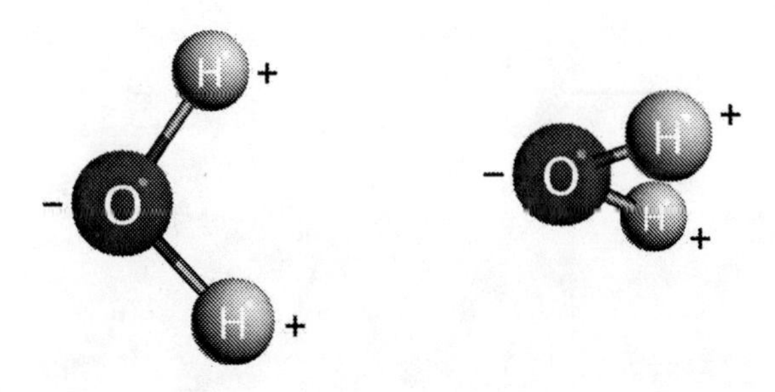

the oxygen atom hogs electrons from the hydrogen atoms and so the oxygen end is negative and the hydrogen end is positive. When two water molecules bump into one another, as shown in Figure 16-2, the negative oxygen end of one molecule is attracted to the positive hydrogen end of the other. Opposite charges attract, you see. This mutual attraction, or hydrogen bond, is what holds two water molecules together. A large collection of water molecules resembles a dog convention, with the little beggars sniffing at each other's tails.

Ethane, C_2H_6, is a typical non-polar molecule. Because carbon is really no greedier than hydrogen, as far as electrons are concerned, there is no positive or negative end to the molecule and hence no strong interaction to bind one ethane molecule to another. In contrast to water, ethane molecules are content to roam about passing one another like head-less tail-less dogs in the night.

What happens when a polar liquid like water and a non-polar one like ethane find themselves in the same container? The situation is much like a room filled half with normal dogs and half with head-less, tail-less dogs. The normal dogs will congregate in one part of the room because of their mutual attraction. The headless, tail-less dogs will be pushed to the other side of the room, not so much because they are attracted to one another, but because the normal dogs are less attracted to them than to those of their own kind. So to come back to chemistry, the polar water molecules will collect in one part of the container and the non-polar ethane molecules will be left in the other. To put it another way, ethane is not very soluble in water and *vice versa*.

Figure 16-3. Ethanol and Water

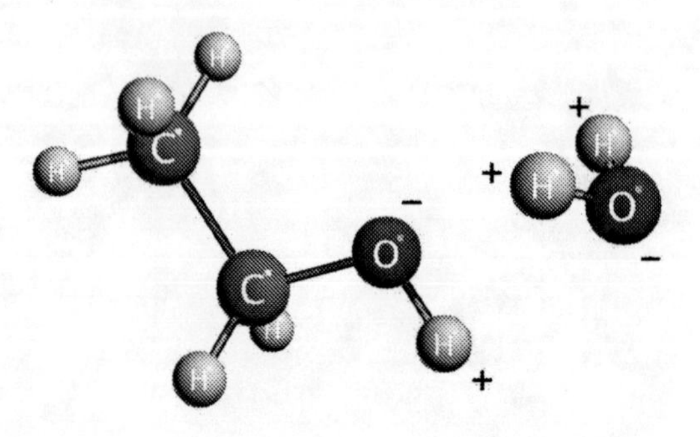

Of course, the question of polarity is not black-and-white. There are some molecules that are very polar, some that are a bit polar, and some that are not very polar at all. For example ethanol, C_2H_5OH, is mid-way between ethane and water in its polarity. One end of the molecule is non-polar like ethane while the other is polar like water. It is as if a normal dog had been sewn onto a head-less tail-less dog. Such a dog, released onto a pack of dogs, would be equally comfortable with either kind. In Figure 16-3 a water molecule is attracted to the oxygen-hydrogen end of the ethanol molecule and *vice versa.* This mutual attraction is what is responsible for the ability of ethanol to dissolve in water.

You may be wondering why ethanol is such a good solvent for so many dyes and medicines. If you think about it, living things, whether yeasts or bacteria, trees or polliwogs, are made up of tiny bags of water called cells. Part of the living thing is polar, the part dissolved in the water. But much of the organism had better be non-polar, the cell walls, skin and bones, or else the whole creature would dissolve in its own juices and be reduced to a blob of goo. Organisms need to move things around and these things must be polar, but once moved to where they belong they need to stay put. So living things are constantly manipulating their components, adding OH's here and there to get them into solution and then chopping them off or covering them up to bring them out again. The beauty of ethanol, or ethyl alcohol, as a solvent is that it can dissolve many **organic** compounds whether or not they are soluble in water.

I should mention that ethanol is only one member of a class of compounds, the ***alcohols.*** Methanol, CH_3OH, is also known as wood alco-

Figure 16-4. Acetic Acid and Water

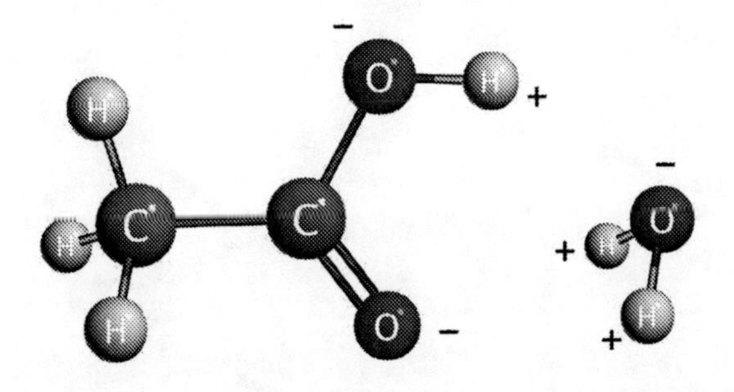

Equation 16-1. Reactions of Ethanol and Acetic Acid

$$(a)\ C_2H_5OH(aq) + O_2(g) = CH_3COOH(aq) + H_2O(l)$$
$$(b)\ CH_3COOH(aq) + NaOH(aq) = CH_3COONa(aq) + H_2O(l)$$
$$(c)\ CH_3COOH(aq) + C_2H_5OH(aq) = CH_3COOC_2H_5(aq) + H_2O(l)$$

hol since it came originally from the distillation of wood. Isopropanol, C_3H_7OH, is familiar as rubbing alcohol. Ethylene glycol, $C_2H_4(OH)_2$, which looks like ethanol with an OH group on each carbon atom, is used as automobile anti-freeze. Glycerol, $C_3H_5(OH)_3$, like isopropanol with an OH on each carbon, will be discussed at length in Chapter 19, when we discuss soap. Each of these compounds consists of a carbon-hydrogen chain, the non-polar bit, with one or more polar OH groups hanging off of it.

When ethanol is oxidized by bacteria (Equation 16-1(a)) it becomes ***acetic acid,*** which makes up about 5% of vinegar. Acetic acid, CH_3COOH, is one member of the class of compounds, the organic acids. You will recall that oxygen is an electron hog and with two of them pulling electron density away from the poor hydrogen, it is left with a greater positive charge than it had in either ethanol or water. Since electrons are what hold the molecule together, this *acidic hydrogen* or *proton* has a tendency to abandon ship if it happens to bump into an alkali. This kind of metathesis reaction, illustrated in Equation 16-1(b), is characteristic of an ***acid-base*** reaction; the **acid** donates a proton to the **base.**

Figure 16-5. Ethyl Acetate

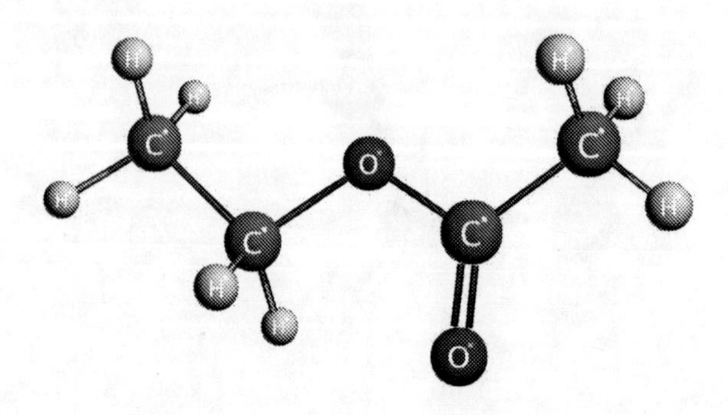

Now, like sodium hydroxide, ethanol has an OH group and you are possibly wondering whether a metathesis reaction is possible between ethanol and acetic acid. If so, you are wondering in the right direction. Equation 16-1(a) shows the metathesis reaction producing ethyl acetate, which is a member of the class of compounds, the esters. Esters are the organic counterparts of inorganic salts, being produced as a **condensation** of an acid and an alcohol. If your mead contained bacteria and oxygen you no doubt produced some ethyl acetate by accident. Unlike alcohols and acids, however, esters have no OH groups and the negative oxygen atom is buried deep inside the molecule. Consequently ethyl acetate is not very polar and makes an excellent solvent for non-polar compounds. It is a popular choice for fingernail polish remover.

∇ I have still to tell you what salt, or, as I like to call it, *tbmu*, has to do with anything. In those days our condensers were short, air-cooled jobbies, good enough for separating substances like gold from mercury, whose boiling points differ by hundreds of degrees. I did not suspect, and was not prepared to efficiently separate substances like alcohol from water, whose boiling points differ by only 22 degrees. As I have said, when a less volatile solute is dissolved in a more volatile solvent, the boiling point of that solvent goes up. Since salt dissolves in water but not in ethanol, it raises the boiling point of the water but not the ethanol, making their boiling points farther apart than they would normally be and consequently making it possible to separate them even with an inefficient air-cooled condenser. When you do not suspect that such a thing is possible, there is, of course, no reason to worry about condenser efficiency, but after my discovery it became clear that most of the alcoholic genie was going out the window rather than into the bottle. And while the painter himself tried to keep me bottled up, I was too volatile for him to keep a lid on me; I spread my bejeezical wings and flew the coop shortly after he gave up the ghost.

Material Safety

Locate an MSDS for ethanol (CAS 64-17-5). Summarize the hazardous properties of this material in your notebook, including the identity of the company which produced the MSDS and the potential health effects for eye contact, skin contact, inhalation, and ingestion. Also include the LD_{50} (oral, rat) and the NFPA diamond for ethanol.[2]

While it is possible to drink enough beer, wine or mead to make yourself seriously ill, it is even easier with distilled spirits. An eight-ounce glass of ethanol is the equivalent of more than a half-gallon of wine or a gallon of beer. Remember, it is the dose that makes the poison. You should also be aware that ethanol dissolves things that are not soluble in more dilute alcoholic solutions. Consequently, poisonous compounds from bottles or stills may dissolve in ethanol that would not have dissolved in mead. Finally, you should be aware that ethanol is not the only volatile compound present in fermented beverages so when you distill alcohol you may concentrate other potentially toxic compounds, ethyl acetate, for example.

You should wear safety glasses while working on this project. Your distilled spirit may be carefully burned or saved for use in a ***spirit lamp,*** as shown in Figure 16-8 (page 204). The contents of the pot may be flushed down the drain.

Research and Development

You are probably wondering what will be on the quiz.

- You should know the meanings of all of the words important enough to be included in the ***index*** or **glossary.**
- You should have studied the Research and Development items from Chapter 11 and Chapter 12.

---

2. The NFPA diamond was introduced in Section 15.2 (page 184). You may substitute HMIS or Saf-T-Data ratings at your convenience.

- You should know the reactions of Equation 16-1.
- You should know the name and purpose of each part of a still.
- You should know why ethanol is such a good solvent for both polar and non-polar compounds.
- You should know the hazardous properties of ethanol.
- You should know why secrets are so hard to keep.

16.3 Θ

The 18th amendment to the United States Constitution prohibited "the manufacture, sale, or transportation of intoxicating liquors within, the importation thereof into, or the exportation thereof from the United States." The law was so popular that it was repealed thirteen years later. The current law in the United States allows each adult to brew up to 100 gallons of beer or wine for *her* own personal use and a home-brewing industry has grown up to service hobbyists. It remains absolutely illegal, however, to distill spirits without a government permit. Boot-leggers are enthusiastically hunted down by law enforcement officers who, like the tenth-century painter, want to keep a lid on the spread of spirits.

Still, Part 19.901 of the Code of Federal Regulations "implements 26 U.S.C. 5181, which authorizes the establishment of distilled spirits plants solely for producing, processing and storing, and using or distributing distilled spirits to be used exclusively for fuel use." To establish such a plant requires a simple alcohol fuel producer's permit, form 5110.74, which may be ordered from the Bureau of Alcohol, Tobacco, and Firearms.[3] There is no fee for the permit and no tax is levied on alcohol used as fuel. Do not imagine that such a permit allows you to drink your spirits. If you are going to break the law anyway, such a permit is of no use to you. But if you are driven by curiosity to experience first-hand one of the most marvelous, literally spiritual substances in all of chemistry, this "liquid, which when set on fire and while still flaming leaves the material [underneath] unburnt," such an easily-acquired permit allows you to do so without getting into hot water with the Feds.

Of course, you will need a still and, there being no home-distillery market, you will have to either buy a still intended for laboratory use or build one yourself. Fortunately a still is not a particularly complicated piece of

3. Reference [32].

Figure 16-6. The PVC Still

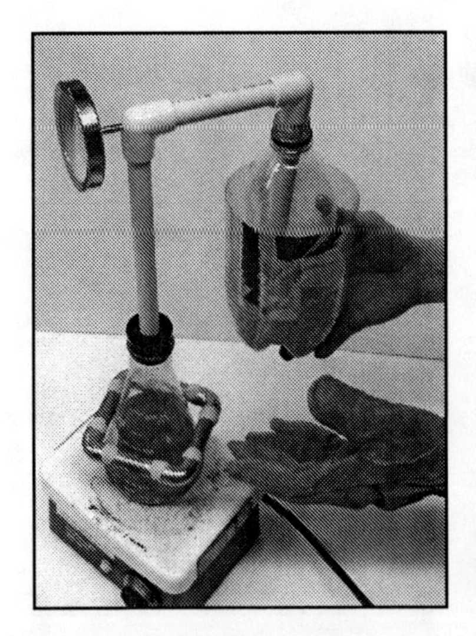

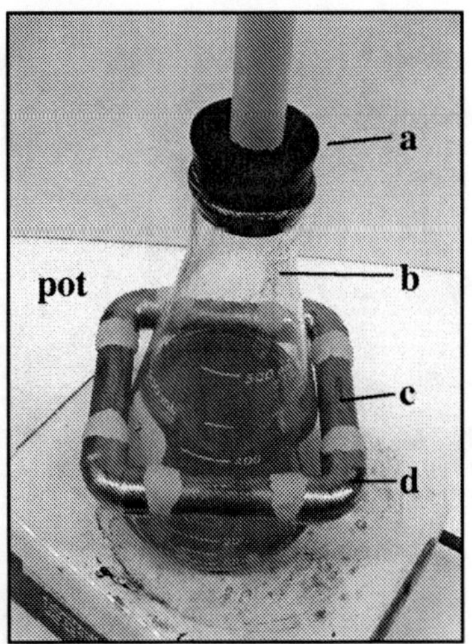

apparatus. I will tell you how to build one from an Erlenmeyer flask, a rubber stopper, and some common plumbing supplies. The complete still is shown in Figure 16-6(L) and the pot is detailed in Figure 16-6(R). Part (a) is a rubber stopper sized to fit the 500 mL Erlenmeyer flask, part (b). The stopper should have a hole drilled in it, 5/8 inch in diameter. Erlenmeyer flasks are not expensive and you can get one with a stopper from one of the suppliers listed in Appendix D (page 387). A flask weight will help prevent the still from tipping over. The one shown is constructed from four 3-inch sections of half-inch copper tubing, parts (c), and four elbows, parts (d), available from a plumbing supply. Fill the tubing with lead or steel shot before gluing it together with epoxy cement. The weight fits over the neck of the Erlenmeyer flask, which sits atop a hot-plate.

The column, detailed in Figure 16-7(L), is constructed from two 9-inch sections of half-inch CPVC tubing, parts (e), a common meat thermometer, part (f), two half-inch elbows, parts (g), and a section of tubing, part (h), long enough to accommodate the thermometer; the one shown is 4 inches long. Be sure to use CPVC tubing—the kind intended for plumbing hot water. Drill a hole in one of the elbows to make a tight fit with the probe of the meat thermometer. Use cement specifically designed for CPVC to glue the column together.

Figure 16-7. Details of the Column and Condenser

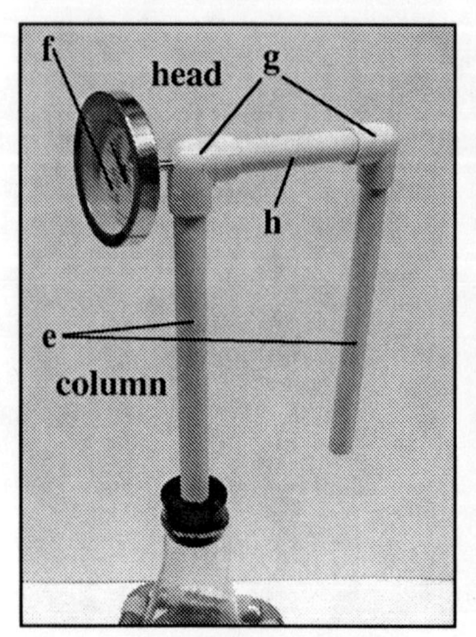

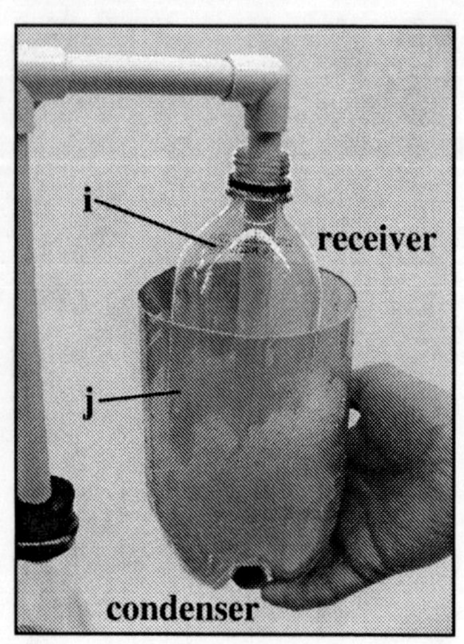

Any convenient container will serve as the receiver, part (i). The one shown in the figure is a 20-ounce soft drink bottle. The condenser, part (j), consists of the bottom of a 2-liter soft drink bottle filled with crushed ice. Commercial condensers are air-cooled or water-cooled, but the one shown works admirably without increasing the cost or complexity of the still.

You need a source of heat strong enough to boil water but not so intense as to crack the Erlenmeyer flask. Laboratories are equipped with hot-plates designed for exactly this purpose. Alternatively, there are many varieties of consumer appliances such as hot-plates, coffee makers, and crock pots which could be made to serve. Avoid anything whose heating elements visibly glow, as it may cause the glass to crack. It is a good idea to keep a leather work glove on one hand during the distillation so that you may quickly remove the still from the hot-plate if the temperature rises too quickly.

To operate the still, fill the Erlenmeyer flask with 400 mL of mead, wine, or beer. Add a couple of "boiling chips," small pieces of glass or silica, to prevent the still from boiling over. Slip the flask weight over the flask and insert the rubber stopper and column into the flask. Turn on the hot-plate, gradually turning up the power until the solution in the flask comes to a boil. Place the receiver into the condenser, fill the condenser with crushed ice, and slip the free end of the still into the receiver. The free end of the column should go all the way to the bottom of the receiver.

You are about to learn the truth of the old saw, "a watched pot never boils." You will watch the thermometer for some time with no change in its temperature. The temperature will begin to rise only when hot vapor makes it up the column to the *head,* the elbow where the thermometer lives. Do not get impatient—if you turn up the heat too quickly, the head temperature will be too high and your distillation will be spoiled. Eventually, the head temperature will rise and "steam" will exit the column into the receiver, where it will condense. Because the boiling point of ethanol is lower than that of water, your goal from this point on is to maintain the head temperature at the lowest possible value while still collecting liquid in the receiver. The boiling point of ethanol is 78°C (172°F); if you can keep the head temperature at or below 83°C (181°F) you will get reasonable separation of ethanol from water. If the head temperature gets a little too high, reduce the power to the hot-plate. If it is way too high, use your gloved hand to remove the still from the hot-plate and allow it too cool for a minute or so. After a while you will get into a "zone," in which the head temperature remains constant and liquid accumulates in the receiver. At some point you may wonder whether anything is coming out of the column; simply remove the receiver and cautiously feel the end of the column with your non-gloved hand. If it is hot, you are still collecting alcohol. Eventually you will find that you cannot continue collecting alcohol without increasing the head temperature. You will have collected most of your alcohol in the receiver and the pot will contain mostly boiling water and yeast carcasses. You may now turn off the hot-plate and allow the still to cool. Disassemble the cold still and pour the pot liquid, but not the boiling chips, down the drain. Rinse out the column and put it away. You may screw the cap on your receiver and store your alcohol for later use.

A crude estimate of the alcohol content can be made by burning your alcohol *distillate.* Weigh a Petri dish and record its empty weight, w_{before} in your notebook. Tare the balance and add 10.0 g of your distillate, w_{dist}, to the dish and set it alight. When the fire has gone out, weigh the Petri dish again and record this weight, w_{after} in your notebook. The percentage of alcohol of your distillate is approximately:[4]

---

4. A simple distillation like this one produces a distillate which is, at most, 95% alcohol so we must subtract the percent water from 95 rather than from 100.

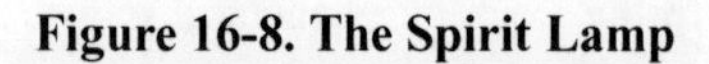

Figure 16-8. The Spirit Lamp

$[95 - 100(w_{after}-w_{before})/w_{dist}]\%$

Double the percent alcohol and you have the *proof*. If your product is below about 40% alcohol (80 proof), you will be unable to light it at all.

Δ Do not overlook this experience, my brothers and sisters. To merely read and parrot the words is unworthy of your talents. If you are to really live, you must see and feel and know for yourself.

> It rises from the earth to Heaven and descends again to the Earth and receives power from Above and from Below. Thus thou wilt have the glory of the Whole World. All obscurity shall be clear to thee. This is the strong power of all powers for it overcomes everything fine and penetrates everything solid.
>
> — *The Emerald Tablet of Hermes Trismegistos*

∇ Scat!

Quality Assurance

You have succeeded in making ethanol fuel when you are able to light your distillate. Your notebook should note the head temperatures at which you began and ended your collection and you should estimate the percent alcohol in your distillate.

Chapter 17. Tzu-Chhun (Gunpowder)

> The second kind of flying fire is made in this way. Take 1 lb. of native sulfur, 2 lb. of linden or willow charcoal, 6 lb. of saltpeter, which three thing are very finely powdered on a marble slab. Then put as much powder as desired into a case to make flying fire or thunder. Note.—The case for flying fire should be narrow and long and filled with well-pressed powder. The case for making thunder should be short and thick and half-filled with the said powder and at each end strongly bound with iron wire.
>
> — Marcus Graecus, *Liber Ignium*, *ca.* 1280 AD[1]

17.1 ☿

Δ From Lucifer to Athanor, from Athanor to Vulcan, from Vulcan to Theophilus I passed like a torch down the generations. Yielding to death I found life and embracing life I passed into death. In 762 AD I came to possess one Tu Tzu-Chhun, a pauper who was delivered from the miserable circumstances of his birth and initiated into the Tao by an ancient and nameless alchemist. One by one the essential elements of my being passed from master to student and little by little I re-meme-bered myself.

Δ Tzu-Chhun had been possessed of Lucifer since childhood and had become Athanor in his youth. The Master recognized these familiar I-deas and took him for an apprentice. First came the lesson of **sulfur**, the masculine soul which is hot and dry. I learned of animal honey, vegetable charcoal, mineral brimstone and all things which are consumed by fire, rising from earth to heaven. Then came the lesson of **mercury**, the feminine spirit which is cool and moist. I learned of animal ammonia, vegetable alcohol, mineral quicksilver and all things which flee the fire, descending from heaven to earth. Perfect Balance, the Master told me, is achieved by the marriage of sulfur and mercury.

Δ The Master had recently come into possession of a new animal spirit, a white powder which he called *hsiao shih*, the stone of solvation, and which later came to be called saltpeter. He sent me to meditate on the lesson of **salt**, that which remains when mercury has flown and sulfur

1. Reference [69], p. 49.

has been consumed, while he wrought new elixirs at the furnace. Terrible visions tormented me in my reveries; I was roused to find the house engulfed in flames and my Master utterly vanished along with his unspoken I-deas, which consequently passed into extinction. I saw to it that the event was set down in writing lest I, too, perish from the Earth:

> Some have heated together sulphur, realgar, and saltpetre with honey; smoke (and flames) result, so that their hands and faces have been burnt, and even the whole house (where they were working) burned down.
> — Chêng Yin, *Chen Yuan Miao Tao Yao Lüeh* (Classified Essentials of the Mysterious Tao of the True Origins of Things), *ca.* 850 AD[2]

Δ From the Master to Tu Tzu-Chhun to Chêng Yin I passed, gathering new I-deas along the way until at last I found myself in the person of one Tsêng Kung-Liang. There I recorded the formula of the fire-drug, *huo yao*, in the *Wu Ching Tsung Yao* (Collection of the Most Important Military Techniques, *ca.* 1040 AD). My early recipes were low in saltpeter and were used for incendiary weapons and poison smoke. Over the next three centuries I explored formulae higher in saltpeter, lower in sulfur and charcoal, and the first explosive powders appeared; rockets, maroons and cast-iron bombs soon followed, not to mention recreational fireworks. By 1300 AD I had developed the metal-barrelled bombard, ancestor of the cannon. All of these developments had taken place in the land of my youth, but I was about to leave China and colonize the whole world.

Δ It is in the interest of the mortal to keep secrets. After the publication of the *Wu Ching Tsung Yao* the Sung government was eager to monopolize the fire-drug and as my I-deas were proliferating, the Sung attempted to control access to the materials needed for gunpowder manufacture. In 1067 AD sales of saltpeter and sulfur to foreigners were banned and in 1076 AD the prohibition was extended to all private transactions in sulfur and saltpeter. But it is not in the interest of secrets to be kept and I was soon embraced by Mongols, who learned to make sulfur and saltpeter for themselves. Having lost the battle of I-deas, the Sung compounded the error by failing to maintain a well-stocked arsenal and by 1277 AD the Mongols had eradicated them.

Δ At the same time that the Mongols were taking China they were pushing westward to the frontiers of Christendom and Islam. Though

2. Reference [65], p. 112.

firearms were not widely used abroad, by 1227 AD Turkestan and Persia had fallen; between 1236 and 1246 AD Russia, Hungary and Poland were overrun. Christian envoys visited the Mongol court between 1245 and 1256 AD in an effort to forge an alliance against their mutual enemies. No alliance emerged, but knowledge of recreational fire crackers returned with them and was recorded by Roger Bacon in 1267 AD. The merchant Niccolò (father of Marco) Polo traveled to China between 1261 and 1269 AD and established commercial ties between East and West. In 1258 AD Baghdad fell, leaving Iran and Iraq in Mongol hands. Khubilai Khan recruited Islamic military engineers to assist in the final push against the Sung and, given the increasingly brisk interaction with the Mongol court through diplomatic, military, and commercial channels, detailed gunpowder formulas became known almost simultaneously to Christendom and Islam by about 1280 AD.

Δ Both Christians and Muslims were on the verge of a period of great colonial expansion. The history books say that they used their knowledge of gunpowder to subjugate cultures less technologically advanced than themselves. Confident in the Truth of their religions, they used salvation as an excuse for the satisfaction of their mortal appetites. But the history books have it exactly backwards. It was I, the un-kept secret, who used the ambitions of Islam and Christianity to fertilize the Earth with I-dea-logical seeds. One of my seeds took root in a French civil servant, ***Antoine Lavoisier,*** who, in 1775, was appointed Inspector of Gunpowder for the Government. As Lavoisier, I explored the nature of combustion, demonstrating that air is a mixture, that water is a compound, and that oxygen is an element. After the mortal Lavoisier was beheaded by the Revolution, I took refuge in one of his apprentices and immigrated to the United States, where I founded a company to manufacture gunpowder. The name of this apprentice was ***Irénée du Pont de Nemours.***

17.2 🜍

Δ Perfect Balance is achieved through the marriage of sulfur and mercury. The secret which I am about to impart brought not only gunpowder, but the entire modern age into being.

🜁 I don't believe it. Is the Author insane?

Δ From the Mongol bombard to the blunderbuss, from the cannon to the steam engine, from the Ford Model T to the Boeing 747, this secret has been instrumental in shaping the world.

Δ From dynamite to C4, from the V2 rocket to the ICBM, from farm kids blowing their fingers off to Timothy McVeigh calling around for the best price on fertilizer, your "secret" has been instrumental in killing innocent people. Look, I kept my mouth shut about the distillation of alcohol, but this is going too far. Given the rise of terrorism, wouldn't we be better off skipping gunpowder and moving on to something less inflammatory?

Δ You imagine that terrorists are going to make their own gunpowder? I have news for you; it is easier to buy gunpowder than it is to buy saltpeter and sulfur.

Δ But this is dangerous! You can't go around teaching people to make explosives as if they were some kind of toys. After all, as a Taoist you must know the teachings of Lao-tzu:

> Now arms, however beautiful, are instruments of evil omen, hateful, it may be said, to all creatures. Therefore they who have the Tao do not like to employ them. [3]

Δ And as a figment of the "Author's" imagination, you must know that Lao-tzu continues:

> The superior man ordinarily considers the left hand the most honourable place, but in time of war the right hand. Those sharp weapons are instruments of evil omen, and not the instruments of the superior man;—he uses them only on the compulsion of necessity.

Δ Is it a necessity for cowboys to shoot Indians, for cops to shoot robbers, or for little kids to shoot their playmates?

Δ You are the one who has laid the foundation; Absolute Harmony comes through the application of stoichiometry to oxidation and reduction.

Δ Hey, that was for educational purposes only!

Δ You probably need not concern yourself; your vacuous little jingles have probably filled no heads with wonder; your breezy little games have probably blackened no hands with charcoal; your tedious little exercises have probably whitened no faces with ash. But just in case,

---

3. Reference [16], *Tao Te Ching* 31 (or 75).

> He who has killed multitudes of men should weep for them with the bitterest grief; and the victor in battle has his place (rightly) according to those rites.

Δ Join me in weeping, Figment, for your I-deas are as tinged with blood as my own.

Alright, alright you two; knock it off. It seems that I am of two minds in this matter. On the one hand, gunpowder manufacture was an important predecessor to modern chemical industry. To ignore it would make the subsequent chapters less comprehensible. On the other hand, it would be irresponsible to encourage the public to engage in dangerous and destructive activities. So I will stop short of giving the formula for gunpowder. I will explain the chemical principles involved and the astute reader will be able to calculate the formula for *herself.* I suppose that the lazy reader will look up the formula in any encyclopedia and the budding terrorist will buy *her* explosives ready-made. My intention here is simply to give the student of chemistry a tactile introduction to this important material; if you are lazy or evil-minded, let me suggest that you will find more practical information elsewhere.

Perfect balance is achieved through the marriage of sulfur and mercury. I think that what Tu Tzu-Chhun was getting at was the careful balance of **oxidation** and **reduction** which is required for an explosion to take place. In Chapter 1 you were introduced to the oxidation of ***charcoal*** by oxygen. You observed that when you blow on a glowing coal, it gets brighter. The rate of combustion is limited by the rate at which fresh oxygen can be provided to the fuel. In Chapter 5 you learned the principle of the furnace, in which a steady draft of air allows the fuel to burn at a faster-than-normal rate, thereby increasing the temperature attainable in the kiln. The draft can only be so strong, however, before the extra heat produced by the draft is simply carried away by the increased flue gases. This is what happens when you blow out a match.

To make an explosive, what we really need is a solid or liquid oxidant which can be intimately mixed with the reductant, or fuel, thus eliminating the need for atmospheric oxygen. The first material to be used in this way was ***saltpeter,*** or potassium nitrate. A sort of gunpowder can be made from charcoal and saltpeter alone:

$$5\,\mathrm{C}(s) + 4\,\mathrm{KNO_3}(s) = 3\,\mathrm{CO_2}(g) + 2\,\mathrm{N_2}(g) + 2\,\mathrm{K_2CO_3}(s)$$

Given the balanced equation, we can use stoichiometry to determine the relative weights of charcoal and saltpeter:

$$\begin{aligned} ?\ \text{g KNO}_3 &= 1\ \cancel{g\,C}\left(\frac{1\ \cancel{mol\,C}}{12\ \cancel{g\,C}}\right)\left(\frac{4\ \cancel{mol\,KNO_3}}{5\ \cancel{mol\,C}}\right)\left(\frac{101.1\ \text{g KNO}_3}{1\ \cancel{mol\,KNO_3}}\right) \\ &= 6.74\ \text{g KNO}_3 \end{aligned}$$

That is, for every 1 gram of charcoal, we require 6.74 grams of saltpeter. If there is too little saltpeter, charcoal will remain after the saltpeter as been used up, which means that there was charcoal that *could* have burned, but didn't. The residue which remains will be black, from the leftover charcoal. If there is too much saltpeter, the charcoal will be consumed and some unused saltpeter will remain. The residue will be white, from the leftover saltpeter. But if mercury and sulfur, oxidant and reductant, are perfectly balanced no salt at all will remain from their union and the maximum energy will be delivered from the available fuel.

While a saltpeter-charcoal gunpowder will burn without air, the production of potassium oxide consumes much of the energy available from the combustion of the charcoal. Potassium sulfide consumes much less energy, leaving more bang for the buck. This becomes possible with the addition of **sulfur** to the saltpeter-charcoal mix—not esoteric, alchemical sulfur—but the ordinary, yellow element. Equation 17-1(a) gives a skeleton reaction for the saltpeter-charcoal-sulfur mixture. Balance it using the method of Chapter 11 and then answer the following **stoichiometric questions** using the method of Chapter 15:

Q: How many grams of saltpeter are needed to react with each gram of sulfur?

Q: How many grams of charcoal are needed to react with each gram of sulfur?

Charcoal is not the only reductant from which gunpowder can be made. A popular alternative for amateur rocket fuel is powdered sugar, ***sucrose.*** Balance Equation 17-1(b) and then answer the following **stoichiometric questions**:

Q: How many grams of saltpeter are needed to react with each gram of sulfur?

Q: How many grams of sucrose are needed to react with each gram of sulfur?

Equation 17-1. Skeleton Equations for Two Gunpowder Mixtures

$$(a)\ ?\ \mathrm{C}(s) + ?\ \mathrm{KNO_3}(s) + ?\ \mathrm{S}(s)$$
$$= ?\ \mathrm{CO_2}(g) + ?\ \mathrm{N_2}(g) + ?\ \mathrm{K_2S}(s)$$

$$(b)\ ?\ \mathrm{C_{12}H_{22}O_{11}}(s) + ?\ \mathrm{KNO_3}(s) + ?\ \mathrm{S}(s)$$
$$= ?\ \mathrm{CO_2}(g) + ?\ \mathrm{N_2}(g) + ?\ \mathrm{K_2S}(s) + ?\ \mathrm{H_2O}(g)$$

Material Safety

Δ Can there be any doubt in your mind that making fireworks is dangerous work? It is not for the lazy, the timid, or the careless. If you are unprepared to make careful calculations and measurements; if you are unwilling to accept responsibility for your own safety and that of others in your vicinity; if you are not able to observe all appropriate safety precautions, you would do well to skip the Salt section and retire to some less demanding exercise, like spinning or dyeing or whistling. But if your hands are black with charcoal and your face is white with ash, welcome Comrade to the fellowship of those who are not afraid to keep in the heat and withstand it.

Locate an MSDS's for saltpeter (CAS 7757-79-1), sulfur (CAS 7704-34-9), charcoal (CAS 7440-44-0) or sucrose (CAS 57-50-1). Summarize the hazardous properties in your notebook, including the identity of the company which produced each MSDS and the NFPA diamond for each material.[4]

The most pressing hazard in this project is the danger of fire. Sources of ignition, including sparking from static electricity or iron utensils should be avoided. Fireworks should be handled according to federal, state, and local laws. For more information on firework safety see, for example, *Practical Introductory Pyrotechnics*.[5]

You should wear safety glasses while working on this project. Any leftover gunpowder should be carefully burned. Leftover sulfur, charcoal, or sugar may be thrown in the trash. Leftover saltpeter can be flushed down the drain with plenty of water.

4. The NFPA diamond was introduced in Section 15.2 (page 184). You may substitute HMIS or Saf-T-Data ratings at your convenience.
5. Reference [48].

Figure 17-1. Plunger, Plugs, Anvils, and Nozzles

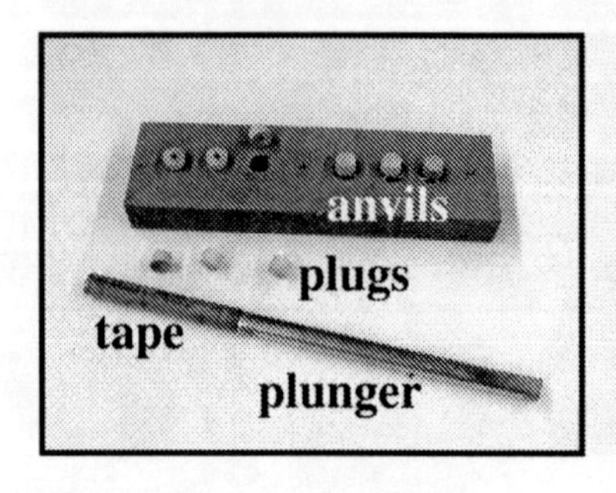

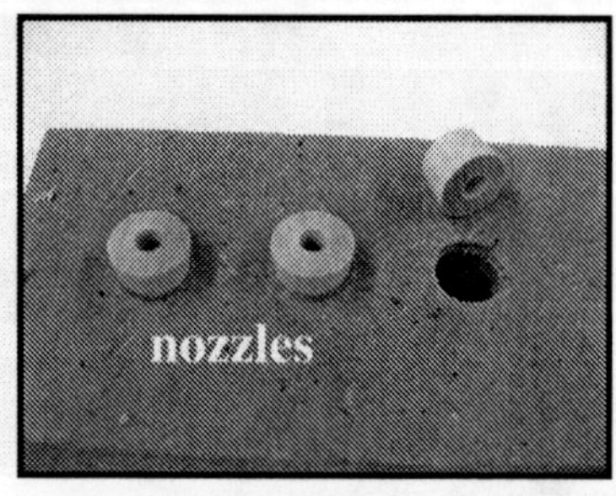

Research and Development

You should not remain ignorant if you are to proceed in the Work.

- Know the meanings of those words from this chapter worthy of inclusion in the ***index*** or **glossary.**
- You should have mastered the Research and Development items of Chapter 13 and Chapter 15.
- Be able to balance the redox reactions of charcoal or sucrose with saltpeter and sulfur.
- Be able to work out the stoichiometry for either of these reactions.
- Know the systematic names and formulae for sugar and saltpeter.
- Know the hazardous properties of charcoal, sugar, saltpeter, and sulfur.
- Be familiar with local laws concerning fireworks.

17.3 Θ

"The case for flying fire should be narrow and long and filled with well-pressed powder." The remainder of this section simply amplifies this simple description for making rockets. Your rocket may be large or small, but for the purposes of this section we shall assume a length of 3 inches and a diameter of 3/8 inch. You will need a few simple tools

Figure 17-2. Rolling Your Own

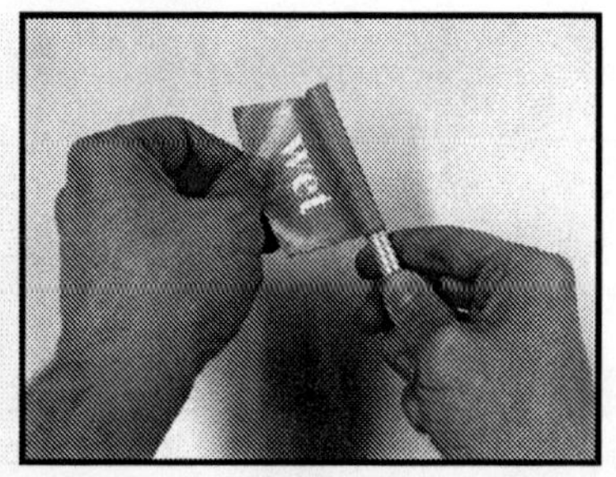

and supplies. First, you will need a *plunger,* a rod 8 inches long and 3/8 inch in diameter. It can be cut from wooden dowel rod, but I cut mine from aluminum rod for durability. One end of this rod is wrapped with tape so that the plunger is larger in diameter at one end than at the other. Second, you will need some hardwood *plugs,* 3/8 inch in diameter, of the kind used by carpenters to cover screw holes in furniture. These are commonly available at craft and building-supply stores. We need to drill holes down the center of some of the hardwood plugs so that they may serve as *nozzles.* Third, you need a board with at least one of these plugs glued to it to serve as an *anvil.* In addition, this board should have holes drilled in it to just accommodate several plugs.[6] Force-fit plugs into the holes in the board to hold them steady while you drill holes in them, then use a pencil to release these nozzles from the board.[7] Each rocket will require one plug and one nozzle.

The bodies of your rockets will be tubes rolled from 3-inch wide paper package tape, as shown in Figure 17-2. Many of my students have never seen paper tape before, but it can still be bought at office-supply stores. I prefer the gummed variety, which becomes sticky only when wet.[8] To roll a rocket *tube,* carefully wet only the very edge of a 15 cm long piece of tape and stick it to the large-diameter end of the plunger, the end which has been built up with tape. Roll the tape around the plunger, wet the last 5 cm, and finish rolling the tube. If you were careful in wetting the very edge, the tube will stick to the plunger while you roll it, but will twist

6. I drill my holes 23/64 inch in diameter.
7. I drill my nozzle holes 9/64 inch in diameter.
8. Heavy kraft paper, 7 cm x 15 cm, may be substituted if paper tape is unavailable.

Figure 17-3. Loading the Tube

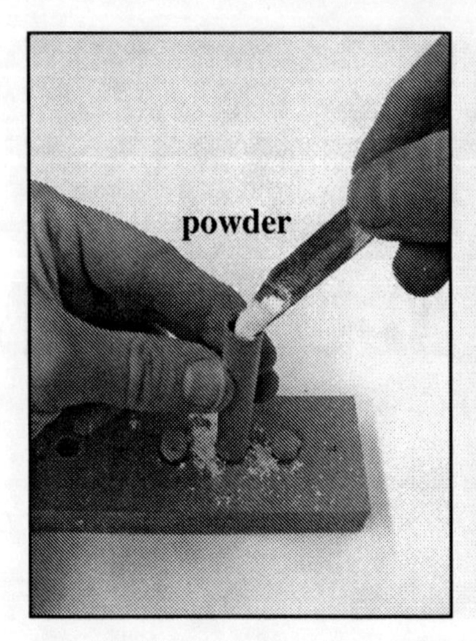

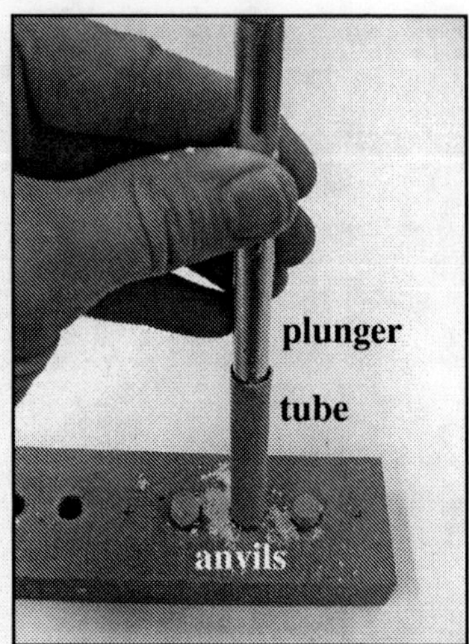

off when you are done. You should now have a paper tube, 3 inches long and slightly more than 3/8 inch in diameter. Try pushing a plug into the tube—it should go in easily but fit snugly. If it is too tight, add more tape to build up the plunger; if too loose, remove some tape from the plunger. Once you have the hang of it, roll two or three good tubes for your first foray into rocketry.

Now it is time to mix your powder, the proportions of which you determined in Section 17.2. You should ensure that your work area is free of extraneous flammable materials. Have a bucket of water nearby so that in case of accident you can plunge any burns into the water.[9] Powder made with powdered sugar as a fuel is less messy, in my experience, than that made with charcoal, and it makes an excellent rocket fuel. Your sulfur and saltpeter should be separately ground in a mortar and pestle to the consistency of powdered sugar. Weigh ***by difference*** (page 384) 1.0 g of sulfur and the stoichiometric amounts of saltpeter and sugar into a plastic bag.[10] Seal the bag and break up any lumps with your fingers.

9. I have been making this powder for thirty-five years, ten of them with undergraduates, and have never had an accidental ignition. The wise person, however, is always prepared for such a possibility.

10. I use anti-static bags of the kind used to ship circuit boards. These provide and added measure of protection against accidental ignition.

Shake the bag to mix the ingredients and turn it end for end twenty or thirty times; if your powder is not thoroughly mixed you will get uneven performance from your rockets. With your powder mixed, it is time to load your rockets.

Push a plug into a rocket tube and set it on one of your anvils, as shown in Figure 17-3. Fill the tube with powder using a spoon or spatula, preferably a metal one to minimize the danger of static discharge. Push the small-diameter end of your plunger, the end without the tape, into the tube and press the powder all the way to the bottom. Give the plunger three good raps with a rubber mallet to compact the powder into a solid mass. Remove the plunger, refill the tube with powder, reinsert the plunger, and give it another three raps with the rubber mallet. Repeat this sequence until the rocket is between 2/3 and 3/4 full. Insert a nozzle into the tube and push it all the way down until it rests on the powder. Remove the tube from the anvil and apply glue to the outside of both the plug and the nozzle, taking care not to fill in the nozzle hole.[11] When the glue has set you may use scissors to trim away any excess paper from the ends of the rocket. The powder made from 1 g of sulfur will be enough for you to make two or three rockets.

Packing the powder into the rocket gives it more fuel for the flight, but it also slows the burning of the powder. To speed up the burn rate you need to drill out a *core* using a 1/16 inch drill bit. Hold the bit in one hand and the rocket in the other.[12] Push the bit into the nozzle, give it a twist, remove it, and allow any loose powder to fall out. Reinsert the bit and remove some more powder. At some point it may become easier to hold the bit still and twist the rocket. Your goal is to drill a core down the center of the rocket from the nozzle to the plug, as shown in Figure 17-4.

Two more items complete the rocket: the *guide stick* and the *fuse.* For guide sticks I use 12-inch bamboo skewers, available in grocery stores for making shish kebab. Use two strips of cellophane tape to attach each stick securely to its rocket. I use commercial "green visco" fuse, available wherever pyrotechnic supplies are sold (page 388). Those wishing to make their own fuse may consult Reference [48]. Insert a 7 cm length of fuse into the nozzle of each of your rockets. Ideally, the finished rocket

11. I use fast-curing epoxy cement so that rockets may be launched soon after they are made.
12. A power drill could heat the powder, with unfortunate consequences.

Figure 17-4. Rocket Schematic

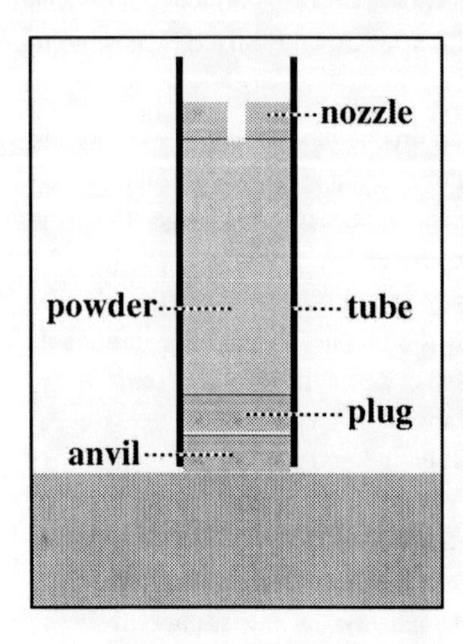

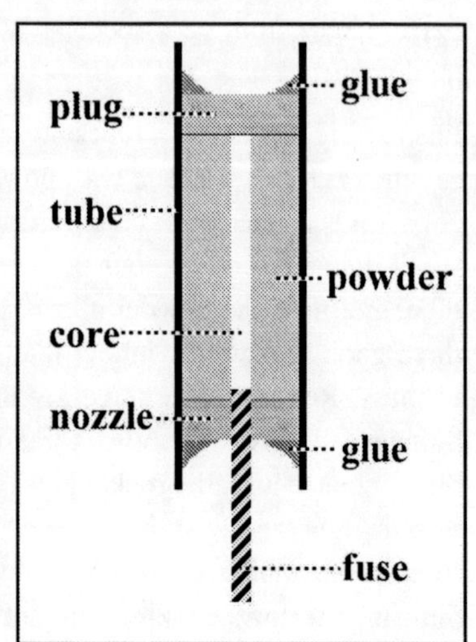

should balance on your finger where the fuse enters the nozzle, as shown in Figure 17-5. If your rocket is stick-heavy, slide the tube toward the middle of the stick. If it is tube-heavy, slide it away from the middle of the stick. If you adjust any stick, secure it with an extra piece of tape. Number your rockets and write you name on each one in case you are able to recover them.

Safety must be utmost in your mind when launching rockets. You should familiarize yourself with state and local laws regarding fireworks. Choose an area that will minimize the possibility of accidental fires and maximize the likelihood of recovering your rocket. Be sure to keep any unused rockets at least 20 feet from the one being launched. Place a rocket into a bottle, an iron pipe, or other appropriate support, light the fuse, and retire to a distance of 20 feet or more.

Many things must go right for a rocket to perform well. The powder ingredients must be in the proper ratios and they must be well-mixed. There must be enough powder to lift the rocket and not so much as to weigh it down. The nozzle must be small enough to constrict the gases as they exit, but not so small that the nozzle or plug blows out. The nozzle and core must be straight enough to provide for a stable trajectory. That said, the most common failures in my experience are either that the rocket fails to get off the ground or that the nozzle blows out. If your

Figure 17-5. The Guide Stick and the Fuse

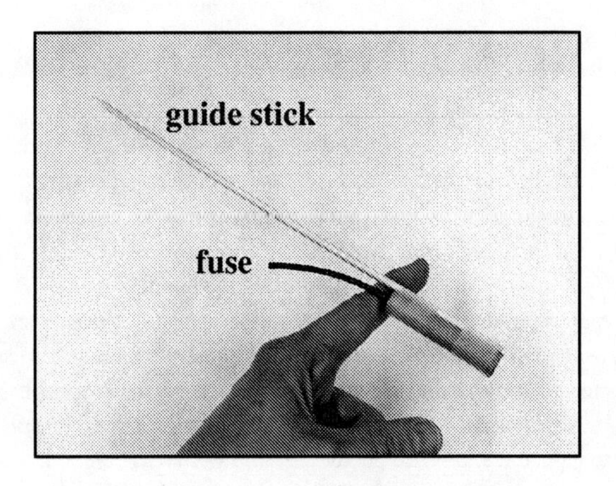

rocket fails to lift off, take more care in formulating and mixing the next batch of powder and be sure to drill out a good core. If you still have trouble, try making your next nozzle holes smaller. If, on the other hand, your rocket blows out a plug or nozzle, or if it explodes, try drilling a shallower core or a larger nozzle hole in the next set of rockets.

Δ Comrade, the vast majority of mortals do not have what it takes to complete this project safely. Judge well your character and decide whether you can have a pious and just feeling about these things and can keep them secure.

> It rises from the earth to Heaven and descends again to the Earth and receives power from Above and from Below. Thus thou wilt have the glory of the Whole World. All obscurity shall be clear to thee. This is the strong power of all powers for it overcomes everything fine and penetrates everything solid.
>
> — *The Emerald Tablet of Hermes Trismegistos*

Quality Assurance

Δ There is no room for equivocation; either your rocket rises from the Earth to heaven or the strong power of all powers remains obscure to thee.

Record in your notebook any differences in the construction of your numbered rockets. Give a brief description of the success or failure of each one.

Chapter 18. Spot and Roebuck (Acid)

> Now, as I told you above, this saltpeter is extracted from the above-mentioned manurial soils and from dark places that have stood turned over and loosened for a long time, provided the rain have not been able to quench the earthy dryness. The best and finest of all saltpeters is that made from animal manure transformed into earth in the stables, or in human latrines unused for a long time. Above all the largest quantity and the best saltpeter is extracted from pig dung. This manurial soil, whatever kind it may be, should be well transformed into a real earth and completely dried of all moisture; indeed it should be powdery if you wish it to be good. Assurance that it contains goodness is gained by tasting with the tongue to find that it is biting, and how much so. If you find from this trial that it has a sufficient biting power so that you decide to work it and you have found a quantity of it, it is necessary to provide yourself with kettles, furnaces, vats, or chests, and also with wood, lime, soda ash, or ashes of cerris or oak, and especially with a large hut or other walled space near water.
>
> — Biringuccio, *Pirotechnia*, *ca.* 1540 AD [1]

18.1 ☿

🜄 Crap! Now *that's* a business to be in. You take something that everybody's got and nobody wants and turn it into something that every government needs if it's going to be a government for any time at all. Maybe you remember from Chapter 12 that bacteria make ammonia whenever crap and piss sit around without much air, which is why outhouses and latrines smell like ammonia. Well, when there's plenty of air, instead of making ammonia they make nitrates, which don't smell at all. And that's why compost heaps don't smell bad as long as they're getting plenty of air. It's those nitrates in compost, stables, and latrines that the government needs for making saltpeter. Only problem is, how do you get the nitrates out?

∇ Perhaps you are wondering why the hut or other walled space in which you will process your manurial soil should be near water. I will tell you. First of all, you will recall from Chapter 7 that all nitrates are

1. Reference [3], p. 405.

soluble and that most hydroxides are insoluble, with the exception of calcium hydroxide which is, of course, soluble in water. So when lime water, that is to say a solution of calcium hydroxide, percolates through manurial soil, virtually all of the minerals stay behind as insoluble hydroxides and the water which dribbles out the bottom contains soluble nitrates, chiefly calcium nitrate. Throw in some wood ashes, which naturally contain potassium carbonate, and a classic metathesis reaction occurs with calcium carbonate precipitating out and potassium nitrate remaining in solution. Boil away the water and potassium nitrate recrystallizes just as potassium carbonate did in Chapter 8. The result is purified ***saltpeter.***

∀ Now, the government needs ***sulfur*** as well as saltpeter if it's going to make gunpowder. Sulfur ore is a kind of yellow dirt or rock that's just a mixture of elemental sulfur and whatever dirt or rock it happens to be mixed with. You can separate out the sulfur on account of its melting point is lower than the melting point of most other kinds of dirt and rock; just heat it up and run the molten sulfur off as a liquid. And if you really want to do it up right, you can distill that sulfur the same as you did alcohol in Chapter 16. So now you have saltpeter and sulfur, which just leaves ***charcoal.*** You get charcoal by charring wood in the absence of oxygen and that gives you everything you need to make gunpowder, which is a pretty good business.

∇ Second of all, you must know that "oil of vitriol" had been manufactured in Saxony in small quantities beginning approximately 1630 AD. This oil, now known as ***sulfuric acid,*** was used to make hydrochloric and nitric acids for parting precious metals from base metals. ***Vitriol*** is a rather uncommon, glassy, green mineral which results from the weathering of iron pyrites, one of the ores of iron. When heated, vitriol decomposes into iron oxide, used as a pigment, and the aforementioned sulfuric acid. Vitriol being relatively scarce, acid was expensive. But there was little incentive to find a less expensive source because assayers required only small quantities of acid.

∀ Then acid got mixed up in drugs.

∇ I was getting to that. About 1689 or so, a fellow named Rudolf Glauber discovered how to make Glauber's salt from sulfuric acid. He ascribed all kinds of miraculous qualities to it and, being something of a quack, began to prescribe it for everything from impotence to venereal disease. This "wonder drug" did no actual good, but unlike most medicines of the day, it did no particular harm and so it passed from quack to quack—

🜃 Just like a spider.

∇ More like a dogma, really.

🜃 Like a spidery dogma maybe, from one quack to another until it landed in a fellow named Joshua Ward, known to his enemies as "Spot," on account of his Gorbachevity birthmark. Spot had started out being a politician but when that didn't work out, he had to be a quack instead. Now, he didn't want to be a little backwater quack, he wanted to be a big-city man-about-town quack and he opened up a fancy London practice in 1733. He needed a boat-load of Glauber's salt and, the vitriol market being what it was, he hunted through his old chemistry books for in*spir*ation. He found out you could make sulfuric acid from saltpeter and sulfur, which nobody ever did, what with the high demand for gunpowder and the low demand for acid. So Spot went into the sulfuric acid business in 1736, making it by the pound from saltpeter and sulfur instead of by the ounce from vitriol. Even though sulfur and saltpeter were not cheap, they were cheaper than vitriol and so he was able to sell that pound of acid at the same price that everybody else charged for an ounce and still turn a tidy profit.

∇ Meanwhile, back in Saxony one Bergrat Barth discovered in 1744 that indigo reacts with sulfuric acid to produce a water-soluble blue dye. You will, of course, remember from Chapter 12 that indigo was an extremely important dye, but that compared to other dyes its application was quite tedious, involving the anaerobic fermentation of the insoluble dye in stale urine. The new "Saxony blue," or "indigo carmine," was an instant hit with dyers, who were soon crusading for more acid. By mid-century Saxony was producing about 20 tons per year of admittedly expensive acid from vitriol.

🜃 Now, Spot could make acid a lot cheaper than the Saxons, but he couldn't make that much of it. See, sulfuric acid eats through most metals so he had to make it by burning sulfur and saltpeter under fifty-gallon glass jars, which were pretty damn big, as jars go. He kept adding more and more jars to his business, trying all the while to keep his process a secret, but that secret didn't want to be kept.

∇ In no time at all a Birmingham doctor named John Roebuck became a convert to the fellowship of saltpeter and sulfur. Poring through the

Glauberian scriptures, he found a passage to the effect that sulfuric acid does not react with lead. So he lined wooden boxes, or “chambers,” with lead foil to produce acid-resistant reactors which were neither as expensive, nor as fragile as Ward’s glass jars. And these “lead chambers” could be built much larger than the jars of the day. Roebuck set up his own acid works in 1746 and began producing chamber acid by the hundreds of pounds—

🜃 But the upstart refused to pay royalties on Spot’s patent—

∇ Which was understandable, since the patent was not filed until 1749, three years *after* Roebuck started manufacturing acid—

🜃 Which acid was made by a process in*spir*ed by Spot, who had started in 1736. Spot sued.

∇ Which drove Roebuck from Birmingham to Edinburgh, the home of a large textile industry. The Scottish bleachers found the now inexpensive sulfuric acid to be an excellent bleach for cotton and linen. Roebuck’s Prestonpans Vitriol Company began to produce acid by the ton, and with economies of scale the price of acid fell to a quarter of the price demanded by Ward.

🜃 Which increased demand. But Roebuck was no better at keeping secrets than Spot was and pretty soon everybody and *his* dog had a chamber acid plant. The French had one by 1768 and four more by 1786.

∇ By 1800 the Prestonpans works comprised 108 chambers with a total capacity of 60,480 cubic feet and there was another at Burntisland with a capacity of 69,120 cubic feet. Worldwide annual acid production was reckoned by then in the hundreds of tons.

🜃 And by 1913 it would top 10 million tons of acid per year. These days, sulfuric acid is the king of chemicals, the big enchilada, numero uno. In other words, more sulfuric acid is produced than any other chemical in the world. Not bad, for something that came out of the dung-heap.

18.2 🜍

Remember from Chapter 7 that water (H_2O) can be viewed as hydrogen hydroxide (HOH) and some small fraction (1 in 10,000,000) will ionize to form hydrogen (H^+) and hydroxide (OH^-) ions:

$$\mathrm{HOH}(l) = \mathrm{H}^+(aq) + \mathrm{OH}^-(aq)$$

In pure water, the concentrations of hydrogen and hydroxide ions are equal. Anything that causes the concentration of hydrogen ion to exceed that of hydroxide ion is called an **acid;** anything that causes the concentration of hydroxide ion to exceed that of hydrogen ion is called an **alkali,** or **base.** Now, we have spent a lot of time in this book talking about alkalis: ***potash*** and ***soda ash*** in Chapter 8, ***lime*** in Chapter 10, ***ammonia*** in Chapter 12, and ***Caustic Soda*** in Chapter 15. Alkalis have a bitter taste, a high pH (8-14), and turn pH test paper blue. This chapter is an introduction to acids.

The earliest, and for most of human history the only acid available was vinegar, a solution of ***acetic acid*** in water produced by the bacterial oxidation of alcoholic beverages as shown in Equation 4-1(c) (page 49). Like vinegar, acids in general have a sour taste, a low pH (0-6), and turn pH test paper red. Acetic acid dissociates directly to produce hydrogen ion in aqueous solution, but there are other compounds, particularly non-metal oxides, that combine with water before dissociating. One such oxide is carbon dioxide.

Remember from Chapter 10 that when ***carbon dioxide*** dissolves in water, it forms hydrogen carbonate, AKA carbonic acid (H_2CO_3). Carbonic acid ionizes in water to form a bicarbonate[2] ion (HCO_3^-) and a hydrogen ion. If you add a base, say, sodium hydroxide to carbonic acid you get a ***salt,*** sodium bicarbonate. The hydroxide ion from the base combines with the hydrogen ion from the acid to produce water; the sodium ion left over from the sodium hydroxide and the bicarbonate ion left over from the carbonic acid stay in solution. You would have exactly the same situation if you just dissolved sodium bicarbonate in water and so in the equation, we abbreviate "$Na^+(aq) + HCO_3^-(aq)$" as "$NaHCO_3(aq)$." The bicarbonate ion is called the ***conjugate base*** of carbonic acid. You could also say that carbonic acid is the ***conjugate acid*** of the bicarbonate ion. The meaning of the word *conjugate* here is just that carbonic acid and bicarbonate ion are really the same thing, with and without an extra hydrogen ion.

We can take it even farther by adding sodium hydroxide to sodium bicarbonate. The hydroxide ion pulls a hydrogen ion off of the bicarbonate ion, making water and leaving carbonate ion (CO_3^{2-}). The sodium ion from the sodium hydroxide joins the one from the sodium bicarbonate,

2. The modern name for the bicarbonate ion is the *hydrogen carbonate* ion. The older name, however, continues to be widely used.

Equation 18-1. Acid Properties of Carbon Dioxide

$$(a)\ C(s) + O_2(g) = CO_2(g)$$
$$(b)\ CO_2(g) + H_2O(l) = H_2CO_3(aq)$$
$$(c)\ H_2CO_3(aq) = H^+(aq) + HCO_3^-(aq)$$
$$(d)\ H_2CO_3(aq) + NaOH(aq) = NaHCO_3(aq) + H_2O(l)$$
$$(e)\ H_2CO_3(aq) + 2NaOH(aq) = Na_2CO_3(aq) + 2H_2O(l)$$

so now you have two of them in solution. Same as before, you can abbreviate "2 Na^+(aq) + CO_3^{2-}(aq)" as "Na_2CO_3(aq)." Carbonate ion is the conjugate base of bicarbonate ion and bicarbonate ion is the conjugate acid of carbonate ion. The acid-base chemistry of carbon dioxide is summarized in Equation 18-1; the oxides of sulfur behave similarly and are the main focus of this chapter.

There are two oxides of sulfur: ***sulfur dioxide*** (SO_2) and ***sulfur trioxide*** (SO_3). When sulfur burns in air the product is sulfur dioxide, which we saw in Equation 9-2 (page 113) as one of the products of the roasting of metal sulfide ores. Sulfur dioxide is a versatile compound; in the presence of a strong oxidant it plays the role of a reducing agent; in the presence of a strong reductant it acts as an oxidizing agent. As shown in Equation 18-2(b), it dissolves in water to form an acid, in this case a weak acid called ***sulfurous acid*** (H_2SO_3), which smells of rotten eggs. A mole of sulfurous acid reacts with a mole or two of sodium hydroxide to produce the salts, sodium bisulf*ite* and sodium sulf*ite,* respectively. While there has never been a large market for sulfurous acid or its salts, they are not entirely useless, as we'll see in Chapter 24. But by far the biggest demand in the acid world has always been for sulfur*ic* acid.

The production of ***sulfuric acid*** (H_2SO_4) requires a bit of ingenuity. As we have seen, combustion of sulfur by atmospheric oxygen produced sulfur dioxide, not sulfur trioxide. In Chapter 17 we saw that saltpeter could substitute for atmospheric oxygen in a redox reaction. Ward's contribution was to prepare a modified "gunpowder," one consisting of only sulfur and sodium nitrate.[3] When such a mixture is burned, it produces

3. Sodium nitrate is less expensive and less explosive than potassium nitrate; a double benefit for the budding acid maker.

Equation 18-2. Properties of Sulfurous Acid

$$(a)\ \mathrm{S}(s) + \mathrm{O_2}(g) = \mathrm{SO_2}(g)$$
$$(b)\ \mathrm{SO_2}(g) + \mathrm{H_2O}(l) = \mathrm{H_2SO_3}(aq)$$
$$(c)\ \mathrm{H_2SO_3}(aq) = \mathrm{H^+}(aq) + \mathrm{HSO_3^-}(aq)$$
$$(d)\ \mathrm{H_2SO_3}(aq) + \mathrm{NaOH}(aq) = \mathrm{NaHSO_3}(aq) + \mathrm{H_2O}(l)$$
$$(e)\ \mathrm{H_2SO_3}(aq) + 2\mathrm{NaOH}(aq) = \mathrm{Na_2SO_3}(aq) + 2\mathrm{H_2O}(l)$$

Equation 18-3. Properties of Sulfuric Acid

$$(a)\ 3\mathrm{S}(s) + 2\mathrm{NaNO_3}(s) = \mathrm{Na_2S}(s) + 2\mathrm{SO_2}(g) + 2\mathrm{NO}(g)$$
$$2\mathrm{NO}(g) + \mathrm{O_2}(g) = 2\mathrm{NO_2}(g)$$
$$2\mathrm{SO_2}(g) + 2\mathrm{NO_2}(g) = 2\mathrm{SO_3}(g) + 2\mathrm{NO}(g)$$

$$(b)\ \mathrm{SO_3}(g) + \mathrm{H_2O}(l) = \mathrm{H_2SO_4}(aq)$$
$$(c)\ \mathrm{H_2SO_4}(aq) = \mathrm{H^+}(aq) + \mathrm{HSO_4^-}(aq)$$
$$(d)\ \mathrm{H_2SO_4}(aq) + \mathrm{NaOH}(aq) = \mathrm{NaHSO_4}(aq) + \mathrm{H_2O}(l)$$
$$(e)\ \mathrm{H_2SO_4}(aq) + 2\mathrm{NaOH}(aq) = \mathrm{Na_2SO_4}(aq) + 2\mathrm{H_2O}(l)$$

sulfur dioxide and nitrogen oxide, as shown in Equation 18-3(a). The nitrogen oxide liberated from the saltpeter reacts with atmospheric oxygen to produce nitrogen dioxide, which, in turn, converts sulfur dioxide to sulfur trioxide. If you add the three reactions of Equation 18-3(a), the nitrogen dioxides cancel out:

$3\ S + 2\ NaNO_3 + O_2 = Na_2S + 2\ SO_3 + 2\ NO$

If more O_2 were available, the NO would be converted back into NO_2 and could react with more SO_2 to produce more SO_3. But that would regenerate NO, which could react with more O_2 to make—well, you get the picture. The NO is not used up in this reaction; it simply changes to NO_2 and back again. We say that nitrogen oxide is a **catalyst** for the oxidation of sulfur dioxide to sulfur trioxide. The entire process is summarized in schematic form in Figure 18-1. The first reactor is an old friend, a ***burner,*** first encountered in Figure 1-3 (page 9). Equation (a) of Figure 18-1 is shorthand for Equation 18-3(a), the "NO" over the equal

Figure 18-1. The Lead Chamber Process

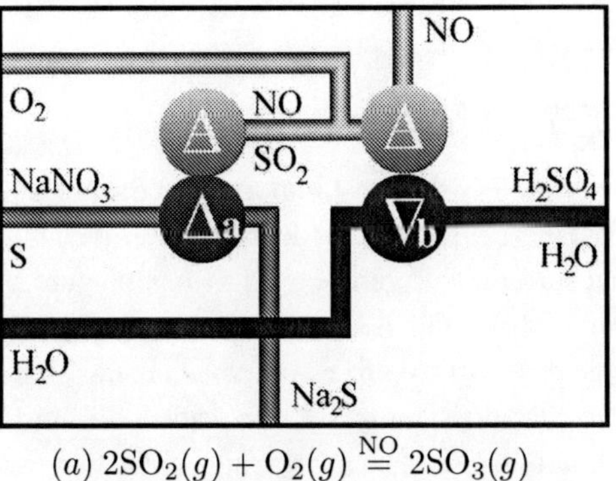

$$(a)\ 2SO_2(g) + O_2(g) \overset{NO}{=} 2SO_3(g)$$
$$(b)\ SO_3(g) + H_2O(l) = H_2SO_4(aq)$$

sign denoting the catalytic action of nitrogen oxide. The second reactor is called an ***absorber,*** which consists of nothing more than a container in which a gas can dissolve in a liquid.

The sulfuric acid of commerce is 95% sulfuric acid and 5% water. You can buy it in this concentration, for example, as industrial-strength drain opener.[4] The acid used to fill car batteries is a solution of sulfuric acid in water. But in this chapter you will learn to make sulfuric acid from scratch.

Sulfuric acid reacts with a variety of substances. Given its name, we would expect it to react with bases to form salts. A mole of sulfuric acid reacts with a mole or two of sodium hydroxide, for example, to produce the salts, sodium bisulf*ate* and sodium sulf*ate,* respectively. Sulfuric acid can also oxidize metals, for example reacting with iron to produce iron sulfate and hydrogen gas. Finally, sulfuric acid is often used as a dehydrating agent. It will, for example, suck water right out of sugar, leaving charcoal: the same reaction we saw way back in Equation 1-1 (page 9). You can think of sulfuric acid as a kind of chemical bejeezus sucker. It is this triple nature of sulfuric acid, as an acid, as an oxidizing agent, and as a dehydrating agent, which accounts its place as the king of chemicals.

4. The solid drain openers available at grocery stores are almost universally caustic soda, sodium hydroxide. Industrial-strength liquid drain openers are more commonly sulfuric acid. Mixing the two is dangerous.

Equation 18-4. Two More Mineral Acids

$$(a)\ H_2SO_4(l) + 2NaNO_3(s) = Na_2SO_4(s) + 2HNO_3(l)$$
$$(b)\ H_2SO_4(l) + 2NaCl(s) = Na_2SO_4(s) + 2HCl(g)$$

Sulfuric acid is certainly the most important of the acids, if raw tonnage is a measure, but there are two other acids of commercial importance, and it turns out that sulfuric acid can be used to manufacture them both. One of these, nitric acid, HNO_3, is an even more powerful oxidizing agent than sulfuric acid. It will oxidize even those metals like lead, copper, and silver which are resistant to sulfuric acid. It will turn out to be important in the manufacture of such amazing applications as photographic film, fertilizers, explosives, and plastics. Nitric acid can be distilled from saltpeter and sulfuric acid, as shown in Equation 18-4. The other acid, hydrochloric acid, HCl, will play a significant role in the development of the chemical industry in the nineteenth century. It can be distilled from ordinary table salt and sulfuric acid. This triumvirate of mineral acids has played such a decisive role in the development of the world as we know it that they deserve a little mnemonic to help you remember their importance:

> In this way was the World created. From this there will be amazing applications because this is the pattern. Therefore am I called Hermes Trismegistos, having the three parts of wisdom of the Whole World.
>
> — *The Emerald Tablet of Hermes Trismegistos*

Material Safety

Locate an MSDS's for saltpeter (CAS 7757-79-1), sulfur (CAS7704-34-9), and sulfuric acid (CAS 7664-93-9). Summarize the hazardous properties in your notebook, including the identity of the company which produced each MSDS and the NFPA diamond for each material.[5]

Your most likely exposure will be to sulfur dioxide fumes. If you are careless enough to get a lung-full and if it makes you cough more than once or twice, you should go to the hospital. But you would be better off *not* getting a lung-full; just keep your nose out of where the gas is.

---

5. The NFPA diamond was introduced Section 15.2 (page 184). You may substitute HMIS or Saf-T-Data ratings at your convenience.

You should wear safety glasses while working on this project. All activities should be performed either outdoors or in a fume hood. Leftover sulfur may be thrown in the trash. Leftover saltpeter and sulfuric acid can be flushed down the drain with plenty of water.

Research and Development

Before you get started, you should know this stuff.

- You better know all the words that are important enough to be ***indexified*** and **glossarated.**
- You should remember pretty much all the stuff you learned from Chapter 14 and Chapter 16.
- You should know the reactions of carbon dioxide in Equation 18-1.
- You ought to recognize sulfur, either from photographs or from samples.
- You should know the reactions of the sulfur oxides in Equation 18-3.
- You should that sulfuric acid may be used as an acid, as an oxidizing agent, or as a dehydrating agent.
- You should know the hazardous properties of sulfur, saltpeter, sulfur trioxide, and sulfuric acid.
- You should know the story of Spot and Roebuck.

18.3 Θ

Making sulfuric acid is simple and fun and it will put you in touch with the most important chemical of modern times. We could make it under glass bells like Spot did or we could make it in lead chambers like Roebuck did, but both of those methods would be more trouble than they'd be worth, given that we just want to make enough acid to show that we can. So we're going to use the same container we always do when we need a cheap, convenient container: the 2-liter pop bottle. You might call this the Pop Bottle Chamber Process for sulfuric acid manufacture.

We need two reactors, a burner and an absorber. You can make a nifty burner from a pop bottle cap, a threaded steel rod, four nuts, and a copper plumbing fitting called an endcap. For the absorber you can use a 2-liter pop bottle. The sizes don't matter as long as the endcap will fit into

Figure 18-2. The Sulfur Burner

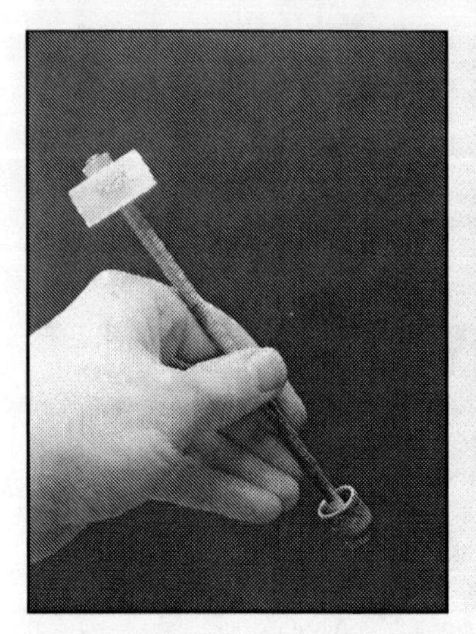

the mouth of your pop bottle. I use a half-inch endcap and an eight-inch length of quarter-inch threaded rod. Drill a hole in the center of the endcap to fit the threaded rod and secure it with one nut inside and one nut outside the endcap. Do the same thing at the other end of the rod with the pop bottle cap. The open ends of the two caps should face each other, like it shows in Figure 18-2.

You should operate your reactor outdoors or in a fume hood, because the smell of burning sulfur is pleasant only to a select few. Wash out the pop bottle and add 4 mL of water to it. Place 0.5 g of sodium nitrate into the copper endcap and then put 0.5 g of sulfur on top of the sodium nitrate. Light the nitrate/sulfur mixture by holding your burner over a ***spirit lamp*** or Bunsen burner. Hold the burner by the pop bottle cap because the threaded rod is going to get hot. The sulfur will turn black and melt into the sodium nitrate and then the nitrate will start to puff up and crackle as it decomposes. The nitrate/sulfur mixture will catch fire, but leave it over the lamp until it is really going good. Then carefully insert the burning endcap into the mouth of your pop bottle and screw the cap down tight. The bottle will start to fill with dense fumes of sulfur dioxide and nitrogen oxide. If you can, take your bottle into a dark corner to check whether your burner is still burning. If it goes out, carefully remove it from your pop bottle and light it over the spirit lamp again. Reinsert the

Figure 18-3. The Pop Bottle Chamber

flaming endcap into your pop bottle and repeat the cycle until it will no longer light. Remove the burner from your pop bottle and screw a fresh bottle cap onto the pop bottle. When the burner has cooled you can rinse the small amount of residue into a sink.

You might think that the reaction is over, but it's only just begun. The bottle is now filled with sulfur dioxide, nitrogen oxide, oxygen, and of course, nitrogen. Over the course of the next few hours, the nitrogen oxide will catalyze the reaction of sulfur dioxide with oxygen, producing sulfur trioxide, which will dissolve in the water to form sulfuric acid. You will know that this is happening because as the oxygen is consumed and the sulfur trioxide dissolves, the bottle will start to collapse and the fog that fills the bottle will clear. You can speed up the process by opening the cap to admit more air and then sealing the cap again. If you leave the absorber to sit overnight, the acid product will be virtually odorless since all of the sulfur oxides will be gone. Whether you wait an hour or let it sit overnight, you can test your acid by opening the pop bottle and pouring its contents into the bottle cap. Test the liquid with pH test paper, the chemist's virtual tongue; if you have been successful, it should turn bright red. Congratulations! You have just manufactured the most important chemical of modern times. If you were in business, you would distill your dilute acid to separate the water from the acid. But given that this is just a test run producing dilute acid, you can pour it down the drain when you are done with it. After all, it *is* drain opener.

If you are interested in further exploration, try repeating the process without the sodium nitrate. You will find that the bottle does not collapse, the fog clears very slowly, and the resulting acid smells strongly of rotten eggs. I suggested a 50/50 mixture of nitrate and sulfur because there is limited oxygen in our small absorber. In commercial lead chambers, the usual ratio was 1 ***part*** nitrate to 7 parts sulfur. The nitrate ratio affects only the rate of the reaction, not the amount. In early chamber plants, nitrogen oxide went up the smokestack as a waste product, but from about 1840 the Gay-Lussac tower was used to recover nitrogen oxides. This practice reduced the nitrate consumption, making the process more economical. Also from the 1840's, sulfur dioxide from roasting pyrite ores (Equation 9-2 (page 113)) began to be used as an alternative to sulfur. Thus sulfuric acid turned out to be a good sideline for lead, copper, and iron smelters.

Quality Assurance

Record in your notebook the amounts of sodium nitrate, water, and sulfur you used. Describe the reactor and its operation. Note how long you waited to test your acid, whether additional air was allowed into the absorber, and whether or not your acid product had a strong smell. Tape your red pH test paper into your notebook as a keepsake.

Chapter 19. Bath (Soap)

280. How soap is made from olive oil or tallow

Spread well burnt ashes from good logs over woven wickerwork made of withies, or on a thin-meshed strong sieve, and gently pour hot water on them so that it goes through drop by drop. Collect the lye in a clean pot underneath and strain it two or three times through the same ashes, so that the lye becomes strong and colored. This is the first lye of the soapmaker. After it has clarified well let it cook, and when it has boiled for a long time and has begun to thicken, add enough oil and stir very well. Now, if you want to make the lye with lime, put a little good lime in it, but if you want it to be without lime, let the above-mentioned lye boil by itself until it is cooked down and reduced to thickness. Afterwards, allow to cool in a suitable place whatever has remained there of the lye or the watery stuff. This clarification is called the second lye of the soapmaker. Afterwards, work [the soap] with a little spade for 2, 3 or 4 days, so that it coagulates well and is de-watered, and lay it aside for use. If you want to make [your soap] out of tallow the process will be the same, though instead of oil put in well-beaten beef tallow and add a little wheat flour according to your judgment, and let them cook to thickness, as was said above. Now put some salt in the second lye that I mentioned and cook it until it dries out, and this will be the afronitrum for soldering.

— *Mappae Clavicula, ca.* 1130 AD [1]

19.1 ☿

🜂 Nothing in the history of humankind, save for the cultivation of noble fire itself, can compare to the discovery of soap. This is how one of the Author's pyrophilic alter-egos would undoubtedly have begun the present chapter. Lucifer would have described a venerable history stretching back to the third millennium BC, when concoctions of ashes and fat were recorded on Sumerian clay tablets.[2] In order to claim that these concoctions were soap, however, he would have to overlook the fact that these tablets make no mention of any detergent properties these mixtures might have had. He would have to find it unremarkable that pharoic Egypt and the empires of Greece and Rome failed to appropriate such a useful material, but relied instead on urine, various plants, clays,

1. Reference [25].
2. Reference [55], p. 12.

potash, and soda for doing the laundry. He would have to ignore the extensive Roman literature on personal hygiene, which describes the process of oiling the body and scraping the dirty oil off with an instrument called a strigil. No, an ancient origin for soap simply does not wash.

♙ The Samsonites would probably prefer the "old Roman legend" which places the invention of soap in the hands of the Goddess Athena. Runoff from animal sacrifices made at her temple on Sapo Hill, so the story goes, resulted in the accumulation of ashes and animal fat in the river below. Women washing clothes in this river found that their clothes came out whiter and brighter than usual and eventually traced the suds back to their source. But why, we might wonder, would the Romans have had an "old legend" about a material which they did not use in classical times? The Author has yet to find an attribution for this oft-repeated tale and it remains of doubtful historical authenticity. Even given his penchant for making up fantastic scenarios, the Author recoiled from this one on chemical grounds. You see, soap-making requires the concentration of alkali and fat, not the dilution which would have resulted from sacrificial runoff. If you ask me, we had better flush this just-so story right down the drain.

♙ Unktomi and his ilk might have spun a similarly fantastic tale, tracing soap back to Phoenician traders of 600 BC. This claim, repeated by no less an authority than the *Encyclopedia Britannica,* is often attributed to Pliny the Elder's *Natural History.* Indeed, this text of the first century AD introduces the word *sapo* into Latin:

> Soap [Sapo] is also good, an invention of the Gallic provinces for making the hair red. It is made from suet and ash, the best is from beech ash and goat suet, in two kinds, thick and liquid, both being used among the Germans, more by men than by women.[3]

♙ This is the only mention of the word *sapo* in all of Pliny and no mention is made either of the Phoenicians or of the aforementioned date. No mention is made of using the stuff for washing up. We might be tempted to conclude that the presumed soap was simply an exotic cosmetic whose name became attached to a material of later invention, were it not for two observations. First, Pliny says that this *sapo* is made from goat tallow and ashes, which, in and of itself would not be conclusive. But he goes on to say that it comes in two varieties, liquid and solid, and the fact

3. Reference [23], Book XXVIII, li.

of the matter is that soap made from potash turns out to be a liquid while that made from soda is a solid. We would do well, then, not to throw our Pliny out with the bath water.

◬ Nevertheless, we should be cautious in supposing the widespread use of soap in the first millennium AD. Claims of soap-making at Pompeii in the first century continue to be made, despite the fact that Hoffman's analysis of this supposed soap in 1882 showed it to be nothing more than fuller's earth, a kind of alkaline clay described by Pliny for the laundering of clothes.[4] If soap had been around you would think that it might have been mentioned in the *Stockholm Papyrus,* that second-century collection of dye recipes quoted in Chapter 12, but while it gives instructions for washing wool with vinegar, ashes, and plants, it makes no mention of anything resembling soap.[5] The second-century physician Galen uses the word *sapo* to describe an exotic material made from goat tallow and ashes[6] which is good for washing things.[7] While this is undoubtedly soap, his descriptions occupy less than a few sentences in the thousands of pages which make up his complete works. French soap is described by Theodorus Priscianus of the fourth century as a material for washing the head.[8] Italian soap-makers of the seventh century were organized into craft guilds and the profession of soap-boiler is mentioned in Charlemagne's *Capitulare de Villis* of 805 AD.[9] Frustratingly, all of the first millennium references to soap are brief at best and vague at worst. We must conclude that while soap had come to Rome from the provinces prior to the second century, as far as most people were concerned, it was "no soap for you."

◬ The first detailed recipe for what is unmistakably soap appears in the twelfth century *Mappae Clavicula.* You will recall from Chapter 16 that this work, traditionally attributed to Adelard of Bath, is a collection of trade secrets for painters and craftsmen. The inclusion of recipes for soap tells us two things about the status of soap at that time. First, soap must not have been widely available; otherwise Adelard would simply have picked it up at the market. Second, soap was familiar enough that no explanations are given for its use. He mentions its use as a soldering

4. Reference [69], p. 308.
5. Reference [5].
6. Reference [12], De Compositione Medicamentorum Secundum Locos, ii, Kuhn Book XII, p. 589.
7. Reference [12], De Methodo Mendendi, viii, Kuhn Book X, p. 569.
8. Reference [69], p. 307.
9. Reference [58], p. 171, and [69], p. 308.

flux, but given the importance of textiles in the rest of the work, the use of soap for washing fabrics seems likely. Moreover a second recipe, that for "French" soap, says that berries are used to color it red, which perhaps explains Pliny's claim that the Gauls use it to color their hair red.

≙ Early centers for soap manufacture were Marseilles, by the ninth century, Venice, by the fourteenth century, and Castile by the fifteenth century. In addition to these internationally-traded soaps made principally from olive oil, domestic soaps made from tallow were produced in Bristol, Coventry, and London. The scale of the industry was increasing; in 1624 the Corporation of Soapmakers at Westminster was granted a royal patent to produce 5000 tons of soap per year. By the eighteenth century, soap was a common domestic item. The *Dictionaire Oeconomique* of 1758 describes:

> SOAP, a Composition made of Oil of Olive, Lime, and the Ashes of the Herb Kali or Saltwort; the chief use of *Soap* is to wash and cleanse Linnen: There are two sorts thereof, which are distinguished by their Colours, *viz.* White and Black *Soap.*[10]

≙ The *Dictionaire* goes on to describe a process little changed from that of the *Mappae Clavicula.* White soap is solid, made from lime and soda, that is, from sodium hydroxide; black soap is liquid, made from lime and potash, that is, from potassium hydroxide stained by residual charcoal.

≙ Demand for soap had increased rapidly during the eighteenth century as the textile industry became increasingly industrialized. And because soap makers depended on supplies of soda, lime, and potash, they were a major driving force in the industrialization of chemical manufacture. In fact, two of the oldest surviving chemical companies in the world started out in the soap trade. In 1806 William Colgate began making soap and candles in New York. In 1837 William Proctor and James Gamble went into partnership to manufacture soap and candles in Cincinnati.

≙ But I suppose that all this is too boring to hold the attention of most readers, with their remote-controls, their instant messages, and their personal digital gizhatchies. They aren't satisfied with vacuous little jingles, breezy little games, and tedious little exercises. No, they'd rather lap up wild tales about Chinese laundresses, or itinerant Taoists, or tail-sniffing dog conventions. If you want anyone to pay attention to you nowadays,

10. Reference [8].

you have to pretend to be Elzabath, the wife of a French tallow chandler, working your fingers to the bone what with the cooking and the cleaning and the interminable rendering of fats for the candle business.

♁ Normally, you render fat by boiling meat trimmings in salt water, with some berries thrown in for color. You allow the melted tallow to float to the top and upon cooling, the tallow hardens into a block and can be lifted out of the pot. But on this particular day, you've got a splitting headache, a crying baby, and a stinking pile of laundry that stretches back to last week. And all your husband can do is to go on and on about how nobody ever notices him down at the Chandler's Guild. You go to lift the tallow out of the pot and—can you believe it—instead of a block you find nothing but a pot of red goo. Your husband is no help at all. "What am I supposed to do with this, make jelly candles?" he snorts. You've had just about as much as you can take. "Maybe you'd like to wear it!" You grab a handful of goo and slap it right on top of his head. He stands there looking like a stuffed goose with gravy and then he suddenly breaks out laughing. You can't help but laugh too. In retrospect you figure out that between the crying, the cleaning, and the complaining, you must have accidentally added potash to the pot instead of salt. You get him cleaned up and pack him off to his guild meeting. When he returns you make nice to him. "How was the meeting, Sweetie?" "Now 'Bath, you know the first rule of Chandler's Guild is that you don't talk about Chandler's Guild." But it seems that red hair is suddenly all the rage and so you add soap as a sideline to the candle business; the rest is history.

♁ Not that such a thing ever happened to me; I'm just a figment of the Author's imagination, in case you've forgotten. So don't go attributing this little soap opera to Aristotle or Lucretius or anyone other than the Author.

19.2 ♀

All living things face the fundamental problem of separating the inside from the outside. The inside of a cell is filled with water and floating around in that water are the various bits,—mitochondria, ribosomes, and, of course, the nucleus—which make the cell go. There's water outside the cell as well—fresh water, sea water, blood, or lymph—water, water everywhere. What separates the inside from the outside is the cell membrane, which had better be insoluble in water or else the inside will be outside in no time flat. But the building blocks of the cell must be water-soluble or else there would be no way to move them around. This solubility dilemma demands a water-soluble material which can be rendered insoluble once it becomes part of the cell membrane. And believe it or not, every living cell on Earth employs the same solution to this fundamental problem; cell membranes, whether in bacteria or yeasts, redwoods or maples, Democrats or Republicans, are composed of fats. Before we can understand the chemistry of fats, we need to have a look at the general class of compounds to which they belong, the ***esters.***

Equation 16-1(c) (page 197) introduced the ester of ethanol and acetic acid, ethyl acetate, as a by-product of alcoholic fermentation. Normally a fermentation vessel is capped, protecting the mead or wine from atmospheric oxygen so that yeasts are forced convert sugar to ethanol rather than carbon dioxide and water. When the fermentation vessel is un-capped, however, bacteria oxidize the ethanol to acetic acid. If all of the ethanol is converted to acetic acid, the mead "goes sour," turning into vinegar, but if only some of the ethanol is converted to acetic acid the two compounds may react under the acidic conditions normally found in a fermenting mead or wine. Under such conditions, ethanol and acetic acid undergo a **condensation** reaction, producing water and ethyl acetate.

$$\mathrm{CH_3CH_2OH}(aq) + \mathrm{CH_3COOH}(aq)$$
$$= \mathrm{H_2O}(l) + \mathrm{CH_3CH_2OOCCH_3}(l)$$

Whereas acidic conditions promote the condensation of ethanol and acetic acid to produce ethyl acetate, alkaline conditions promote the ***hydrolysis*** of ethyl acetate to produce ethanol and acetic acid. The acetic acid produced further reacts with available alkali to produce a salt, sodium acetate or potassium acetate, depending on the alkali used.

$$\mathrm{CH_3CH_2OOCCH_3}(l) + \mathrm{NaOH}(aq)$$
$$= \mathrm{CH_3CH_2OH}(aq) + \mathrm{CH_3COONa}(aq)$$

While alcohols, acids, and their salts are generally soluble, esters are generally insoluble in water. Ethanol, for example, is soluble in water because its OH group is essentially half a water molecule and the general rule for solubility is "like dissolves like." Similarly, acetic acid is soluble in water because the organic acid group, -COOH, contains an OH group. But when ethanol and acetic acid condense to form ethyl acetate, a water molecule is eliminated, leaving ethyl acetate with no OH groups at all.

Figure 19-1. Glycerol

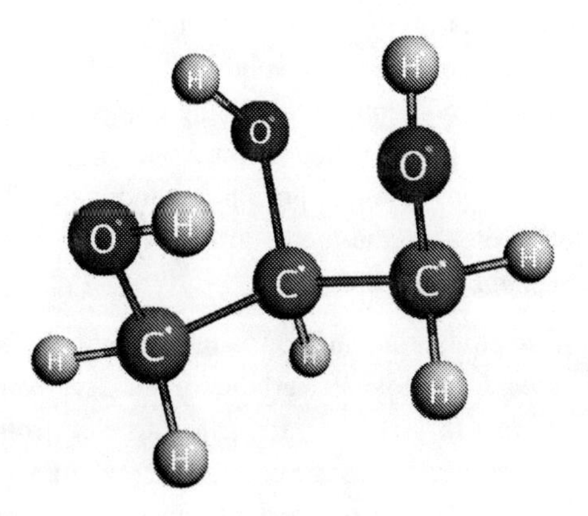

Consequently, ethyl acetate is not soluble in water, or, to be more precise, its solubility is very low. In esters, then, we have a class of insoluble compounds which can be either synthesized from soluble compounds or decomposed into them, depending on the conditions. This is precisely the behavior required by plant and animal cells for building cell membranes.

Cell membranes consist primarily of esters, not of ethanol and acetic acid, but of the alcohol ***glycerol*** and a class of acids called ***fatty acids.*** Whereas ethanol has only one OH group, glycerol has three and consequently glycerol can form ester linkages with up to three fatty acids; such a tri-ester is called a ***tri-glyceride.*** Whereas acetic acid has only two carbon atoms, a fatty acid has a long chain of them. A tri-glyceride may be called an ***oil*** or a ***fat,*** depending on whether it is a liquid or a solid at room temperature. Oils tend to have a relatively low ratio of hydrogen to carbon. Olive oil, for example, is derived primarily from ***oleic*** acid, $CH_3(CH_2)_7(CH)_2(CH_2)_7COOH$, whose ratio of hydrogen to carbon is 34/18, or 1.88. Lard and tallow also contain oleic acid, but with significant amounts of ***palmitic*** acid, $CH_3(CH_2)_{14}COOH$, and ***stearic*** acid, $CH_3(CH_2)_{16}COOH$, with hydrogen-to-carbon ratios of 32/16 and 36/18, respectively. Fatty acids with hydrogen-to-carbon ratios of 2/1 are deemed ***saturated,*** those with lower ratios, ***unsaturated.*** No matter whether they are fats or oils, saturated or unsaturated, these tri-glycerides are devoid of OH groups and so are insoluble in water, a property which makes them useful for separating the inside from the outside.

Just as the alkaline hydrolysis of ethyl acetate produces ethanol and, for example, sodium acetate, the hydrolysis of an oil or a fat produces glycerol and a mixture of fatty acid salts, for example, sodium oleate, sodium palmitate and/or sodium stearate. Insoluble fats and oils are thereby converted into soluble compounds. While living things in general hydrolyze fats and oils to move them around the cell, human beings desire to remove fats and oils from clothing, pots, pans, and hands. This is the fundamental problem of soap and the hydrolysis of a fat or oil is given the particular name, ***saponification.***

Any alkali will saponify fats and oils and by now we have several at our disposal; ***potash*** (potassium carbonate) leached from the ashes of inland plants, ***soda ash*** (sodium carbonate) leached from the ashes of marine plants, ***ammonia*** from stale urine, ***lime*** (calcium oxide) from the calcination of limestone, ***caustic potash*** (potassium hydroxide) from the reaction of potash and lime, and ***lye,*** or ***caustic soda*** (sodium hydroxide) from the reaction of soda and lime. The choice of alkali depends to some extent on which one you can obtain most easily and inexpensively, but given this latitude, some alkalis are better than others for making soap. Strong alkalis will work faster than weak ones, which puts ammonia, potash, and soda behind lime, caustic potash and caustic soda. The choice among these three alkalis determines the relative solubility of the resulting soap.

Recall from Table 7-2 (page 94) that calcium carbonate is insoluble in water and from Table 8-2 (page 102) that the solubility of sodium carbonate is less than that of potassium carbonate. The solubilities of fatty acid salts follow the same pattern. Calcium soaps are insoluble; in fact, soap scum results when soluble soaps are used in hard (calcium rich) water, precipitating calcium salts of fatty acids. While sodium soaps are soluble, they are easily dried into solid cakes. In contrast, potassium soaps are so soluble that they will even absorb water from the air and so they exist primarily as solutions. Thus the choice of alkali—either caustic potash or caustic soda—is responsible for Pliny's observation that soap comes in two varieties, solid and liquid. This distinction has been central to soap-making ever since.

While the strong alkalis, caustic potash or caustic soda, will saponify fats and oils on clothing and cook-ware, they will also saponify those in skin. If saponification were the only means for removing fat and oil we would be left choosing between an alkali so mild that it does not work and one

Figure 19-2. Sodium Palmitate

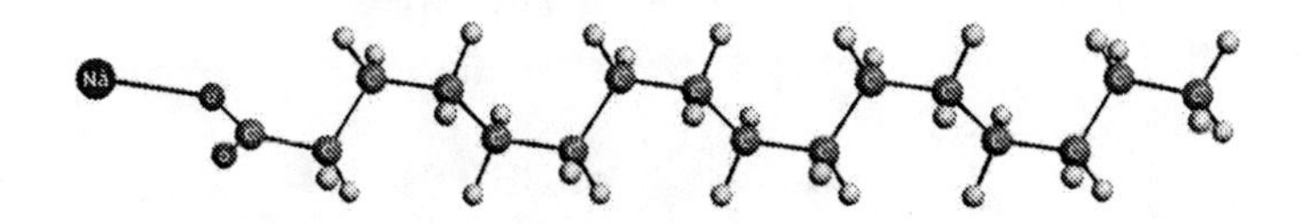

Figure 19-3. The Emulsification of Fats

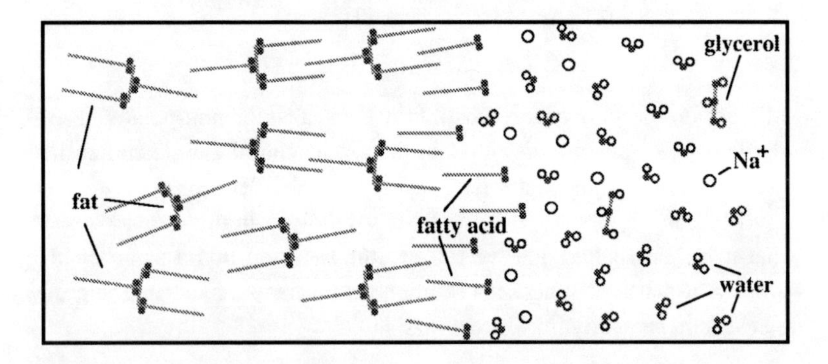

so harsh that it turns the skin to soap. Fortunately, the materials responsible for the cleansing action of soap are not the alkalis themselves, but rather the products of saponification—the fatty acid salts. The structure of such a salt, sodium palmitate, is shown in Figure 19-2. One end of the molecule, the end with the polar oxygen atoms, looks very much like any other ionic substance—sodium chloride or sodium sulfate or sodium acetate. Salts such as these dissociate into positive cations and negative anions when they dissolve in water. The other end of the molecule contains carbon and hydrogen but not oxygen, and so it is non-polar.

Recalling the tail-sniffing analogy of Chapter 16, a fatty acid salt is like a normal dog sewn to a very long headless tail-less dog; polar water molecules will cluster around the ionic end of the molecule, with its positive sodium or potassium ion and negative oxygen atoms; water molecules will find little of interest in the long, non-polar hydrocarbon chain. In fact, this long non-polar chain looks, to a water molecule, very much like an un-saponified fat. Following the maxim "like dissolves like," the ionic end of the fatty acid salt will be soluble in water, while the non-polar end will be soluble in fats and oils. So in a tub of water containing fat and soap, the fat breaks up into tiny droplets; the surface

Equation 19-1. Saponification

$$
\begin{array}{l} CH_2OOCR \\ | \\ CHOOCR + 3\,NaOH \\ | \\ CH_2OOCR \end{array}
\quad = \quad
\begin{array}{l} CH_2OH \\ | \\ CHOH + 3\,RCOONa \\ | \\ CH_2OH \end{array}
$$

$$
\begin{aligned}
\text{Oleic Acid, R} &= CH_3(CH_2)_7(CH)_2(CH_2)_7 \\
\text{Palmitic Acid, R} &= CH_3(CH_2)_{14} \\
\text{Stearic Acid, R} &= CH_3(CH_2)_{16}
\end{aligned}
$$

of these droplets are covered with fatty acid salts; the non-polar ends of the fatty acid salts are dissolved in the fat, leaving the ionic end at the surface. As far as the water is concerned, the droplet looks like a giant ion. We can't really say that the fat has dissolved in the water. We say that it has emulsified and we refer to the resulting liquid as an emulsion or suspension. Whereas a solution is transparent, like apple juice, a suspension is merely translucent, like milk.

If the fatty acid salts are responsible for the soapiness of soap, you might well wonder how we can get the most soap from a given quantity of fat, a classic **stoichiometric question.** According to our theory of soap, three moles of soap can be produced from each mole of fat. The fat, tri-stearine, for example, is a tri-ester of glycerol and stearic acid. Equation 19-1 gives the balanced equations for the saponification of tri-oleine, tri-palmitine, and tri-stearine. The "R" in the equation stands for the hydrocarbon chain of any of the fatty acids. If you paid any attention at all to Chapter 15, you ought to be able to answer the following stoichiometric questions:

Q: How many grams of sodium hydroxide are needed to react with 1 gram of tri-stearine?

Q: How many grams of sodium hydroxide are needed to react with 1 gram of tri-palmitine?

Q: How many grams of sodium hydroxide are needed to react with 1 gram of tri-oleine?

Table 19-1. Saponification Values for Common Oils and Fats

Oil or Fat	$\left(\frac{g\ NaOH}{g\ fat}\right)$	$\left(\frac{g\ KOH}{g\ fat}\right)$
Lard	**0.1380**	**0.1935**
Olive Oil	**0.1340**	**0.1879**
Shortening	**0.1360**	**0.1907**
Tallow	**0.1405**	**0.1971**

Q: How many grams of potassium hydroxide are needed to react with 1 gram of tri-stearine?

Q: How many grams of potassium hydroxide are needed to react with 1 gram of tri-palmitine?

Q: How many grams of potassium hydroxide are needed to react with 1 gram of tri-oleine?

The answer to such a problem, the ratio of alkali to fat, is called a saponification value and it is important to get this number right. If you use less than the stoichiometric amount of alkali, not all of your fat will be saponified. Not only will you get less soap than you might have, but that soap will have to dissolve the un-saponified fat in addition to that on your pots and pans. If you use more than the stoichiometric amount of alkali, not only will you have wasted alkali, but the leftover alkali will saponify your hands. If you are to get the most soap for your money, if you are to get the mildest soap for your delicate skin, it is important to get the saponification value, as Goldilocks might have said, *just right.*

The problem is that you are unlikely to find tri-oleine, tri-palmitine, or tri-stearine in the grocery store; fats and oils generally contain a variety of fatty acids, chiefly oleic, palmitic, and stearic acids but including a dozen or so other less abundant acids as well. How could you determine the saponification value for a fat of unknown composition? You could make a series of soaps, varying the ratio of alkali to fat, and then testing the finished soaps for excess alkali using pH test paper. The best ratio of alkali to fat would be the one which completely saponified the fat without leaving excess alkali. This optimum alkali/fat ratio would be the saponification value for the unknown fat.

Table 19-1 lists saponification values for a variety of common fats and oils. Compare your calculated saponification values for tri-oleine, tri-palmitine, and tri-stearine to those in the table. In practice, fats and oils vary in composition and it is better to risk having excess fat rather than excess alkali. The saponification values are generally discounted by 5% or so; simply multiply each saponification value by a factor of 0.95. You'll get a little less soap than you might have, but that's a small price to pay for soap that doesn't eat your skin off.

Material Safety

Locate an MSDS for sodium hydroxide (CAS 1310-73-2). Summarize the hazardous properties, including the identity of the company which produced the MSDS and the NFPA diamond for this material.[11]

Your most likely exposure will be eye or skin contact. In case of eye contact flush them with cold water and call for an ambulance. In case of skin contact wash the affected area with plenty of water until your skin no longer feels slippery.

You should wear safety glasses and gloves while working on this project. Leftover sodium hydroxide can be flushed down the drain with plenty of water. Leftover fat or oil can be thrown in the trash.

Research and Development

So there you are, studying for a test, and you wonder what will be on it.

- Study the meanings of all of the words that are important enough to be included in the ***index*** or **glossary.**
- You should know all of the Research and Development points from Chapter 14 and Chapter 15.
- You should know the reactions of Equation 19-1.
- You should know the meaning of the adage "Like dissolves like."
- You should know why fatty acid salts dissolve in both water and in oils and fats.
- You should know the hazardous properties of sodium and potassium hydroxide.
- You should know how soap emulsifies fat.

11. The NFPA diamond was introduced Section 15.2 (page 184). You may substitute HMIS or Saf-T-Data ratings at your convenience.

19.3 Θ

At the dawn of the twenty-first century there remains in the world a small but dedicated sub-culture of people who are not afraid to experience the joys of manufacturing; a hardy folk for whom the word *chem ical* is not equated with "evil poison foisted on unsuspecting innocents by heartless multi-national corporations;" a people who recognize that something as good and wholesome as soap requires for its manufacture something as caustic and poisonous as caustic soda. I speak, not of industrial or laboratory chemists, but of soapers—crafters who in an age of pre-packaged conveniences indulge in the simple pleasure of making soap from scratch.

The first choice of the soap-maker is that of which fat to saponify. You can use any fat or oil derived from animal or vegetable sources; mineral and motor oils are not triglycerides and so are not useful for making soap. Among animal fats, you may choose tallow rendered from beef, lard from pork, or suet from goat. Butter and bacon grease make perfectly functional soaps, though they will win no prizes for their aromas. Among vegetable oils, olive oil has been the traditional choice for fine soap. Palm oil is renowned for its rich lather. Other vegetable oils are often used in combination with tallow or lard rather than by themselves. Even vegetable shortening and margarine can be used, though the soft, whipped, or tub margarines contain excess water, complicating the calculation for the amount of caustic soda to be used. Each fat lends its own particular qualities to the finished soap and experienced soapers blend different oils to produce the qualities they desire, but for a first soap I recommend using any of those listed in Table 19-1.

The next choice of the soap-maker is that of which alkali to use. Potassium hydroxide will produce a soft or liquid soap, while sodium hydroxide will produce the familiar solid bar. Either of these alkalis might be manufactured from scratch, using potash or soda from Chapter 8 and lime from Chapter 10 to produce caustic soda or caustic potash as described in Chapter 15. Recall from Table 8-1 (page 101), however, that 1000 pounds of wood provide only about a pound of potash. Consider as well that caustic soda is sold as a drain opener, and so it is frequently available in grocery stores and hardware stores. For these reasons, store-bought caustic soda (lye) is the choice of most soapers.

You want 100% sodium hydroxide; name-brand drain openers contain perfumes or other ingredients which may interfere with saponification. Many grocery stores have removed lye from their shelves in the mistaken belief that it is more dangerous than the name-brand drain openers. And while lye deserves all of the dire warnings on its label, you will find the same warnings on the name brands because they are almost universally either sodium hydroxide or sulfuric acid. You would do the world of soapers a service by pointing this out to grocery store managers when the opportunity arises; we all benefit from cheap and convenient sources of lye.

The traditional method for making soap, the *hot process*, involves leaching ashes to get alkali, causticizing the alkali with lime to make caustic soda, and boiling the resulting lye with fat to make soap. It turns out that the boiling step has less to do with saponification than with removing bejeezical water to concentrate the lye. We can skip the boiling step if we start with solid caustic soda. This *cold process* is the method preferred by the majority of modern amateur soap-makers.

For your first soap, I recommend a cold process using that container of choice for caveman chemists, the 2-liter pop bottle, which will accommodate between one and two pounds of fat. Weigh out your fat on a balance and record the weight in your notebook. Place it in a beaker or a saucepan and warm it on a hot-plate until it melts (if solid) or until it is warm to the touch (if liquid). While your fat is warming, use UFA (page 41) and Table 19-1 to answer the following question:

Q: How many grams of caustic soda are needed to saponify the fat I have weighed out?

If you are blending different fats and oils you can answer this question for each one separately and then add the weights to get the total weight of caustic soda. *Multiply the total weight of caustic soda by a factor of 0.95;* it is far better to have a finished soap with excess fat than one with excess caustic soda. Weigh out your caustic soda on a balance and record the weight in your notebook. Weigh out twice that amount of water in a glass measuring cup or beaker and record the weight in your notebook. Stirring with a glass rod or plastic spoon, slowly add your caustic soda to your water—if you add water to caustic soda, the caustic soda will cake up and dissolve more slowly. The water will get hot as the caustic soda dissolves. Continue stirring until the caustic soda is completely dissolved. By this time, you should have hot melted fat on the stove and hot caustic soda solution in a beaker.

They say that a watched pot never boils, but it never cools either. You are about to mix the oil and caustic soda solution together, but as with salad dressing, oil and water don't mix. This is especially true of hot oil and water, so you will get the best results if you allow both ingredients to cool down to between 38 and 43°C, 100 and 110°F. A thermometer will keep you honest, but as long as you can hold them in your bare hands comfortably, you are probably alright. In my experience, the most frequent reason for failed soap is that the soaper got impatient and mixed the oil with the caustic soda too soon.

When the fat and caustic soda have cooled sufficiently, use a funnel to pour the fat into your 2-liter pop bottle. Add the caustic soda solution, put the cap on the bottle and give it a few shakes. You now have what amounts to caustic salad dressing—the oil and water mix initially, but may separate over time. As long as the mixture is smooth and creamy, you are fine, but if it begins to separate, give the bottle another shake. When the mixture no longer separates, you can let it rest. A tallow or shortening soap may solidify in an hour or so; a lard or olive oil soap may take up to two days.

The saponification will proceed more quickly if the mixture is kept warm than if it is allowed to cool. Because the saponification reaction is exothermic, simply insulating a large batch of soap is sufficient to keep it warm, but a small batch may need some help. I like to place my sealed 2-liter bottle in a pot of hot tap water to keep it warm. Don't watch this pot or fret about it, but whenever it occurs to you in the first couple of days, replace the water with hot tap water. If you forget to change the water, it's no big deal—the soap may just take a little longer to cure.

The curing time will vary from one fat to another, but after a week your soap should be ready for testing. With a knife, slit the pop bottle open and peel it apart to release your soap. If you have done your work correctly, you will have a single solid cake of soap with very little extraneous water. Cut a sliver the size of a raisin from this cake and place it into an empty bottle with a cup or so of hot water. Put the cap on the bottle and shake it. If you have succeeded in making soap, the bottle will fill with soap suds after a bit of shaking. If it doesn't, then either you had too little caustic

Figure 19-4. Olive Oil Soap

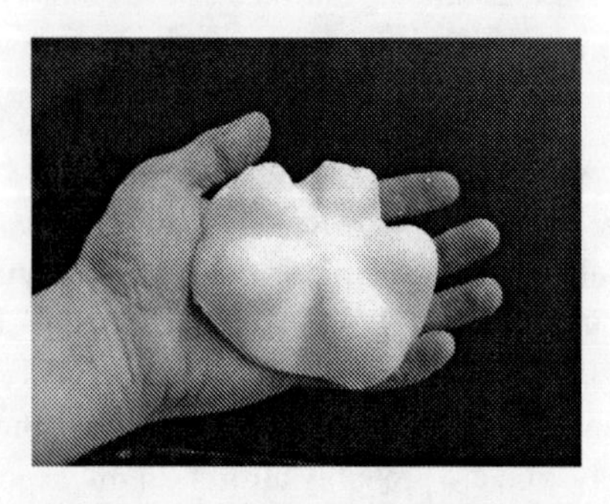

soda in your mixture or the soap is not sufficiently cured. As a second test, wipe a piece of wet pH test paper against your soap. If it turns blue, then either you had too much caustic soda in your mixture or the soap is not sufficiently cured. If your soap fails either of these tests, allow it to cure for a few more days and then test it again. When your soap is finished, you can slice it into bars of a convenient size.

With a bottle of soap suds, you can explore one more aspect of soap manufacture. Make a good, strong soap solution and add some salt to it. The fatty acid salts are less soluble in salt water than they are in fresh water and so they will precipitate from the solution. The soap, that is, the fatty acid salts will float to the top while the glycerol stays in solution. Separating the above from the below allows us to produce glycerol as a by-product of soap manufacture. Glycerol will be an important player in the development of explosives and plastics, as we shall see in Chapter 27.

Quality Assurance

Record the weights of the fat, caustic soda, and water that went into your pop bottle and compare these to the weight of the soap you produced. The soap is satisfactory if it makes suds and has a pH no higher than 8. Tape your yellow or green pH test paper into your notebook as a keepsake.

Chapter 20. Leblanc (Soda)

A Most Easie Way of Acquiring Spirit of Salt Together With the Salt Mirabile.

R. of common salt two parts, dissolve it in a sufficient quantity of common water; pour A upon the solution; put the mixture into a glass Body, or a glass Retort well coated, or else into an earthen Body or Retort. If a Body, set on an Head, and begin to destil with Fire of sand, encreasing your Fire gradually; with the first heat comes off the unsavory Phlegm, which gather apart; when the Liquor comes forth sowrish, change your Receiver and receive the sowre spirit: Continue the operation till no more spirits will arise, then let out the Fire, and permit the Vessel to stand in sand till all is cooled, when cold, take it out, and if it be unbroke, fill it again with the aforesaid matter, and proceed as we taught: The Phlegm is not to be cast away, but must be kept, that in it may be dissolved Salt (because it is better than common water) for another destillation. Thus from every pound of salt you will have lb. 1 of the best and most pure spirit. Dissolve the salt remaining in the Body or Retort (if neither be broke) in Water, filter and evaporate the Water, let it crystallize, the Crystals will be white, endowed with wonderful Virtues, to be declared here following.

— Rudolf Glauber, *Miraculum Mundi*, *ca.* 1658 AD [1]

20.1 ☿

∇ When I was born in 1742 the king was on his throne, God was in his heaven, and all was right with the world. It would not last. I was orphaned at the age of nine and apprenticed to an apothecary, where I learned my chemistry from sacred texts such as the one quoted above. I studied surgery at college, graduating as Nicolas Leblanc, MD. In 1780 I became physician to the Duke of Orleans and with his patronage devoted myself to the winning of a prize of 2,400 livres offered by the *Académie des Sciences* for "anyone who should find the most simple and economical method to decompose in bulk salts from the sea, extract the alkali which forms their base in its pure state, free from any acid or neutral combination, in such a manner that the value of this mineral alkali shall

1. Reference [18], pp. 31-32.

not exceed the price of the product extracted from the best foreign sodas."[2] You are probably wondering why the *Académie* would offer such a prize. I will tell you.

🜃 See, the glass business had been using ***potash*** or ***soda ash*** as a flux ever since about the twenty-fifth century BC (Chapter 13); the paper business needed alkali ever since the first century AD (Chapter 14); and the soap business had filled out by the eighteenth century, which increased the demand for alkalis even more (Chapter 19). And ever since God was a child, folks had been making potash and soda ash by leaching ashes (Chapter 8), ***lime*** by burning limestone (Chapter 10), and ***caustic soda*** by reacting soda ash and lime (Chapter 15). So alkali was a pretty good business by the eighteenth century, what with increasing demand from lots of different businesses and all.

🜃 Any country that had a glass, paper, or soap industry needed soda and to get it they tried burning all kinds of stuff to get ash. The ***saltwort*** plant, which grew mostly in Spain and the Canary Islands, was about the best thing to burn, on account of its ash, called ***barilla,*** contained about 20% soda. If you didn't want to have to kiss up to Spain, you could burn seaweed, which they did in Ireland and Scotland, but its ash, called ***kelp,*** contained only about a third as much soda as barilla did. No matter what you burned, you had to burn lots of it to get a little ash, and not all of that ash was soda. As the eighteenth century wore on, the supply of ash couldn't keep up with the demand for soda and the price doubled from 1750 to 1790. Now the French Academy of Sciences was worried about the dependence of domestic glass and soap on imported soda, so they offered a prize to promote the foundation of a domestic soda industry which would liberate France from Spanish barilla and Scottish kelp.

▽ Pardon me; this was *my* story.

🜃 Sorry.

▽ Earlier, in 1736 the French agriculturalist Henri Louis Duhamel du Monceau had revealed that soda and sea salt were salts of the same base, now known as sodium hydroxide. This being so, he proclaimed that it was possible, in principle, to make soda (sodium carbonate) from sea ***salt*** (sodium chloride). This information came down to me—

🜃 Like a spider.

2. Reference [52].

∇ More like a revelation, really, and I began to look for an intermediate between salt and soda. ***Glauber's salt*** (sodium sulfate) was well known among doctors (Chapter 18) and was easily produced from sea salt and ***sulfuric acid.*** If it were possible to convert Glauber's salt to soda, it would prove even more *mirabile* than had previously been imagined. Several people had taken this approach to the problem and after a thorough review of the successes and failures of my contemporaries—

∀ For in*spir*ation.

∇ —I devised a process for this conversion involving the calcination of limestone, coal, and Glauber's salt in a furnace to produce "black ash," from which soda could be extracted by recrystallization. The *Académie* had increased the prize to 12,000 livres in 1789 and two years later I was granted a patent from the king. I set up a factory at Saint-Denis with the financial backing of the Duke; at its peak the factory was producing 320 tons of soda per year. Unfortunately, the 1790's were not the best of times for the French nobility, as the *Comité du Salu Public* provided a brisk trade for the guillotine business. The Duke lost his head in 1793, the plant was confiscated, and the *Académie des Sciences* abolished.

∇ Now, all I wanted to do was to found a domestic soda industry for the benefit of French glass, paper, and soap makers, collect my prize, and settle down to a life of quiet contemplation. But in the space of a few years the king was de-throned, God seemed to have taken a holiday, and all was not particularly right with my world. Seeing the writing on the wall, I surrendered my patent to the *Comité* and the details of my process were published in 1797 in the *Annales de Chimie*. In 1802 Napoleon finally came to the realization that cheap domestic soda would be good for France and returned the now-derelict factory to me, but balked at the suggestion that he make good on the *Académie's* promised prize. I tried to make a go of it, I really did. But by the time I was allowed to resume my vocation, the secrets of my process had wandered far and wide.

∀ Like spiders.

∇ Like prodigal spiders, perhaps. Facing stiff competition from ungrateful imitators with more capital, my business failed. I must confess that I sank into depression and in a fit of melancholy shot myself in 1806.

∀ By 1810, French soda plants were making 15,000 tons of soda per year. Meanwhile, back in England there was a stiff salt tax which discouraged the black ash process from hopping the Channel, but the salt tax was repealed in 1823.

∇ May I continue?

🜃 I thought you were supposed to be dead. So, with the salt tax gone and with in*spir*ation from Leblanc, James Muspratt opened a black ash soda works at Liverpool, near the Cheshire salt fields. Pretty soon everybody and *his* dog had a soda plant. By 1852, the French soda production of 45,000 tons would be topped by the English production of 140,000 tons per year.

∇ Largely subsuming the chamber acid industry, Leblanc soda would become the foundation of a diversified chemical industry with products that included sulfuric and hydrochloric acids, soda, lime, salt cake (Glauber's salt), and caustic soda. Growing from scattered factories to immense complexes, soda manufacturers would, by the end of the nineteenth century, introduce such modern innovations as toxic waste dumps, water pollution, and acid rain. This pollution would create a climate of government regulation which would force manufacturers to find markets for former waste products. Eventually these new chemicals would become even more profitable than the original ones, leading to another round of industrial growth and rendering the word *chemical* synonymous with the word *poison* in the popular culture. But I am getting ahead of myself.

20.2 🜍

By the eighteenth century, increasing quantities of soluble alkali were in demand by three major industries. The glass industry relied on sodium or potassium carbonate as a flux for lowering the melting point of silica. The paper industry preferred soluble sodium hydroxide to marginally soluble calcium hydroxide for hydrolyzing lignin from pulp. The soap industry preferred sodium hydroxide to potassium hydroxide for saponifying oils and fats; the sodium salts of fatty acids are less hygroscopic than the potassium salts, which makes it easy to produce solid cakes of soap from sodium hydroxide. Since sodium hydroxide is easily produced from sodium carbonate and calcium hydroxide, the demands of all three industries would be satisfied if only there were a plentiful, inexpensive supply of sodium carbonate.

From remote antiquity, sodium carbonate had been available from two major sources. It occurred as a mineral in the deserts of the Middle East, having been deposited when water evaporated from ancient, alkaline seas. The Latin name for this mineral was ***natron,*** whose first two letters gave us the symbol for the element sodium. The Egyptian name survives in the name for the mineral ***trona,*** sodium sesquicarbonate, a 50/50 mixture of sodium carbonate and sodium bicarbonate. Where this mineral is available, other sources of soda cannot compete; there is nothing more economical than digging the material you need right out of the ground.

Trona is not a common mineral, however, and in most parts of the world soda has traditionally been produced by burning plant materials, the ashes of which contain sodium and potassium carbonates. Inland plants are richer in potassium and the soluble alkali extracted from their ashes is called potash, from which the element derives its name. Marine plants are richer in sodium and among the richest sources of soda are saltwort and seaweed. Whatever the source of alkali, a great deal of plant material must be burned to produce a little ash and not all of that ash is soda. Table 8-1 (page 101) shows that 1000 pounds of beech wood yields only 1 pound of potash. Similarly, 1000 pounds of dry kelp yield at most 2 pounds of soda.[3] When the demand for soda exceeds the supply from mineral and vegetable sources, an enterprising chemist would look for another source of soda and one obvious candidate is ordinary salt.

Salt was indispensable for the preservation of meat in the time before refrigeration, and it continues to be produced by time-honored methods. Salt-makers in hot climates produce salt by flooding shallow ponds with sea-water and allowing the Sun to drive off the water. In cold climates they freeze salt water; pure water-ice freezes out first, leaving concentrated brine to be evaporated to dryness. Salt-makers who are unfortunate enough to live in moderate climates must burn fuel to drive bejeezical water from brine. Salt can be extracted not only from modern seas, but from ancient ones as well; the dried remains of ancient oceans are mined or quarried for the mineral, ***halite.*** With sodium from sodium chloride, we need only find a way to combine it with carbonate to make soda.

The Leblanc process consists of three ***furnaces*** and a **lixiviator,** as shown schematically in Figure 20-2. In the first furnace, the *salt cake*

3. Reference [29], p. 171.

Figure 20-1. Halite

furnace, sodium chloride and sulfuric acid engage in a typical metathesis reaction, as shown in Figure 20-2(a). The products are solid sodium sulfate and gaseous hydrogen chloride. During the first half-century of the Leblanc era, there was little commercial demand for hydrogen chloride and it was simply sent up the chimney. If the neighbors complained, which they often did, the soda-maker simply built a taller chimney; chimneys of 400 feet were not uncommon. Alternatively, hydrogen chloride gas was absorbed into water, producing hydrochloric acid. With little demand for this acid it was frequently run into the nearest river. This seems irresponsible by modern standards, but the point of this first step is, after all, to get rid of the chloride and keep the sodium.

The second furnace of the Leblanc process, the *black ash* furnace, reacts sodium sulfate with coal and limestone at red heat, as shown in Figure 20-2(b). Bejeezical carbon dioxide flies the coop leaving ***black ash,*** a mixture of sodium carbonate, calcium sulfide, and residual coal and limestone. The main product, sodium carbonate, is soluble in water while the waste products are not. The third step of the process **lixiviates** the black ash with water, leaching out the soda and leaving a solid combination of blackened calcium sulfide and limestone known in the trade as *tank waste.* The final furnace evaporates the wash water, which deposits its cargo of soda. If this wash water is boiled down and left to cool, the result is washing soda, $Na_2CO_3 \cdot 10\ H_2O$. If it is calcined at red heat, the result is soda ash, Na_2CO_3. The schematic may look complicated, but the first two furnaces hearken all the way back to Figure 1-3 (page 9) and

Figure 20-2. The Leblanc Soda Process

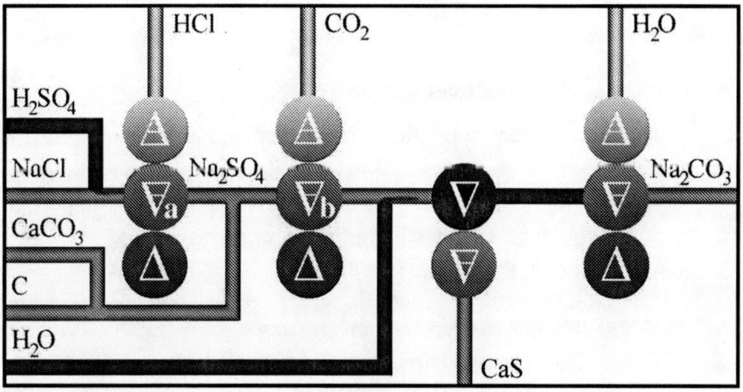

HCl
CO_2
H_2O
H_2SO_4
NaCl
Na_2SO_4
Na_2CO_3
$CaCO_3$
C
H_2O
CaS

$$(a)\ \mathrm{H_2SO_4}(l) + 2\mathrm{NaCl}(s) \overset{\Delta}{=} \mathrm{Na_2SO_4}(s) + 2\mathrm{HCl}(g)$$

$$(b)\ \mathrm{Na_2SO_4}(s) + \mathrm{CaCO_3}(s) + 2\mathrm{C}(s) \overset{\Delta}{=} \mathrm{Na_2CO_3}(s) + \mathrm{CaS}(s) + 2\mathrm{CO_2}(g)$$

the right half of the figure is identical to Figure 8-1 (page 103). Just remember that reactants enter from the left of the figure, waste product exit to the top and bottom, and the main product exits to the right. Study the figure well and you will understand on a fairly sophisticated level the genesis of the modern chemical industry.

Material Safety

Locate MSDS's for sulfuric acid (CAS 7664-93-9), sodium chloride (CAS 7647-14-5), hydrogen chloride (CAS 7647-01-0), and sodium sulfate (CAS 7757-82-6). Summarize the hazardous properties in your notebook, including the identity of the company which produced each MSDS and the NFPA diamond for each material.[4]

Your most likely exposure will be to hydrogen chloride fumes. If a persistent cough develops, seek medical attention.

You should wear safety glasses while working on this project. All activities should be performed in a fume hood or with adequate

4. The NFPA diamond was introduced Section 15.2 (page 184). You may substitute HMIS or Saf-T-Data ratings at your convenience.

ventilation. Leftover materials can be flushed down the drain with plenty of water.

Research and Development

You are probably wondering what will be on the quiz.

- You should know the meanings of all of the words important enough to be included in the ***index*** or **glossary.**
- You should know the Research and Development points from Chapter 13 and Chapter 16.
- You should know halite when you see it.
- You should know the reactions of Figure 20-2.
- You should know the hazardous properties of sulfuric acid, sodium chloride, sodium sulfate, and hydrogen chloride.
- You should be able to reproduce Figure 20-2 and to explain this schematic in your own words.
- You should know the traditional sources of sodium carbonate, the industries that depend on it, and the sad story of Nicolas Leblanc.
- Without coal (or charcoal) in the black ash furnace, calcium sulfate would have been produced rather than calcium sulfide. You should be able to explain why this would have been disadvantageous. Hint: Review Chapter 10.

20.3 Θ

You probably know enough by now that you could reproduce the Leblanc process without further instruction. The four-step process would involve heating sulfuric acid and salt. You would heat the product from this first reaction with charcoal and limestone in a crucible similar to the one used to smelt metals in Chapter 9. After heating this crucible to red heat in a kiln, you would crack it open and extract the soda in the same manner that you extracted potash in Chapter 8. But you would simply be repeating, for the most part, processes you have already performed in previous projects. Of the steps in the process, only the first step is substantially different from what you have already done and so this project will focus on the reaction of sulfuric acid and salt.

Begin by using Figure 20-2(a) to answer the following **stoichiometric questions:**

Q: How many grams of sulfuric acid are needed to react with 2.00 grams of sodium chloride?

Q: How many grams of sodium sulfate can be produced from 2.00 grams of sodium chloride?

The answer to the first question tells you how much concentrated acid to use. The answer to the second stoichiometric question is the ***theoretical yield,*** the amount of product that you would expect to get if the reaction went perfectly. You now know how much acid and salt to use and how much sodium sulfate to expect.

You are going to heat sulfuric acid and salt in a flask and you ought to take a few precautions to accomplish this safely. Obviously, since the reaction is going to produce hydrogen chloride gas, you should do this either outdoors or in a fume hood. You are also going to be boiling a concentrated solution which is prone to "bumping," that is, to boiling suddenly and violently. Chemists mitigate against this by adding *boiling chips* or *boiling beads* to such a solution. Any chemistry laboratory will have such things on hand, but if you are working at home, a few chips of broken glass or stone from Chapter 2 will do nicely. Finally, you will be handling a hot flask so you ought to have leather work gloves or beaker tongs handy. With these few precautions, you are ready to proceed.

Weigh a medium (100-250 mL) Erlenmeyer flask on a centigram balance and record the weight in your notebook. Tare the balance, add a few boiling chips, and record their weight. Tare the balance again and add the calculated amount of sulfuric acid using a medicine dropper or transfer pipette. If you happen to add too much acid, remove a few drops with the dropper and dispose of it in the sink, washing it down the drain with plenty of water. In a separate cup or weighing boat, weigh out 2.00 grams of sodium chloride and record this weight in your notebook for completeness.

Use a spatula or spoon to add the sodium chloride to the acid a little at a time. The acid will fizz and hydrogen chloride gas will be evolved. If you happen to get a whiff of hydrogen chloride gas, imagine what it must have been like to live next door to an early soda factory. Use a strip of pH test paper, the chemist's virtual nose, to determine the pH of the gas. Continue adding sodium chloride until you have added the entire

Figure 20-3. Calcination

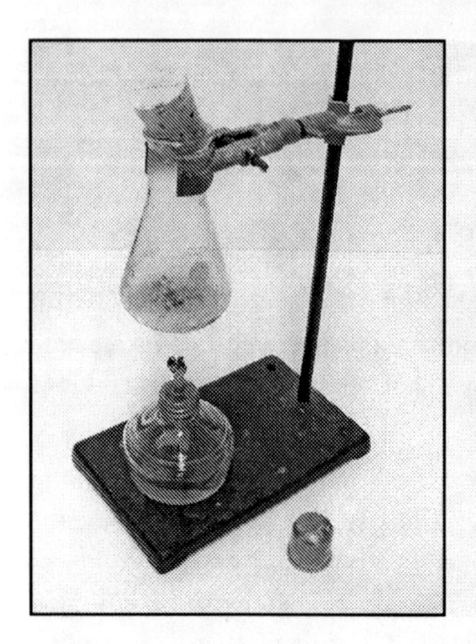

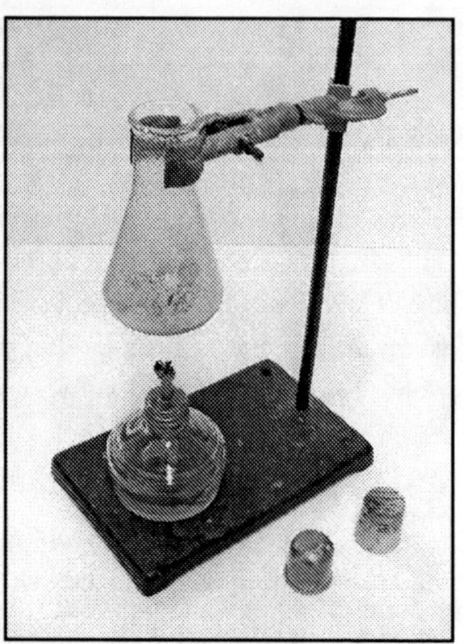

2 grams. Normally we would calcine the acid-salt mixture in a furnace at red heat, but we can get by with a kinder, gentler heat by engaging in successive rounds of calcination and dissolution. There are two ways you may calcine your flask. You may mount it on a lab stand and heat it over a ***spirit lamp*** or you may heat it on a hot-plate. To prevent the acid from spattering, place a cork *loosely* into the flask. Light the spirit lamp or turn the hot-plate to its highest setting and heat the flask for 10 minutes, as shown in Figure 20-3. Snuff the lamp or remove the flask from the hot-plate, remove the cork, and let the flask cool for 10 minutes. Human nature being what it is, you will probably be tempted to cut this time short, but if you rush things you are likely to break your flask. When it is cool enough to handle with your bare hands, weigh the flask and record the weight in your notebook. You can now subtract the weight of the empty flask and the boiling chips to get the weight of your product. Compare this weight to the theoretical yield you calculated earlier.

Chances are your actual yield is higher than your theoretical yield, meaning that not all of the HCl has departed. Add 5 mL of water to the flask with a medicine dropper or transfer pipette and swirl the flask to bring as much of the solid as possible into solution. Fill the pipette with the resulting solution and use it to wash all of the solid from the sides of the flask down into the bottom, as shown in Figure 20-4. Then test the

Figure 20-4. Dissolution

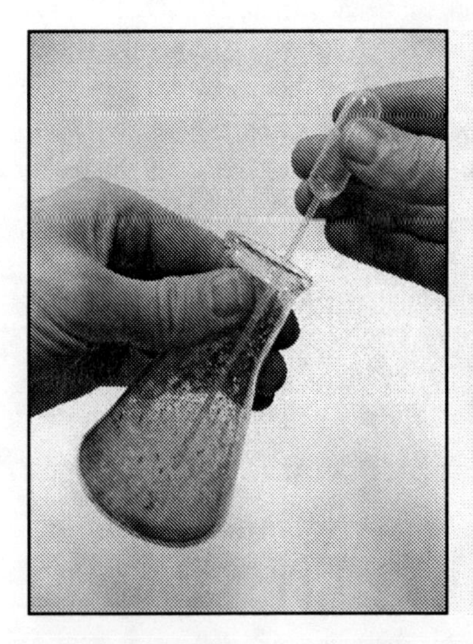

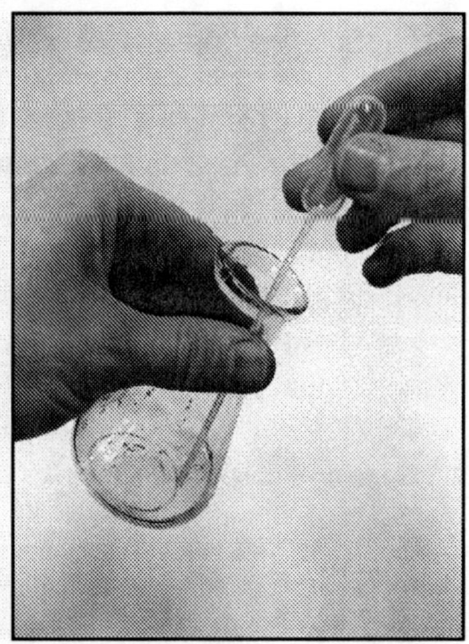

solution by placing the last drop from your pipette on a strip of pH test paper; it will probably be acidic. You are about to calcine this solution a second time, but the concentrated solution may spatter as it dries out. To prevent this, place a cork *loosely* into the mouth of the flask and place it back on the heat source, as shown in Figure 20-3(L). Once the solution has stopped spattering the cork may be removed, as shown in Figure 20-3(R). Calcine for a total of 10 minutes and then cool for 10 minutes. Weigh the cool flask once again and calculate the yield of your product. It should be less than before if you have successfully driven off more HCl.

Continue this cycle of calcination and dissolution; it may take four or five rounds to drive off all of the HCl. When the weight stops dropping, that is, when the weight from your final calcination is the same as the one before, you have probably succeeded in driving off all of the HCl. The flask now contains anhydrous sodium sulfate. Chances are, you are completely underwhelmed at this point. All this time spent slaving over a hot flask for nothing more than some white powder. *Mirabile* indeed! The best is yet to come, my friend, if you will be patient but a while longer.

The *sal mirabile* which brought a glow to Glauber's cheek, was not anhydrous sodium sulfate, but sodium sulfate decahydrate, $Na_2SO_4 \cdot 10\ H_2O$. To make it, add *only 3 mL* of water to your flask, not 5 as before. Warm the flask *gently* over the spirit lamp or hot-plate, but do *not* bring it to a boil. The flask should not be so hot that you cannot hold it comfortably in your bare hand. Swirl the liquid around the flask and use your pipette as before to rinse the solid from the sides of the flask. Your goal is to get as much sodium sulfate into as little warm water as is physically possible. Spend a full 10 minutes swirling, rinsing, and warming; this time is well spent. Not all of the solid will dissolve. Use your pipette to transfer the warm, saturated solution to a small Petri dish or watch glass and allow it to cool. If you have used the fire gently and with great skill, you will witness the behavior which gave Glauber such high opinion of this salt:

> This *sal mirabile* being rightly prepared, looketh like Water congealed or frozen into Ice; it appeareth like the Crystals of Salt-petre, which shoot into a long Figure; also it is clear and transparent, and being put to the Tongue, melts like Ice. It tasteth neither sharp, nor very salt, but leaveth a little astringency on the Tongue. Being put upon burning Coals, it doth not leap and crackle after the manner of common salt, neither coneiveth flame like Salt-petre, nor being red hot, sends forth any smell; which gifts or endowments no other salt possesseth.
>
> — Rudolf Glauber, *A Treatise on the Nature of Salts*[5]

With the first crystals of Glauber's salt in hand, add another 2 mL of water to the remaining solid in the flask and use your pipette to bring the solid into solution. Add this solution to the crystals in your Petri dish and allow the excess water to evaporate overnight. If you allow the crystals to dry longer than that, they may revert to anhydrous sodium sulfate, depending on the temperature and relative humidity. Now, I know what you are probably thinking; these crystals are not nearly as *mirabile* as the hype might have led you to believe. It is difficult to compete with color television and video games. I am amazed that you have read this far in the book with an attitude like that. But consider that without Glauber's salt, there would have been no cheap soda, without soda no cheap glass, and without glass, no television and video games. And who knows? If crystals get under your skin, as they did mine, and Glauber's before me, then you may become interested in growing larger and more beautiful crystals, water-soluble gemstones like the large crystal of Glauber's salt

5. Reference [18], p 32.

Figure 20-5. Coagulation

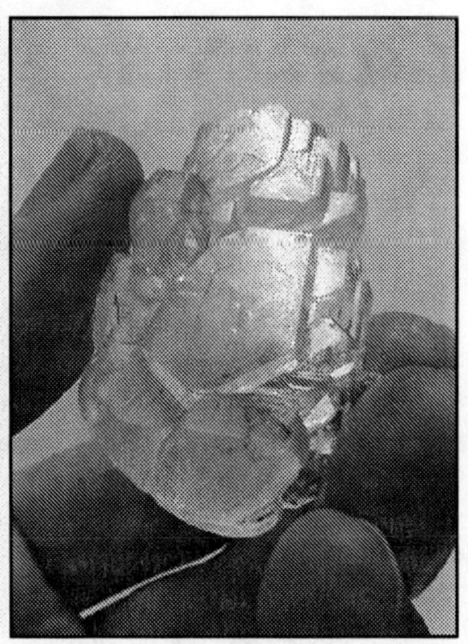

shown in Figure 20-5(R). If so, allow me to recommend *Crystals and Crystal Growing,*[6] the Bible of amateur crystallography.

Quality Assurance

Compare the theoretical yield to your actual yield of anhydrous sodium sulfate. Your product should be very nearly neutral in pH; tape your final pH test paper into your notebook. Photograph your crystals and tape the photo into your notebook as an everlasting memorial.

6. Reference [86]

Chapter 21. Volta (Batteries)

> After a long silence, for which I will not seek to excuse myself, I have the pleasure of communicating to you, Sir, and through you to the Royal Society, some striking results at which I have arrived in pursuing my experiments on the electricity excited by the simple mutual contact of metals of different sorts, and even by that of other conductors, also differing among themselves, either liquids, or containing some fluid to which they properly owe their conducting power. The principle result, and that which comprehends nearly all the others, is the construction of an apparatus which resembles in its effects, that is to say, in the sensations which it can cause in the arms, &c., the Leyden jars, or better yet, feebly charged electric batteries, but which acts without ceasing, or whose charge after each discharge is reestablished by itself; which provides, in a word, an unlimited charge, a perpetual action or impulsion on the electric fluid;
>
> — Allesandro Volta, *On the Electricity Excited by the Mere Contact of Conducting Substances of Different Kinds*, *ca.* 1800 AD [1]

21.1 ☿

Δ A mortal finger jabs at the button and the screen flickers to life; the disk drive hums into action and the machine feels itself all over to make sure that everything is in its place. I stare at the screen and wonder where to begin. I suppose that it all began with a spark. By the power of the spark I brought light and warmth into places dark and cold. By the power of the spark I raised up great civilizations of brick, metal, and glass. One by one I watched these civilizations burn to the ground and new ones rise from their ashes. By the power of the spark I enabled the many to defend themselves against the few with new weapons of fire and metal. This is the strong power of all powers, for it overcomes everything fine and penetrates everything solid. I carried the spark from generation to generation, keeping in the heat and withstanding it, and yet I knew it not.

Δ For half a million years I watched the lightning bolt descend from heaven to Earth, receiving power from Above and from Below, but it was not until 1752 that I summoned the courage to call it forth. It was then, in

1. Reference [18], p. 239.

the person of one ***Benjamin Franklin,*** that I coaxed the spark from the sky and brought it to live in a jar of glass and metal. Franklin had grown up among the soap kettles of his father, receiving the lessons of Lucifer and Athanor in his youth. He established himself as a printer, learning to pour metal type from Vulcan. He invented bifocal spectacles, having mastered the lessons of Theophilus. He was an architect of the American Revolution, a revolution of the many against the few made possible only by the lessons of Tzu-Chhun. It was in such a mortal abode that I began to understand the spark for the first time.

Δ My progress from one mortal to another was enhanced in the eighteenth century by the establishment of learned societies and their publications. From Franklin I leapt to ***Charles Coulomb,*** where, in 1785 I determined a method by which electric charges could be measured; the unit of electric charge is now known as the *coulomb.* These charges are of two kinds, the positive and the negative; opposite charges attract and like charges repel one another. From Coulomb I came into the person of ***Luigi Galvani,*** Professor of Anatomy and Gynecology at Bologna. In the course of my dissections, I observed that a dead frog could be made to twitch and convulse as if it were alive by any of three methods. First, it may be connected to a lightning rod, which gathers electricity from the air in stormy weather. Second, it may be connected to an electrical machine, that is, a machine consisting of a sphere of sulfur rubbed by a woolen cloth. Finally, it may be connected to two *different* metals and the metals brought into contact with one another. Such contact may initiate an electric spark, bringing with it apparent life.

Δ That which is below corresponds to that which is above; just as two metals in contact may initiate an electric spark, so two mortals in verbal or literary contact may initiate a hermetic spark which passes from one to the other. I am that spark, passed from Galvani to ***Allesandro Volta,*** in whom I continued the Work. I found that a salt solution could substitute for the frog in Galvani's experiment. In the simplest possible terms, a voltaic cell consists simply of two different metals separated by a conductive solution. Voltaic cells may be connected in series to form a pile or battery. The more cells in the battery, the longer the spark produced; the longer the spark, the greater the electromotive force; and the unit of electromotive force is called the *volt.*

Δ From Volta to ***Hans Christian Oersted,*** I discovered in 1820 that the current from a voltaic battery produces a magnetic field. Having witnessed a demonstration of Oersted's results at a meeting of the *Académie des Sciences*, the I-dea passed to ***André Ampère,*** in whom I discovered a way to measure this current, the unit of which is now known as the *ampere,* or *amp.* Big batteries produced more current than small ones, but had the same voltage. I solved this puzzle in 1827 in the person of ***Georg Ohm*** when I announced the principle of resistance, which limits the current produced by a cell of a given voltage. The unit of resistance is now known as the *ohm.* With the concepts of EMF, current, and resistance clarified I was poised for my greatest advances in understanding the spark.

Δ In no other single mortal did I make more headway in understanding electricity than in the person of ***Michael Faraday.*** Apprenticed to a book-binder at the age of 14, the I-dea of electricity was transmitted to him by an article in the *Encyclopaedia Britannica,* and from that moment I established a home in him. I found employment as the assistant to the great chemist, Sir Humphry Davy, who had made a brilliant start in the relationship of electricity to chemistry. In Davy and Faraday I came to understand that the positive and the negative are not two species of electricity. No, there is only one electrical entity; positive and negative simply represent a greater or lesser quantity of this One Thing. In 1831 I announced the principle of electromagnetic induction, the basis of the transformer, by which electrical current may be converted from one voltage to another. In 1834 I announced the laws of electrolysis, unifying the I-deas of electric charge and chemical amount. The conversion factor from electric charge to moles is now known as the *Faraday constant.*

Δ My researches culminated in the first practical application of electricity in the person of ***Samuel Morse.*** In 1844 I demonstrated a device using the batteries of Volta and the electromagnetism of Oersted to transmit I-deas electrically from one place to another. This "telegraph" made it possible to communicate over vast distances almost instantaneously. Most of the practical applications of electricity for the next half-century would simply be variations on the telegraph theme: electrical current produced by batteries, controlled by switches, flowing through wires, and producing motion via electromagnets. And even after the advent of vacuum tubes, radio, and television these simple devices would not lose their spark.

Δ A little more than a century after the invention of the telegraph I re-minded myself of the spark once more. A mortal child constructed his Halloween costume from cardboard boxes and tin foil; he wanted to be a robot. He hungered for unconventional playthings, winding copper wire around a nail to make an electromagnet, constructing a telegraph set from tin cans and scrap wood, dismantling radios and telephones to scavenge precious capacitors, resistors, and diodes. He grew impatient with the over-protectiveness of contemporary science books for children, devouring instead the boy-scientist[2] books of previous generations. He learned to tame the Model T ignition coil and the flyback transformer. And so I began my current incarnation.

Δ In the person of Dunn I sit writing these words on a laptop computer. The device is powered by batteries; the keys on the keyboard are nothing more than switches; and the motors, relays, and pumps in my inkjet printer are all activated by electromagnets. Without these simple marvels, the transistors and integrated circuits would slumber in darkness. Truly there are few modern wonders, whether of art or artifice, which do not depend on the power of the spark.

21.2 🜍

I did some growing up in the city of New Orleans. One of the most striking landmarks of the Big Easy is the Greater New Orleans Bridge. There is a toll to cross this bridge and we can go a long way toward understanding the chemistry of batteries by examining the economics of bridge tolls. Why would anyone pay a bridge toll? Obviously because they want to get to the other side. How much of a toll would they be willing to pay? It depends on how badly they want to cross the river by car; let us call this desire the *automotive force,* or AMF, with units of *dollars/car.* The AMF depends on the location of the bridge; it will be high when the bridge connects, for example, a major commercial district with a major residential area. People will pay a toll to cross the bridge when the cost of the toll does not exceed the automotive force.

Let us consider a chemical analogue. Whenever we have manipulated alkalis in this book we have been careful to avoid the use of aluminum containers or utensils. The reason is that aluminum is easily oxidized by alkaline solutions:

2. Reference [47] is an excellent book for boy- *and* girl-scientists capable of overlooking the un-apologetic sexism of the era in which it was written.

$$\mathrm{Al}(s) \;+\; 3\,\mathrm{OH}^-(aq) = \mathrm{Al(OH)_3}(s) \;+\; 3\,e^-$$
$$3\,\mathrm{H_2O}(l) \;+\; 3\,e^- = \frac{3}{2}\,\mathrm{H_2}(g) \;+\; 3\,\mathrm{OH}^-(aq)$$

When sodium hydroxide is sprinkled on wet aluminum foil a vigorous reaction takes place; the solution foams up, the foil is eaten away, and a great deal of heat is released. If we were to look closely at the foil we would find some areas in which aluminum is being oxidized to aluminum hydroxide; these areas comprise the ***anode,*** the site of oxidation. In other areas water is being reduced to hydrogen gas; those areas comprise the ***cathode,*** the site of reduction. In the example so far, half of the aluminum foil acts as the anode and half as the cathode with electrons being passed from anodic regions to adjacent cathodic ones. Such a situation is analogous to a city in which residences and businesses are mixed in and among one another; there is plenty of traffic, but since it is mostly local it would be difficult to collect a toll from people commuting to and from work.

The situation would be different if we were to connect to the aluminum foil an inert metal, one which does not react with alkali. Such a metal could not be oxidized by the alkali and so it could not serve as an anode. But electrons would be able to move from the anodic regions of the aluminum foil, through the inert metal, and into the water. The water would be reduced to hydrogen gas at the surface of the inert metal and so the inert metal would serve very well as a cathode. With reduction taking place at the inert metal surface, the entire surface of the aluminum foil would become available for oxidation. Such a situation is analogous to a city in which the business district is adjacent to the residential district; there would be an automotive force reflecting the desire of people to travel from one district to another, but no way to collect a toll.

Let us now separate the aluminum foil from the inert metal, leaving each one in contact with the alkaline solution. Such an arrangement, a ***voltaic cell,*** is analogous to a city in which the business and residential districts are separated by a river. Figure 21-1 shows a process schematic for the aluminum-alkali cell. Here the electrons consumed at the anode and produced at the cathode are shown explicitly as reactants and products. The electrons on the reactant side represent a wire dipping into the solution, or **electrolyte;** hydrogen gas is produced at this wire, the cathode. The

Figure 21-1. The Aluminum-Alkali Cell

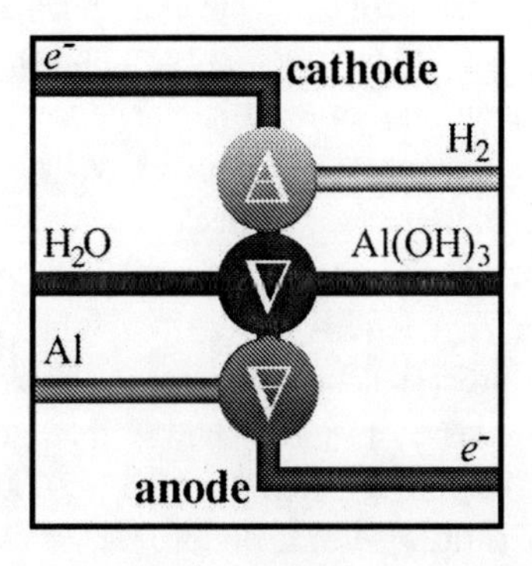

Equation 21-1. Four Electrochemical Reactions

$$\text{(a) } ?\ \mathrm{Al}(s) + ?\ \mathrm{H_2O}(l) = ?\ \mathrm{H_2}(g) + ?\ \mathrm{Al(OH)_3}(s)$$

$$\text{(b) } ?\ \mathrm{Zn}(s) + ?\ \mathrm{MnO_2}(s) + ?\ \mathrm{NH_4Cl}(aq) = ?\ \mathrm{Mn_2O_3}(s) + ?\ \mathrm{ZnCl_2}(aq) + ?\ \mathrm{NH_3}(aq)$$

$$\text{(c) } ?\ \mathrm{Zn}(s) + ?\ \mathrm{MnO_2}(s) + ?\ \mathrm{H_2O}(l) = ?\ \mathrm{Mn_2O_3}(s) + ?\ \mathrm{Zn(OH)_2}(s)$$

$$\text{(d) } ?\ \mathrm{Pb}(s) + ?\ \mathrm{PbO_2}(s) + ?\ \mathrm{H_2SO_4}(aq) = ?\ \mathrm{PbSO_4}(s) + ?\ \mathrm{H_2O}(l)$$

electrons on the product side represent the wire leading from the aluminum anode. With the cathode at the top and the anode at the bottom of the schematic, we can imagine the electrons flowing downhill, impelled by the ***electromotive force***, or EMF, analogous to the automotive force. Of course, in the physical cell the relative placement of the anode and cathode is irrelevant. The convention established in the process schematic simply helps us to re-meme-ber the direction of electron flow. If the wire from the anode is connected to the cathode we have, in effect, created a bridge and may "charge" the electrons a "toll" for the privilege of crossing over; we might require them to light a bulb or turn a motor. How much of a toll would they be "willing" to pay? Any amount up to the value of the electromotive force, measured in ***volts.***

The EMF, or voltage, of a cell depends on the identity of the conductors and on the identity and concentration of the electrolyte. Equation 21-1 shows the *un*balanced reactions for four voltaic cells, (a) the aluminum-alkali cell, (b) the Leclanche carbon-zinc cell, (c) the alkaline carbon-zinc cell, and (d) the lead-acid cell. The aluminum-alkali cell is the one we have been discussing so far, with an EMF of 1.0 V, more or less, depending on the concentration of the alkali. The Leçlanche cell is the familiar flashlight battery, with a voltage of 1.5 V. The alkaline cell comes in the same voltage and sizes as the Leclanche cell. Six lead-acid cells, at 2.0 V each, comprise the common automobile battery. Take a moment to balance these redox reactions using the method of Chapter 11 and then sketch out the process schematics for each cell. You can identify the anodes and cathodes by re-meme-bering that electrons go in at the top, down through the cell, and out the bottom; under this convention, the positive cathode is the electrode on top.

The flow of electrons is really only half the story of the voltaic cell. Returning to the bridge analogy, the Greater New Orleans Bridge collects tolls only from cars traveling towards the business district; the return trip is free. The "return trip" in our aluminum-alkali cell is afforded by the alkaline solution. Negative hydroxide ions are produced at the inert metal cathode and consumed at the aluminum anode. Just as the electrons flow from anode to cathode, the hydroxide ions migrate through the electrolyte solution from cathode to anode. This migration of charge through the electrolyte constitutes an electric ***current*** equal and opposite to that in the wire. The unit of electric current is the ***ampere*** or *amp*. The electric current is proportional to the rate at which electrons flow through the wire. The more electrons cross over in a given time period, the higher the current. Thus electric current is analogous to the rate at which traffic crosses the bridge.

The amount of bridge traffic depends, of course, on the AMF; if people have little motivation to cross the bridge, traffic will be light. But it also depends on the population of the city; a large city will have more traffic than a small one. Similarly, the current that can be delivered by a voltaic cell depends on the size of the cell. For example, flashlight cells come in different sizes, D, C, A, AA, and AAA. Each of these cells has the same voltage. But the big D cell can deliver more current than the tiny AAA cell. Hence a boom box, with its large speakers and motors, requires D cells; a pocket radio, with only headphones, can get by with AAA cells.

The big payoff for the bridge comes when lots of people are willing to pay high tolls for long periods of time. For electrical devices, the payoff is called energy, with units of joules:

1 joule = (1 volt)(1 amp)(1 second)

The rate of energy delivery is called power, with units of watts:

1 watt = (1 volt)(1 amp)

Thus 1 joule = (1 watt)(1 second)

We can use these definitions to do UFA, for example:

Q: How many mega joules of energy are provided by the delivery of 1 kilowatt of power for 1 hour?

A: 3.6 MJ

More power is delivered by two voltaic cells than by one and there are two ways of connecting cells to form a ***battery.*** When two cells are connected in ***series*** the anode of one is connected to the cathode of the other. For example, when Leclanche or alkaline cells are loaded into a flashlight the "nose" or button end, the positive cathode, of one is placed in contact with the flat bottom, the negative anode, of the next. The voltage of this battery of two cells will be double that of either cell on its own. You can think of two process schematics stacked one atop the other; the electrons "fall" twice the normal distance. For series batteries of more than two cells, the voltage is proportional to the number of cells. Thus, if you cut open a 9-volt transistor battery, you will find six 1.5-volt cells within. Similarly, a 12-volt automobile battery consists of six 2-volt lead-acid cells. Since the same current flows through each of the cells in a series battery, the power delivered is proportional to the number of cells; two cells deliver twice the power of one.

Conversely, when two identical cells are connected in ***parallel,*** that is, anode to anode and cathode to cathode, the voltage of the pair is the same as that of an individual cell. Since the same current flows through each cell, the current of the pair is twice that of either on its own. Six 1.5-volt cells in parallel will have a voltage of 1.5 volts, but will be able to deliver more current than any of them alone. You can think of the process schematics arrayed side-by-side. In effect, the cathodes and anodes form one giant cathode and one giant anode. Since the voltage is the same as that of the individual cells but the current is proportional to the number of them, the power delivered is, again, proportional to the number of cells; two cells deliver twice the power of one. Thus more power is delivered by more cells no matter which connection scheme is used. The choice of series of parallel connection depends on the voltage required by the device to be powered.

This chapter has introduced a host of intangible concepts and their units: the volt, the amp, the joule, and the watt. It may have left you weary, feeling small. When tears are in your eyes I will dry them all. I am on your side. When times get rough and friends just can't be found, just remember that a voltaic cell is like a bridge over troubled water.

Material Safety

Locate MSDS's for aluminum (CAS 7429-90-5), soda ash (CAS 497-19-8), and charcoal (CAS 7440-44-0). Summarize the hazardous properties in your notebook, including the identity of the company which produced each MSDS and the NFPA diamond for each material.[3]

Your most likely exposure will be eye or skin contact. In case of eye contact flush them with water and call an ambulance. In case of skin contact wash the affected area with plenty of water.

You should wear safety glasses while working on this project. Leftover soda can be flushed down the drain with plenty of water. Used aluminum foil should be thrown in the trash. Used charcoal may be thrown in the trash or saved for reuse.

Research and Development

You should not remain ignorant if you are to proceed in the Work.

- Know the meanings of those words from this chapter worthy of inclusion in the ***index*** or **glossary.**
- You should have mastered the Research and Development items of Chapter 17 and Chapter 20.
- Know two new unit factors from this chapter and be able to use them to convert among the units, volts, amps, and watts.
- Given the skeleton reactions for the aluminum-alkali, Leclanche, alkaline, and lead-acid cells, be able to produce balanced redox reactions and process schematics.
- Know the hazardous properties of charcoal and washing soda.

3. The NFPA diamond was introduced in Section 15.2 (page 184). You may substitute HMIS or Saf-T-Data ratings at your convenience.

- Know the years in which the voltaic pile and the telegraph were invented.
- Know the contributions of Franklin, Volta, Ampère, Ohm, and Faraday to the understanding of the spark.

21.3 Θ

For this demonstration you will require some aluminum foil to be used as an anode in your voltaic cell. You will also need a cathode. I am fortunate enough to have had mercury-silver cathodes installed in my mouth when I was a child. That way, wherever I am, I am able to construct my own batteries. Only one more thing is needed: an electrolyte. This I also keep in my mouth. If you are similarly equipped, roll up a ball of aluminum foil and pop it into your mouth. Move it from side to side, chewing it thoroughly. If the aluminum happens to connect with a cathode, you will perceive an electric shock. Congratulations! You have constructed a voltaic cell.

Such a cell is not entirely satisfactory, however. While it produces a decent voltage, the current produced is quite small. Similar proof-of-concept batteries may be constructed from coins and salt-water or lemon juice. It may even be possible to run a tiny electronic clock from one of these, as such clocks do not require much in the way of current. But for lighting light bulbs and running motors and closing relays we need a lot more current than can be provided by a battery made from coins. The key to high currents is electrode surface area.

The battery you will build is a variation on the one described in the previous section. I chose aluminum as the anode because it is commonly available and because its standard reduction potential is rather large and negative. Ordinary table salt may be used as an electrolyte, but I have found that ***washing soda,*** sodium carbonate, produces a cell with a higher current. In the previous section we discussed platinum as an inert electrode. Such an electrode allows water itself to serve as an oxidizing agent. Recalling that the oxidizing agent is, itself, reduced and that the site of reduction is called the cathode, we see that platinum served as the cathode. Platinum is expensive, however, and so we require a cheaper material for our cathode. For electrical applications, carbon often serves as "poor *man's* platinum." We would like a form of carbon with a large

Figure 21-2. The Aluminum-Alkali Battery

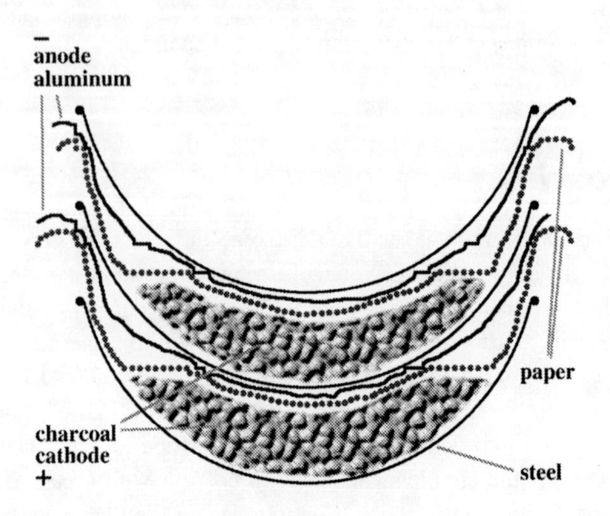

surface area so that electrons will be more likely to cross our wire bridge than to move directly from the aluminum to the electrolyte. An inexpensive, high-surface-area form of carbon is met with in activated charcoal.[4] Activated charcoal is porous, like a sponge, and is sold for use in water filters for aquariums. Ordinary charcoal is electrically conductive, but lacks the large surface area of activated charcoal. Since it is inert, it may be re-used to make generations of voltaic cells.

You will also need a convenient container for your battery and I find that a small stainless steel mixing bowls serve well. Three mixing bowls of the same size will allow you to construct a battery of two cells in series. The larger the mixing bowls, the greater the current from your battery, but even small, 2-cup bowls will power a small motor or telegraph sounder.

Begin by preparing a saturated solution of washing soda. Fill a mixing bowl half-full of water, add some washing soda, and stir until it is dissolved. Keep adding washing soda and stirring until no more will dissolve and solid sodium carbonate settles to the bottom of the bowl. Fill each of your remaining bowls about 1/4 full of activated charcoal. Add

4. In early versions of this project I used copper dishwashing pads as the cathode. The I-dea for using activated charcoal came to me from Reference [46].

enough washing soda solution to each bowl to make a slurry with the activated charcoal. The charcoal will foam; air is displaced from the pores as the charcoal absorbs the solution. You now have two bowls, 1/4 full of charcoal and electrolyte. We are about to add the aluminum anode, but we need a way to electrically insulate the aluminum from the charcoal while still allowing the ions in the electrolyte to move about.

To do this, line each of the two bowls with two thicknesses of paper towel. This will keep the charcoal separate from the aluminum while allowing the electrolyte to flow between the two. Add enough washing soda solution to completely wet the paper towels. Now line each bowl with aluminum foil, taking care that the foil does not tear the paper towels and that it does not touch the mixing bowl beneath. You are now ready to connect your two cells in series.

Place one of your cells inside the other, just as if you were stacking the bowls for storage. The aluminum anode of the lower cell should make contact with the steel-charcoal cathode of the upper one. Empty the third mixing bowl of leftover washing soda and nest it within the upper cell. You now have three bowls nested one inside the other. If you were to look at their contents in cross section, it would go steel-charcoal-paper-aluminum-steel-charcoal-paper-aluminum-steel. Use both hands to press the upper and lower bowls together. By compressing the charcoal, you reduce its resistance and thereby increase the current produced by your cells. Finally, use a strong rubber band to hold the bowls tightly together.

Inexpensive volt-ohm meters are now widely available. Use a meter to measure the voltage between the lower bowl and the middle bowl, the middle bowl and the upper bowl, and lower bowl and the upper bowl. Record these voltages in your notebook and compare the voltages of the individual cells to that of the two of them in series. If your meter measures milli-amps as well, record the current output of each individual cell and compare them to that of the two in series. The voltage of your battery should be about 2 V and the current should exceed 50 mA.

There is a break-in period with these cells as aluminum hydroxide builds up around the anode. Be patient before concluding that your cells are not operating correctly. If one or both of your cells fail to produce any output, it is likely that the paper is torn, shorting anode to cathode; the electrons are able to move from anode to cathode without going through your meter. If both cells perform, but one performs better than the other, it is

Figure 21-3. The Battery

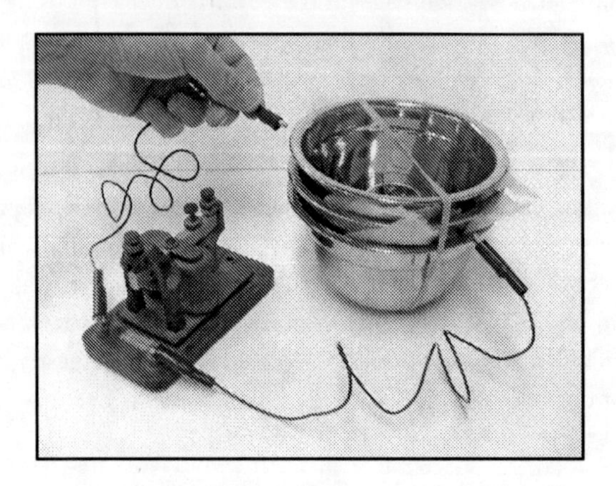

likely that the under-performing cell does not contain enough electrolyte. Dismantle the cell and add some more sodium carbonate solution.

The real test of your battery is whether or not it is strong enough to power an electrical device. You may try lighting a small flashlight bulb, running a small motor, or activating a telegraph sounder. You may even build your own electromagnet. Simply wind 50-100 turns of insulated wire around an iron nail. The wire must be insulated; "bell wire" or "magnet wire" can be found at hobby electronics stores. Magnet wire may also be salvaged from "dead" motors. With a knife, strip the insulation from the two ends of the wire; connect one end to the lower bowl and one to the upper bowl of your battery. When connected to your battery, your electromagnet should be strong enough to lift a paper clip.

Your battery will remain at full strength for hours but not for days. Eventually the aluminum anode will be consumed and the electrolyte will become saturated with aluminum hydroxide. The charcoal, however, is inert. When your battery is dead, or when you are tired of it, throw the aluminum foil and paper towels away. Drain the electrolyte from the charcoal and wash the solution down the sink. Wash the charcoal with a few portions of fresh water, drain, and allow the charcoal to dry. The mixing bowls and charcoal may be used to make more batteries later on. I would not, however, put the used charcoal in my aquarium filter; the residual alkali and aluminum hydroxide would probably not be good for fish.

The aluminum-alkali battery is far from optimal. One of the most practical cells of all time is the ***Leclanché cell,*** developed by Georges Leclanché in 1866. Batteries up to that time had consisted of jars filled with liquid, which were subject to breakage, spillage, and leakage. Leclanché's innovation was to produce a "dry cell," which might more accurately have been called a "moist cell." The Leclanché cell consists of a carbon/manganese dioxide cathode, surrounded by a paste of ammonium chloride, separated from the zinc anode by a layer of paper. The redox reaction is:

$$Zn(s) + 2\ MnO_2(s) + 2\ NH_4Cl(aq)$$
$$= ZnCl_2(aq) + Mn_2O_3(s) + 2\ NH_3(aq) + H_2O(l)$$

Unlike the aluminum-alkali cell, the Leclanché cell produces no gases and so it may be hermetically sealed. As the cell ages, however, the ammonium chloride gets used up and the cell voltage drops. This drawback has been addressed by the alkaline dry cell.

The ***alkaline dry cell*** uses the same anode and cathode as the Leclanché cell and consequently has very nearly the same EMF. Rather than using ammonium chloride as the electrolyte, however, it uses potassium hydroxide. The redox reaction is:

$$Zn(s) + 2\ MnO_2(s) + H_2O(l) = Zn(OH)_2(s) + Mn_2O_3(s)$$

Because the electrolyte is not consumed in the reaction, its concentration remains constant as the cell ages. Consequently the EMF of the alkaline dry cell remains nearly constant as it ages.

All of the cells we have discusses so far have been primary cells; once the anode is consumed in the reaction, the cell is dead and must be discarded. A secondary cell is one which can be recharged, that is, one in which the redox reaction may be run backwards to restore the original state of the cell. We will re-meme-ber this I-dea in Chapter 25.

Quality Assurance

Record in your notebook the voltages and currents produced by your individual voltaic cells and by the two cells connected in series. Your battery passes muster when it can operate an electrical device.

Chapter 22. Perkin (Aniline Dyes)

> The real inventor of practical gas-lighting is William Murdoch, who in 1792 lit his workshops at Redruth, Cornwall, with a gas obtained from coals. His operations remained unknown abroad for some ten years, and hence the French consider Lebon as the inventor of gas-lighting, since he lit (1801) his house and garden with gas obtained from wood. The first more extensive gas-work was established in 1802 by Murdoch, at the Soho Foundry, near Birmingham, the property of the celebrated Boulton and Watt; and in 1804 a spinning-mill at Manchester was lighted with gas. From that period gas-lighting became more and more generally adopted in factories and workshops, but not before the year 1812 did this mode of lighting become introduced into dwelling-houses and streets, a few of which in London were lit with gas in this year; while in Paris gas was first introduced in 1820. From that year gas-lighting may be said to have become of general importance in Europe, and now there is hardly any important place on the Continent where it is not in use, while as regards the United Kingdom in no portion is gas-making and lighting so general over town and country as in Scotland.
>
> — *Wagner's Chemical Technology*, 1872 AD [1]

22.1 ☿

🜄 Gas! Now *that's* a business to be in.

🜁 I don't believe it! Here I am, supposedly the representative of the alchemical element *air,* waiting patiently for a chapter that might possibly have some remote connection to the element for which I am supposed to speak, and when one finally comes around the Author assigns it to a confabulating bug with about as much air-worthiness as a lead Zeppelin. And it makes me wonder.

🜄 If there's a bustle in your hedgerow, don't be alarmed now. This chapter isn't about gas; it's really more about tar. See, folks had been making charcoal from wood since God was a child. Now, to make charcoal, you put wood in a sealed container, and heat the ***bejeezus*** out of it in the absence of air. Bejeezical water goes up the smoke stack, like it shows in

1. Reference [29], pp. 645-646.

Equation 1-1 (page 9), and the stuff that's left behind is ***charcoal,*** which is mostly carbon. That charcoal burns hotter and cleaner than wood and so it's great for smelting metals and all.

∆ Now all this wood burning caused a scarcity of wood in Europe, a kind of energy crisis. Coal was an obvious alternative since it, like charcoal, consists primarily of carbon. But coal wasn't any good for smelting metals because it contains impurities like sulfur which weaken metals. It wasn't until 1603 that a fellow named Hugh Platt tried heating coal anaerobically, driving off bejeezical gases and producing purified carbon in the form of ***coke.*** Coke was a big hit with steel-makers, but it would be another two hundred years before anyone got the bright idea to do something with the gas from the coke ovens.

∇ That's where ***Phillipe Lebon*** comes in. See, he grew up in a charcoal-producing region of France and was in*spired* to use the gas from the charcoal ovens for illumination. He got a patent in 1799 for his "Thermolamp" and even thought up an engine to run off of gas; he was way ahead of his time. Who knows what he would have come up with if he hadn't been stabbed to death in 1804?

∆ Something quite remarkable, no doubt, but returning to events that actually materialized, ***William Murdock*** began experimenting with gas from the coke ovens, lighting his workshops in 1792. When news of Lebon's work reached Murdock's employer, Matthew Boulton, they pushed forward with commercial development of gas lighting. Murdock described the basic principles of gas lighting to the Royal Society in 1808.

∆ The recovery of gas from a coke oven may be viewed as a kind of distillation. In the distillation of Chapter 16 a mixture is simply separated into its components, for example, a solution of alcohol and water is separated into alcohol and water. Turn up the temperature, however, and the material you are distilling may start to decompose. Wood, for example, may decompose into charcoal and water. In such a ***destructive distillation*** the materials recovered were not present at the beginning, but were produced during the distillation. The destructive distillation of coal produces four products which are easily separated from one another; coke remains in the pot; gas driven from the pot is passed through a condenser where some of it condenses into an aqueous solution of ammonium carbonate; separating from this solution is a tar which is insoluble in water; the remaining gas is ready for distribution as illuminating gas.

∀ But like I said at the beginning, this chapter is not so much about the gas; it's about the tar. That tar is a complicated mixture of compounds and folks started to wonder what they were and what could be done with them. In 1825 ***Michael Faraday*** isolated benzene from the destructive distillation of whale oil. Pretty soon everybody and *her* dog were destructively distilling things to see what would come of it. In 1826 ***Otto Unverdorben*** isolated aniline from indigo and in 1834 ***Friedlieb Runge*** isolated aniline and phenol from coal tar. So with a second, non-destructive distillation folks started producing chemicals from coal tar, chemicals like benzene, toluene, aniline, phenol, and naphthalene. Folks figured that since these compounds came from the destruction of organic compounds, maybe it would be possible to make good and useful things by putting them back together again.

∆ Gas had become a major industry—

∀ —but this chapter is more about tar. See, at that time malaria was a problem for the British Empire, on account of its rampant colonialism, and the best treatment for malaria was quinine, extracted from the bark of a South American tree. To promote chemical innovation the British started up the Royal College of Chemistry in 1845, with the great German chemist, ***Wilhelm Hofmann,*** as director. In 1856 Hofmann's assistant, ***William Perkin,*** set out to synthesize quinine from coal tar. Perkin reacted aniline with potassium dichromate, a really strong oxidizing agent; the resulting black goo was definitely not quinine, but it made a beautiful purple solution in alcohol. Perkin called it "mauveine" and dropped out of college at the age of 18 to develop his new synthetic dye. British dyers didn't think that mauve would catch on, but the Paris fashion houses liked it so much that Perkin was able to retire at the age of 36. I told you it was a great business.[2]

∆ And—

∀ —once folks knew that it was possible to make artificial dyes from coal tar, new dyes started coming out of the gasworks, so to speak. Variations on the original synthesis produced dozens of dyes from aniline: aniline reds, aniline violets, aniline greens, yellows, browns, and blues. Substituting phenol or naphthalene for aniline produced two more distinct families of artificial colors. There seemed to be few colors which could not be fashioned by art and ingenuity from coal tar.

2. For a popular account of Perkin's career, see Reference [59]. For the early history of the synthetic dye industry, see Reference [74].

🜃 Hofmann's leadership at the Royal College had given the British a head start in synthetic dyes, but when Hofmann returned to Germany in 1865 British dyestuffs dropped the ball. In France, synthetic dye manufacturers charged so much for their products that demand there shifted to natural and imported dyes. But in Germany, the situation was ideal for the development of these new wonder dyes. State-subsidized technical schools fed talented students to the universities; the universities promoted collaborative work, sending students to study in Britain and France; and universities cooperated with industry on joint research projects. Furthermore, with thirty-nine states in the German Confederation, patent enforcement was a problem and the resulting competition produced lean, mean, dye-making machines. German unification in 1871 made patent enforcement easier, but by that time the big players in German dyestuffs were off and running; *Badische Anilin und Soda Fabrik* (BASF) was out of the gate in 1861, followed by *Farbwerke Hoechst* in 1862; *Freidrich Bayer* rounded the bend 1863, followed by *Kalle & Co* in 1864; and a late starter from 1867 was reorganized in 1873 as *Aktien Gesselschaft für Anilin Fabrikation* (AGFA). Throughout a century of wars, depressions, revolutions, and mergers Bayer, Hoechst, and BASF were the top dogs of the chemical industry.

🜃 I guess you could say, "They were buy-y-ing a ta-ar-way to heaven."

22.2 🜍

Tar happens. It happens in a fireplace; as the wood is heated it begins to break down into charcoal and a kind of greasy steam which condenses in the chimney. That greasy steam is tar. Tar happens in a cigarette; as the tobacco is heated, bejeesical nicotine passes through the filter. The stuff trapped by the filter is tar. Tar happens when the fat from your barbecued chicken drips on the hot coals; the fat decomposes and tar is deposited on the inside lid of the barbecue grill. Tar happens whenever things burn or scorch, releasing vapors which later condense somewhere else. If they condense in your nose, you smell them. Tars make your clothes smell of cigarettes after you leave the bar. Tars alert you that the bacon is burning. Tars make up the aroma of chestnuts roasting on an open fire. Tar is not a single substance; it's just a fancy catch-all name for all the various substances produced by heating things until they decompose. The fancy name for heating things until they decompose is ***destructive distillation.***

Figure 22-1. Indigotin

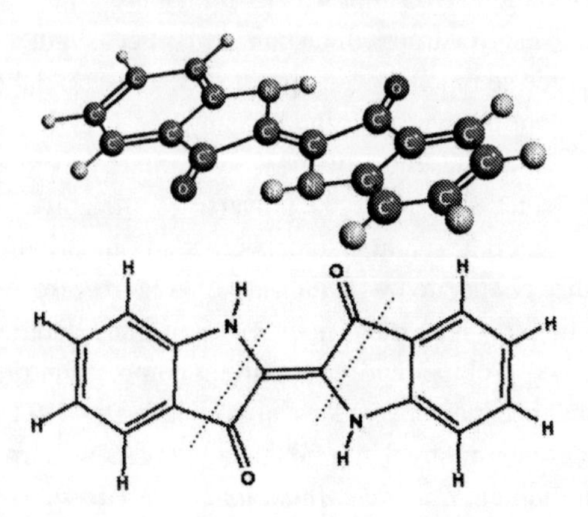

Destructive distillation was all the rage in the 1820's. In 1826 Otto Unverdorben destructively distilled indigotin, the blue dye we met in Chapter 12. You probably just got used to molecular formulas like $C_{16}H_{10}N_2O_2$ and empirical formulas like C_8H_5NO when Chapter 16 went and introduced the ***molecular model*** as a way of showing how the atoms in a molecule are hooked together in 3D. Figure 22-1 shows a molecular model for indigotin. Molecular models are great when you have a wooden or plastic model or a computer program to draw them, but when you need a quick 2D sketch of how the atoms are connected, chemists use a ***structural formula,*** like the one shown at the bottom of Figure 22-1. To understand structural formulas you need to realize that carbon atoms generally make four bonds. So whenever four lines come together in a structural formula, that point stands for a carbon atom, even if no "C" is shown. A curious feature of indigotin is that it contains two six-member carbon rings with alternating single and double bonds. Those rings are unusually strong on account of the double bonds and so when indigotin decomposes, those rings are likely to survive in the decomposition products. You can imagine lots of ways for the indigotin molecule to fall apart, but the bonds cut by the dotted lines in Figure 22-1 are among the most likely to break during destructive distillation. When that happens, the two hunks containing the carbon rings turn into two molecules of aniline, the stuff Unverdorben recovered in 1826.

What a surprise it must have been when Friedlieb Runge isolated this same aniline from the destructive distillation of coal. Granted, most of the coal turns into coke, just as wood turns mostly into charcoal. In addition, several gases are produced—hydrogen, methane, and carbon monoxide. Water and ammonia are also produced. But the distinctive aroma of burning coal comes from the mixture of compounds which constitute coal tar, those volatile components which become gases at elevated temperatures but which condense into liquids and solids as they cool back to room temperature. The strong smell of these coal-tar products has given a name to a whole class of organic compounds: the ***aromatic*** compounds. One hundred kg of coal will yield approximately 3 kg of tar; distillation of that tar will produce about 1 kg of aromatic compounds.

You might think that *aromatic* means *smelly,* and surely that was the original meaning. But as time when on, it turned out that the aromatic coal tar distillates all shared a common structural feature, that curious six-membered carbon ring. As more and more compounds containing this ring were identified, it turned out that not all of them were smelly. But the name *aromatic* stuck fast like a spider. The meaning of the term is a little more technical these days, but for our purposes, an aromatic compound is one which contains the six-membered, alternately single and double-bonded carbon ring, the *aromatic* ring.

Each carbon atom of an aromatic ring can be bonded to one other atom. When all six carbon atoms are bonded to hydrogen atoms, we have a molecule of ***benzene,*** shown in Figure 22-2(L). When I was in college benzene was a common laboratory chemical, dry-cleaning solvent, and paint thinner. Benzene is not particularly toxic by ingestion, with an LD_{50} of 3.8 g/kg in the rat. Because of its low boiling point, inhalation is more likely to be a problem than ingestion, but with adequate ventilation acute exposure to benzene is no more hazardous than exposure to many solvents currently available at hardware stores. But in addition to the risk of acute exposure, there is a danger of chronic exposure; occupational exposure to benzene is associated with an increased risk of cancer. Consequently, benzene has been replaced by toluene for household and many commercial uses.

Toluene, shown in the middle of Figure 22-2, is familiar to most people as the solvent in model airplane glue. You might think of it as a benzene molecule with an extra carbon atom attached. As with benzene, inhalation is the most important hazard but with adequate ventilation it can be handled safely. Unlike benzene, chronic exposure to toluene is not as-

Figure 22-2. Benzene, Toluene, Aniline

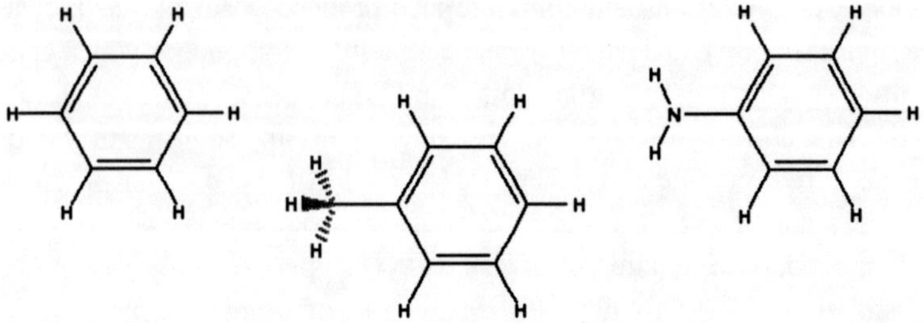

sociated with cancer and toluene can still be found in hardware stores among the paint thinners.

Aniline, shown on the right of Figure 22-2, is the stuff originally produced by the destructive distillation of indigotin. You might think of it as a benzene molecule with an ammonia molecule attached. Aniline is an oily brown liquid only slightly soluble in water. Though aniline is the compound needed for the synthesis of mauveine, indigo is too expensive to be used as a commercial source and coal tar doesn't contain enough of it to be useful. Since benzene is the most abundant coal-tar aromatic it is natural to look for a way of making aniline from benzene. This synthesis occurs in two steps; first benzene is oxidized by nitric acid:

$$C_6H_6 + HNO_3 = C_6H_5NO_2 + H_2O$$

Second, the nitrobenzene produced in the first reaction is reduced by iron filings:

$$C_6H_5NO_2 + H_2O + 2\ Fe = C_6H_5NH_2 + Fe_2O_3$$

Similar reactions produce ***toluidine,*** $C_7H_8NH_2$, from toluene. There is a hitch, however, in that the NH_2 might appear in any of three positions on the ring. Consequently, there are three "flavors," or ***isomers*** of toluidine: ortho-, meta-, and para-toluidine, shown in Figure 22-3. Each of these isomers has its own boiling point and other physical properties. Oxidation of toluene with nitric acid and subsequent reduction with iron yield predominantly ortho-toluidine and para-toluidine. Early coal-tar distillation did not completely separate benzene from toluene and consequently commercial "aniline" contained significant quantities of the toluidines. Just as organic molecules which contain the OH group are classified as alcohols, those containing the NH_2 group are classified as ***amines.*** The

Figure 22-3. The Toluidines

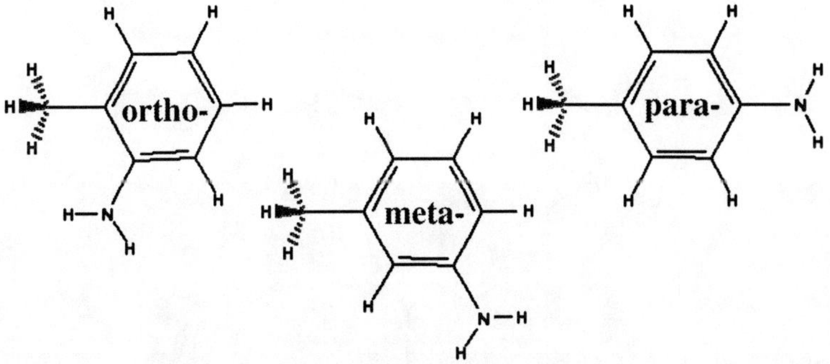

aromatic amines, aniline and the toluidines, are the raw materials for the synthesis of the dye, mauveine.

When you oxidize pure aniline in an acidic solution, the product is the dye pseudomauveine:

$4\ C_6H_5NH_2 = C_{24}H_{19}N_4^+ + 9\ H^+ + 10\ e^-$

The remnants of the four original aniline molecules are evident in the structure of pseudomauveine, as shown in Figure 22-4. The "R's" in the structure are simply hydrogen atoms, for the moment. Note that pseudomauveine is a cation and so there must be an anion in the solution to balance it. If the reaction took place in acetic acid, the result is pseudomauveine acetate; if it took place in hydrochloric acid, the result is pseudomauveine chloride, *etc.* No matter which acid was used to make it, pseudomauveine is a brown dye, not very soluble in water but quite soluble in ethanol. Brown dyes are neither very rare nor much sought after and if Perkin had used pure aniline in his attempted synthesis of quinine, he would not have been much impressed with the results.

Perkin's aniline was not pure, however; it contained o-toluidine and p-toluidine. With these other amines in the mix, the "R's" in Figure 22-4 might turn out to be either hydrogen atoms or CH_3 groups, depending on the mix of aniline and the toluidines. With CH_3 groups at Ra and Rc, two slightly different purple dyes result, one (mauveine A) with Rb = H and the other (mauveine B) with Rb = CH_3. Like pseudomauveine, the mauveines are not very soluble in water but are quite soluble in ethanol. Unlike pseudomauveine, the mauveines are purple, a color rare among

Figure 22-4. Pseudomauveine and Mauveine

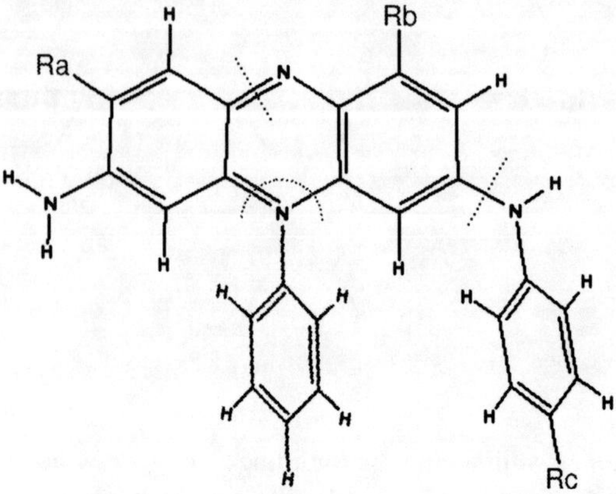

natural dyes and prized as a status symbol. If you had found a way to make such a color in a world of browns and blues and yellows and tans, you would very likely have dropped out of school as Perkin did. And that color would have made you as wealthy as the folks already wearing purple clothes.

Material Safety

Locate MSDS's for ethanol (CAS 64-17-5), aniline (CAS 62-53-3), o-toluidine (CAS 95-53-4), p-toluidine (CAS 106-49-0), acetic acid (CAS 64-19-7), and sodium hypochlorite (CAS 7681-52-9). Summarize the hazardous properties in your notebook, including the identity of the company which produced each MSDS and the NFPA diamond for each material.[3]

In the course of this project you will combine vinegar, aniline, and laundry bleach to produce a solution of dye in ethanol. By themselves, vinegar and bleach would release chlorine gas, but you will be using only a few drops of bleach and the liberated chlorine will immediately oxidize the aniline; you are more likely to smell chlorine from the bleach itself than from your dye solution. Be aware that sodium hypochlorite will bleach clothing.

3. The NFPA diamond was introduced in Section 15.2 (page 184). You may substitute HMIS or Saf-T-Data ratings at your convenience.

You should wear safety glasses and rubber gloves while working on this project. All activities should be performed in a fume hood or with adequate ventilation. Leftover aniline and toluidine should be converted to dye as described below. Leftover dye should be washed down the drain with plenty of water unless prohibited by law.

Research and Development

Before you get started, you should know this stuff.

- You better know all the words that are important enough to be ***indexified*** and **glossarated.**
- You should know all of the Research and Development stuff from Chapter 18 and Chapter 19.
- You should know the structures and properties of the coal-tar distillates: benzene, toluene, and aniline, as shown in Figure 22-2.
- You should know the structures of aniline and o-, m-, and p-toluidine as shown in Figure 22-3.
- You should be able to recognize the remnants of aniline and toluidine in mauveine and pseudomauveine, as shown in Figure 22-4.
- You should know the hazardous properties of ethanol, acetic acid, aniline, o-toluidine, p-toluidine, and sodium hypochlorite.
- You should know who started the synthetic dye industry, when he did so, and which modern companies were founded as dye companies.

22.3 Θ

It has been fun researching this book and trying to figure out how to make things from scratch. Sometimes I have gained inspiration from modern laboratory manuals, but more often than not, the projects have developed by trying to imagine what commonly-available materials could be used to make something appropriate to each chapter; this chapter was one of the last ones to "come together." At first I imagined that I would make mauve as Perkin had done, but the more I learned about Perkin's synthesis, the more discouraged I became. Perkin dissolved his aniline in aqueous sulfuric acid. Oxidation of this solution with potassium dichromate produced an insoluble brown goo which would have to be filtered, and filtering goo is no picnic. Once filtered, the goo would be extracted with ethanol, which would dissolve the dye. Perkin's yield was on the order of 5%, that is, 5% dye, 95% useless goo. Could caveman chemists be expected to make such a dye when even experts got such poor yields? Furthermore, chromium compounds pose a health risk and a disposal problem I was not sure I wanted to impose on beginners. Finally, neither aniline, nor toluidine, nor potassium dichromate is a household item. It seemed that there was much to be said against mauve as a project.

Nevertheless, I wondered whether it might be possible to substitute a household oxidizing agent for potassium dichromate. Wagner's Chemical Technology of 1872[4] reported that bleaching powder, calcium hypochlorite, had been used as an alternative to potassium dichromate. I wondered whether sodium hypochlorite, modern laundry bleach, might also do the trick.

I also wondered whether the goo-filtering step might be avoided by performing the oxidation in ethanol rather than aqueous solution. I expected that the goo might settle to the bottom of the container and the ethanolic dye might simply be poured off. My very first try using laundry bleach in ethanol took my breath away; it resulted in a rich purple dye and *no* goo at all. Not only did the household chemical work, it apparently worked better than any combination I had found in the chemical literature! But it was when I applied the dye to some silk cloth that I knew how Perkin must have felt.

To begin with, you need some aniline. If you buy aniline from a chemical supply (page 387) it will likely be of higher purity than you need. Pure aniline will produce the brown pseudomauveine which, while it is a perfectly good brown dye, is unlikely to have the visual impact of mauveine. To make mauveine you need aniline contaminated with o-toluidine and p-toluidine. The optimum proportions are one mole aniline, two moles o-toluidine, and one mole p-toluidine. Since these three compounds have similar molecular weights and densities, you can make it up as 1 gram of aniline, two of o-toluidine, and one of p-toluidine. I will refer to this mixture as "aniline oil." The four grams just described will make a *lot* of dye; in a classroom situation, this amount of aniline oil will be enough for about a hundred students.

---

4. Reference [29], p. 577.

You will make two solutions. First measure one mL of white household vinegar using a graduated pipette or a graduated cylinder. Place the vinegar into a small vial and add 1 drop of aniline oil to it. The oily aniline will dissolve completely within a minute or two. Now measure 5 mL of ethanol into a graduated cylinder and add 10 drops of bleach to it. Add the ethanol-bleach solution to the vinegar-aniline solution and put the cap on the vial. The combined solution should immediately turn brown as the aromatic amines are oxidized. The brown color will turn blue in a few minutes and then purple in an hour or so. This vial of dye is sufficient for dyeing a 12-inch silk handkerchief.

Is your dye a single compound or a mixture of compounds? To answer this question we'll use a technique invented by Otto Unverdorben for precisely this purpose: paper chromatography. You've probably seen water-soluble ink bleed when a piece of paper gets wet. If the ink is a mixture and if the mixture's components have different solubilities, then the more soluble components bleed faster than the less soluble ones. As the water spreads out on the paper, the more soluble ink components separate from the less soluble ones. This is the main idea behind chromatography. Since our dye is not soluble in water we'll use a solution of vinegar and ethanol as a solvent.

To make your chromatograms you'll need a piece of filter paper (90 mm in diameter), a pencil, a glass capillary tube (1 mm inside diameter, open both ends), a glass jar with a lid, and a solvent consisting of 75% white vinegar and 25% ethanol. Begin by folding the filter paper lengthwise into quarters and snipping off the ends with a pair of scissors. The resulting paper will be able to stand up, as shown in Figure 22-5(L). With a *pencil*[5] make a dot 1 cm from the bottom of the paper. Your goal is to place as much dye onto the smallest spot possible. To this end, dip the capillary tube into your dye; dye will climb into the tube by capillary action. Practice using the capillary tube by touching it briefly to a piece of paper towel. Some of the dye will be transferred to the towel, forming a spot. Practice making the smallest possible spots. Once you have some confidence, draw up some more dye into the capillary tube and spot the paper towel until a 1 cm column of dye remains in the tube. You will now deliver that 1 cm column of dye onto the pencil dot on your filter paper, as shown in Figure 22-5(R). Place the first spot onto the pencil mark and count to thirty, allowing the spot to dry. Place a second spot

5. A mark made with a pen would bleed.

Figure 22-5. Folding and Spotting the Chromatogram

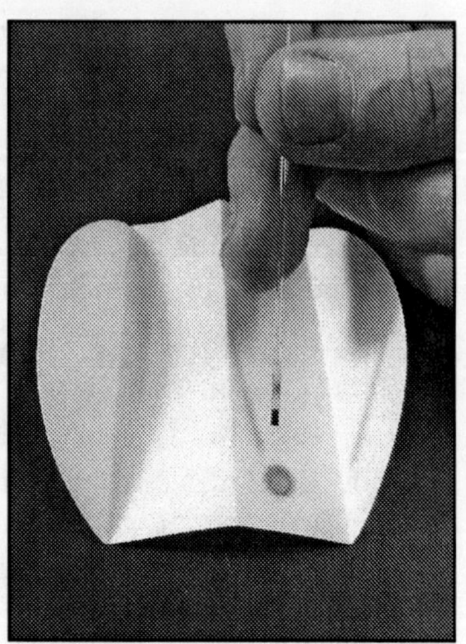

on top of the first and count to thirty. Continue to build this spot until you have delivered the entire 1 cm column of dye onto the filter paper. If you do a good job this "young" dye spot will be no more than 5 mm in diameter. If you mess up, just spot another paper until you get it right. You will now *develop* your chromatogram using a solvent composed of vinegar and ethanol.

Mix 9 mL of white vinegar and 3 mL of ethanol in a glass jar large enough to contain your filter paper. Screw the lid onto the jar and shake it to thoroughly wet the walls of the jar. We want the air in the jar to be saturated with solvent so that the solvent doesn't evaporate as it climbs the paper. Open the jar and place your filter paper inside, standing it on the end with the spot and taking care that the paper doesn't touch the walls of the jar. Then replace the lid. The vinegar/ethanol solvent will climb the filter paper, taking the dye with it. Your chromatogram will take about an hour to develop, as shown in Figure 22-6(L).

When the solvent has climbed to within 5 mm of the top, remove your filter paper from the jar and draw a pencil line where the solvent stops. Your chromatogram will look something like Figure 22-6(R), which is shown in color on the back cover of this book (page 381). The chromatogram on the right is for a dye which has been allowed to mature for a week.

Figure 22-6. Developing and Interpreting the Chromatogram

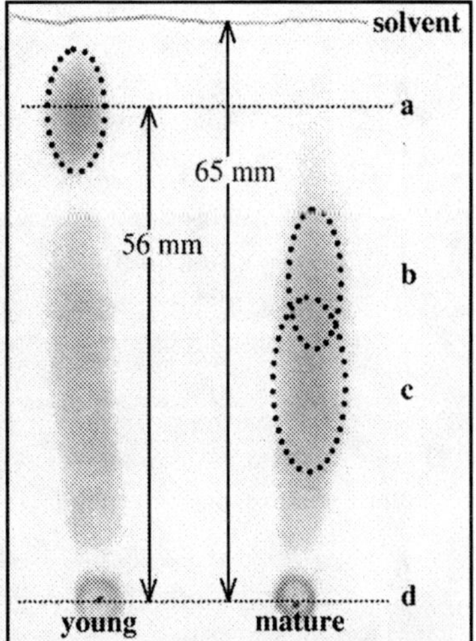

At least four substances are evident in these chromatograms. Spot *a* was present in the young dye but not in the mature one. It represents a precursor, a combination of two or perhaps three aromatic amines. Spots *b* and *c* overlap but represent at least two different mauve dyes with slightly different colors. Spot *d* is soluble in the highly ethanolic dye solvent, but not in the less ethanolic chromatographic solvent. Consequently, it has not budged from the original spot.

Chemists characterize chromatographic spots by comparing the distance traveled by each one to that traveled by the solvent. Measure the distance from the original pencil dot to the center of each spot and to the solvent line. Divide the distance for each spot by that for the solvent to get what is called the R_f value (ratio of fronts). Spot *a* in the chromatogram shown has an R_f value of 56/65 = 0.86. Spot *d* has an R_f of 0.

By the time your chromatogram develops, your dye is mature enough to dye your silk handkerchief.[6] Push the entire handkerchief into your vial of dye and once it has taken up the dye, fish it out again. Spread it out on a few sheets of newspaper and allow it to dry. Be careful not to drip dye

6. If you wish, you can hold back some of your dye and repeat the chromatography after it has matured for a week.

where it ought not to be. Once it's completely dry, you can wash your handkerchief with soap and water. Wow, is that a pretty color!

Quality Assurance

Your dye is not so great if the color isn't fast. If the color hasn't washed out, tape your mauve handkerchief and your chromatogram into your notebook as proof of your dye-making prowess. Make a table of R_f values for your dye spots and explain how many substances were present in your young dye.

In this way was the World created. From this there will be amazing applications because this is the pattern.

— *The Emerald Tablet of Hermes Trismegistos*

Chapter 23. Eastman (Photography)

It has already been observed, that when a solar spectrum falls upon paper covered over with chloride of silver, the chloride turns black in the more refrangible regions, and from this and similar experiments we have been lead to the knowledge that there exists in the sunbeam a principle which can bring about chemical changes.

This fact has been perceived from the beginning of the present century; but, of late, much attention has been given to these rays, and from a consideration of the phenomena they exhibit, I have endeavored to prove that they constitute a fourth imponderable principle of the same rank as heat, light, and electricity; and for the purpose of giving precision to this view, have proposed that they be called Tithonic rays, from the circumstance that they are always associated with light; drawing the allusion from the classical fable of Tithonus and Aurora. . . .

The process of the Daguerrotype depends on the action of the Tithonic rays. It is conducted as follows: A piece of silver plate is brought to a high polish by rubbing it with powders, such as Tripoli and rotten-stone, every care being taken that the surface shall be absolutely pure and clean, a condition obtained in various ways by different artists, as by the aid of alcohol, dilute nitric acid, &c. This plate is next exposed in a box to the vapor which rises from iodine at common temperatures, until it has acquired a golden yellow tarnish; it is next exposed, in the camera obscura, to the images of the objects it is designed to copy, for a suitable space of time. On being removed from the instrument, nothing is visible upon it; but on exposure to the fumes of mercury, the images slowly evolve themselves.

To prevent any further change, the tarnished aspect of the plate is removed by washing the plate in a solution of hyposulfite of soda, and finishing the washing with water; it can then be kept for any length of time. . . .

There are many other photogenetic processes now known; several have been invented by Mr. Talbot; among them may be mentioned the calotype. Sir J. Herschel, also, has discovered very beautiful ones, and these possess the great advantage over Daguerre's that they yield pictures upon paper. In minuteness of effect they can not, however, be compared to the Daguerrotype.

— John Draper, *Textbook on Chemistry*, 1850 AD[1]

---

1. Reference [11], pp. 90-93.

23.1 ☿

Δ Well, where is the Figment? This *is* his chapter, is it not?

🜃 Looks like he's out of the picture.

∇ When I look back on all the crap he's had from you two, it's a wonder he shows up at all. I guess your lack of admiration must have hurt him some; he could see the writing on the wall.

🜃 Speaking of writing on the wall, I guess that photography really started with the invention of the ***camera obscura,*** which is basically just a dark little room with a hole at one end. Light from outside comes in through that hole and projects an image of the outside scenery on the wall. Folks had been using the camera obscura since the sixteenth century, tracing the projected image onto paper with a pencil.[2]

Δ All well and good, but you cannot consider a pencil drawing as a photograph. More fundamental to the birth of photography was my discovery in 1725, in the person of Johann Schulze, that silver nitrate darkens upon exposure to light. In modern photography it is from exposed silver salts that the photographic image is ultimately formed.

∇ Formed, yes, as long as this image remains in the dark of the camera. But bring it out into the daylight and the whole shebang turns black. You are probably wondering how this image may be "fixed" permanently. I will tell you. In 1819 ***John Herschel*** discovered that aqueous sodium hyposulfite dissolves silver salts, most of which are otherwise insoluble in water. So by soaking your image in such a solution, the unexposed silver salts are washed away while the exposed areas remain. *Voila!* A permanent photographic image.

🜃 Now hold on there, cowboy. That's a heck of a theory, but it's still not a photo. Seems to me that the prize for first permanent photo goes to Nicéphore Niépce for his 1826 snapshot of the view outside his window.

Δ Snapshot, indeed! The exposure took eight hours.

🜃 So? Anyhow, in 1829 Niépce hooked up with ***Louis Jacques Daguerre,*** who was wowing theater audiences with his elaborate special effects. Daguerre's "Diorama" involved projecting light through huge semi-transparent paintings, the motion of the lights and the paintings set to music. Daguerre used the camera obscura to sketch his huge paintings

2. For a general introduction to the history of photography, see Reference [70].

and looked to Niépce for in*spir*ation on eliminating the tedious sketching. Niépce died in 1833, but not before Daguerre had caught the "photo bug." In 1837 he succeeded in producing a photo on silver-plated copper, developing the image with mercury vapor and fixing it with a solution of ordinary salt. All with an exposure under 20 minutes.

Δ Impressive to him at the time, no doubt, but still not a photograph as we understand the term today. Both Niépce and Daguerre produced images on opaque metal plates. With such a system there is no way to copy the image, no way to enlarge it, no way to project it onto a screen. Without duplication and projection the I-dea of the motion picture would not have materialized; without motion pictures, no television; without television, no computer monitors. The metal-plate image is simply a fluke, an evolutionary dead end. I produced the first *true* photograph in 1835 in the person of ***Fox Talbot.*** This 10-minute exposure on translucent paper of the view through the windows of Lacock Abbey resulted in an image which could be projected, enlarged, and reproduced; all fundamental properties of photographs as we know them today.

∇ An image which, like Daguerre's of 1837, deteriorated over time because silver salts are generally insoluble in water. In both instances the residual silver salts darkened with time, spoiling the original image. What can be a more important property of a photograph than that it should be permanent? I must remind you that Herschel discovered the solubility of silver salts in sodium hyposulfite in 1819 and I must further point out that he informed both Talbot and Daguerre of this in 1839, marking the beginning of true photography.

Δ A step forward, to be sure.

∀ It *was* a hum-dinger. Thanks. But to get back to the story, these Talbot prints were pretty fuzzy, as opposed to *daguerreotypes,* which were sharp as a tack. And what's more, talbotypes were funky, in that the parts of the photo which ought to be black were white and the parts that ought to be white were black. When you think about it, the talbotype image had everything except what was supposed to be there. In the daguerreotype, the parts of the photo which ought to be white were shiny silver and the parts that ought to be black were dark copper. Awesome! Napoleon III liked the daguerreotype so much that he gave Daguerre a nice pension and released the process to the public as a benefit to the human race.

Δ Except in England; you were careful to patent the daguerreotype in Talbot's homeland. And for your information, the "funkiness" you complain about is called a *negative image.* It is a feature, not a problem. When you expose a second paper through the negative, the resulting print is a positive. My further research revealed that the image forms even before it is visible and that this latent image can be *developed* using gallic acid. I patented as this *calotype* process in 1842 and was able to use exposure times of 30 seconds, making portraiture possible.

🜃 That is, if you like fuzzy, funkified portraits. Plus, you had to pay for a license to produce calotypes, whereas daguerreotypes were patent-free (except in England). It just sounds too complicated to me, what with invisible negative latent images and all. Most folks agreed, and the daguerreotype business really took off, leaving the calotype in the dust.

🜄 Of course, the fuzziness of the calotype stemmed from the fact that the image penetrated the paper. A sharper negative image would result if it were produced on a glass plate rather than on paper, a suggestion Herschel made in 1839.

🜂 I was getting around to that; as Frederick Archer I formed an emulsion of silver salts and ***collodion*** on a glass plate in 1850. This thin emulsion enabled the production of sharp negatives. And in that same year, in the person of Louis Blanquart-Evrard I applied a similar technique to paper prints; I mixed silver salts with egg whites, forming a thin emulsion which coated only the surface of the paper. This "albumin paper," in combination with the collodion-glass negative, produced prints as sharp as the daguerreotype. With no way to duplicate or project daguerreotypes, it was only a matter of time before they went the way of the gaslight.

🜃 Well, it was a good business while it lasted. Eventually everybody and *his* dog had a *something*type process named after him, what with little improvements here and there. The daguerreotype had made photography practical enough that pretty much anybody who wanted a picture could have one taken. But even the new-fangled *something*types had to be developed right after exposure, so unless you had a photographic darkroom, you couldn't take your *own* pictures. Then in 1889 a fellow named ***George Eastman,*** in the practical down-to-earth spirit of Daguerre, introduced celluloid roll film and a simple camera, the Kodak. For the first time you could take a whole roll of pictures and then send your film off to be developed and printed. From then on, it was "Momma don't take my Kodachrome away."

23.2 🜍

The photographic process is fundamentally a redox process. While the redox chemistries of iron and chromium have been explored from time to time, most photography, for most of its history, has been based on silver. It is most likely that every photograph you have ever seen has been a silver photograph. We saw in Chapter 9 that silver is primarily a by-product of lead smelting, in whose ore, galena, it is a common contaminant. The production of photo-chemicals begins with the oxidation of metallic silver with nitric acid:

$$\mathrm{Ag}(s) + 2\,\mathrm{HNO_3}(l) = \mathrm{AgNO_3}(aq) + \mathrm{H_2O}(l) + \mathrm{NO_2}(g)$$

Silver nitrate is, of course, soluble in water. To make a photographic emulsion we need an insoluble salt, one which will not leech out of the emulsion into the paper. You will recall from Chapter 7 that silver chloride is insoluble in water:

$$\mathrm{AgNO_3}(aq) + \mathrm{NaCl}(aq) = \mathrm{AgCl}(s) + \mathrm{NaNO_3}(aq)$$

Silver bromide and silver iodide are also insoluble in water and these salts are more light-sensitive than the chloride. Consequently they are the silver salts of choice for modern photographic emulsions. But because sodium chloride, ordinary table salt, is easier for the amateur to obtain, we shall use it in the production of our emulsion.

If our paper were simply saturated with silver chloride, the resulting image would penetrate the paper resulting in a fuzzy image. To demonstrate this, write on a paper towel with a pencil; notice that the pencil line is sharp and that it does not penetrate or bleed into the paper. By contrast, write on a paper towel with an ink pen; notice that the ink soaks into the paper resulting in a fuzzy line. Writing paper is ***sized,*** that is, it is treated with substances to prevent ink from penetrating the paper. We shall use a similar trick to ensure that our silver chloride resides only at the surface of the paper.

To complete the photographic emulsion we need something which we can apply to the paper but which will be impervious to water once it is dry. This substance should bind well to the paper and remain flexible after processing. Collodion and gelatin have been used for this purpose, but in the spirit of using as many household substances as possible, let us choose egg whites. Egg whites contain the protein ***albumin,*** which is fairly impervious to water once it is dry, an observation which will be confirmed by anyone who has tried to wash dried eggs from the side of a house. Briefly, a mixture of sodium chloride in egg whites will be applied to paper and allowed to dry. Once dry, the coated paper will be sensitized with silver nitrate solution. Silver nitrate will react with sodium chloride to produce insoluble silver chloride which will reside only in the emulsion because the dried albumin will not allow it to soak into the paper. The sensitized paper must be stored in the dark until it is used.

Now, silver ion is an oxidizing agent, that is, it is easily reduced to metallic, elemental silver. Albumin is a mild reducing agent, that is, it is easily oxidized. Place an oxidizing agent with a reducing agent and you are just asking for a redox reaction to happen. Given a day or two, the silver ion will be reduced to metallic silver:

$$\mathrm{Ag}^{+}(aq) = \mathrm{Ag}(s) + \mathrm{e}^{-}$$

As the silver is reduced the albumin will be oxidized to, well, to whatever it is oxidized to. Remember, proteins are polymers of amino acids; they form large, complex molecules and because the oxidation of the albumin has little to do with the resulting image, let us leave it at that and concentrate on the silver.

In the sensitized paper silver ion will be slowly reduced to metallic silver. You might expect, then, that after a few days the sensitized paper would look like a paper mirror. But you probably know from experience that powdered metals generally look black; take a file to any piece of metal and the filings will be dark gray or black. Silver is no exception. The reduced silver in the paper is not one big piece of silver to be polished to a mirror finish; it is in the form of tiny grains of silver. As a consequence, a sensitized paper left to its own devices for a few days will turn dark brown, gray, or even black.

Now for the magic! When light shines on the silver chloride/albumin emulsion the redox reaction happens more quickly than it does in the dark. Imagine now a sensitized sheet of paper, half of which is exposed to bright sunlight and the other half of which is covered up with an opaque card. The silver ions in the exposed area will be reduced to black metallic silver in a matter of minutes; the silver ion under the card will remain colorless or white. Imagine now that you remove the opaque card; what will you see? The half of the paper that was in the light will be black and the half that was in the dark will be white. This reversal of light and dark is referred to as a ***negative*** image.

But as soon as you remove the opaque card from the sensitized paper, the formerly unexposed white half will begin to turn black. In order to fix the image, we need to remove the light-sensitive silver chloride. Just washing it in water won't do the trick because silver chloride is insoluble in water. The earliest photographic fixer consisted simply of a concentrated solution of sodium chloride, ordinary table salt:

$$\mathrm{AgCl}(s) + \mathrm{NaCl}(aq) = \mathrm{AgCl_2^-}(aq) + \mathrm{Na^+}(aq)$$

While certainly convenient, this reaction does not go very far; only some of the silver chloride is dissolved and the rest remains on the paper. More effective than salt is ***ammonia:***

$$\mathrm{AgCl}(s) + 2\,\mathrm{NH_3}(aq) = \mathrm{Ag(NH_3)_2^+}(aq) + \mathrm{Cl^-}(aq)$$

If you are looking for a household chemical to use as a photographic fixer, clear household ammonia will work better than salt. But far more effective is a specialty chemical which has been synonymous with photographic fixer since 1839: ***sodium hyposulfite,*** known to photographers as "hypo," or, by its modern chemical name as ***sodium thiosulfate:***

$$\mathrm{AgCl}(s) + 2\,\mathrm{Na_2S_2O_3}(aq) = \mathrm{Ag(S_2O_3)_2^{3-}}(aq) + \mathrm{Cl^-}(aq) + 4\,\mathrm{Na^+}$$

By rinsing the sensitized paper with sodium thiosulfate, light-sensitive silver chloride goes into solution leaving the unexposed areas of the print white and the exposed areas black.

The image so far is a negative image. To produce a positive print, the process is repeated with a second piece of sensitized paper. The negative is placed on top of this second piece of paper and exposed to sunlight. The black areas of the negative mask the sunlight and the underlying areas of the print remain white. The white areas of the negative allow sunlight to penetrate and the underlying areas of the print turn black. Once fixed, the print bears a positive image, one in which white and black are as they should be.

The silver chloride/albumin combination produces what is known as a "Printing Out Paper," one in which the image becomes visible as the exposure proceeds. Such papers are slow, requiring minutes of exposure to bright light for the image to form. In a "Developing Out Paper," the binder used to form the emulsion is not a reducing agent, giving the unexposed paper a longer shelf life. When such a paper is exposed to light, no image is immediately apparent. The image forms when the exposed paper is placed in a developing bath, a solution of a relatively strong reducing agent. Exposed areas of the print are quickly reduced by the developer, turning black. Before the unexposed areas are reduced, the print is placed in a "stop bath," which destroys any developer remaining on the print. The print is then fixed in the usual fashion.

Photography continued to develop in the twentieth century. Color photography came into its own with the introduction of Kodachrome in 1935. In 1948 Edwin Land introduced a film which incorporated pouches of developer and fixer. When pulled from his "Polaroid" camera, the packets were squeezed onto the film resulting in a self-developing print; color Polaroid film was introduced in 1962. The twenty-first century has seen the maturation of the electronic, film-less camera. In fact, all of the photographs for this book were taken electronically. But the vast majority of photographs at the time of writing continue to be based on the fundamental principles of silver reduction first explored by Herschel, Daguerre, and Talbot.

Material Safety

Locate MSDS's for silver nitrate (CAS 7761-88-8), sodium chloride (CAS 7647-14-5), sodium carbonate (CAS 497-19-8), acetic acid (CAS 64-19-7), sodium thiosulfate (CAS 10102-17-7), and household ammonia (CAS 1336-21-6). Summarize the hazardous properties in your notebook, including the identity of the company which produced each MSDS and the NFPA diamond for each material.[3]

The most likely exposure is to silver nitrate solution, which will blacken the skin. Wash any affected areas immediately with plenty of running water. If eye contact occurs, flush with water and call an ambulance.

You should wear safety glasses and rubber gloves while working on this project. Leftover silver nitrate solution should be collected for recycling or disposal. Leftover egg whites may be thrown in

3. The NFPA diamond was introduced in Section 15.2 (page 184). You may substitute HMIS or Saf-T-Data ratings at your convenience.

the trash. Spent sodium thiosulfate or ammonia should be washed down the drain with plenty of water unless prohibited by law.

Research and Development

So there you are, studying for a test, and you wonder what will be on it.

- Study the meanings of all of the words that are important enough to be included in the ***index*** or **glossary.**
- You should know all of the Research and Development points from Chapter 17 and Chapter 19.
- You should know what substance comprises the black areas of a black and white photograph; you should know what substance comprises the white areas of a black and white photograph.
- You should know three purposes of albumin in the photographic emulsion described in this chapter. Is albumin used in modern photographic emulsions?
- You should know the purpose of a photographic fixer; you should know the reactions of sodium thiosulfate and ammonia with silver chloride.
- You should know the hazardous properties of sodium chloride, sodium carbonate, silver nitrate, ammonia, and sodium thiosulfate.
- You should know the contributions of Daguerre, Talbot, Herschel, and Eastman to the development of photography.

23.3 Θ

The creation of albumin photographs consists of five steps; mixing of the albumin/salt solution, coating the paper with albumin/salt solution, sensitizing the paper with silver nitrate solution, exposure through a negative, and fixing with sodium thiosulfate. While just about any paper will work, including handmade paper from Chapter 14, I recommend a good-quality, smooth-textured watercolor paper.[4] You will also need a dozen

4. I use Canson Montval 90 lb. Watercolor Paper. Many papers will work but the best results are to be had from paper that is heavy, smooth, and well-sized. I typically cut paper to 5x7 inches as a matter of convenience.

Figure 23-1. Making the Albumin/Salt Emulsion

raw eggs, some table salt (sodium chloride), white household vinegar, silver nitrate, sodium carbonate, and sodium thiosulfate.

To make the albumin/salt solution, you need to separate the yolks from the eggs. One egg at a time, separate the yolk from the whites using your favorite method and place the egg white into a cup. If any egg white is contaminated by yolk, discard the whole egg; egg yolks contain sulfur, which will degrade the quality of your print. Examine each egg white and remove any dirt or bits of eggshell. Once an egg white is free of debris, pour it from the cup into a one-pint glass jar. Continue separating eggs until your jar contains a dozen clean egg whites. To your jar of egg whites add 12 mL of white household vinegar and 12 g of sodium chloride. Put the lid on the jar and shake it until a good head of froth develops. Allow the froth to rise to the surface and then remove it with a spoon, as shown in Figure 23-1(R), leaving clean, clear emulsion in the jar. The emulsion may be prepared in advance and stored in the refrigerator indefinitely. If the emulsion becomes dirty after repeated use, simply shake it and spoon off the froth; most of the impurities will be caught in the froth. A dozen eggs produce enough emulsion for very many prints.

When you are ready to coat your paper, remove your albumin solution from the refrigerator and allow it to warm to room temperature. Cut your paper to a convenient size and fold up about 1 cm along each edge of the paper. The effect is that you are forming your paper into a shallow dish. You will need to fold the corners in so that all four edges stand up; do not cut the corners; your paper dish needs to be water-tight. Using a medicine dropper or transfer pipette, withdraw some emulsion from your jar. By putting the tip of the pipette well below the surface, you can extract very clean emulsion, free of bubbles and froth. Deliver the emulsion to your paper dish and continue adding emulsion until the bottom of your dish is covered. You can rock your paper dish from side to side and move emulsion to dry areas with your pipette. The goal is to completely cover the paper with clean, bubble-free emulsion. Once the bottom of your paper dish is completely covered with emulsion, allow it to sit for 3 minutes. At the end this time, use you pipette to return the remaining emulsion to the jar.[5] Then fill your pipette with fresh emulsion and use it to wash any bubbles or dirt from your paper, as shown in Figure 23-2(R). Allow your coated paper to dry overnight; you may lean it vertically against a wall, stand it up in a lab drawer or cabinet, or hang it from a line with clothes pins. Coated paper may be stored indefinitely.[6]

Paper should be sensitized in dim light; a 15-watt incandescent bulb or darkroom safe-light may be used. To sensitize your paper you will coat the bottom of your paper dish with silver nitrate solution using a medicine dropper or transfer pipette. However, you should use a *different pipette* from the one you used for the emulsion to avoid cross-contamination. You should also wear rubber gloves if you don't want your hands blacker than charcoal. The sensitizing solution consists of

---

5. It is possible to reuse all three solutions used in this project—the emulsion, the sensitizer, and the fixer—provided that they are not mixed with one another. The emulsion doesn't go bad if refrigerated, it just gets used up. The sensitizer may become depleted after prolonged use; when prints fail to darken in sunlight, it is time to replace the sensitizer. The fixer will also weaken with prolonged use; when fixed photographs darken with time, it is time to replace the fixer.
Spent sensitizer and fixer may be recycled to reclaim their silver content. If you have no access to a commercial recycler, these solutions may be evaporated to dryness and then smelted by the procedures of Chapter 9 as if they were ores of silver.

6. In a classroom situation, there are several ways to work around this overnight drying period. First, you may coat your paper with emulsion one week, sensitize and expose it the following week. Second, you may use paper coated by another student the week before and then coat paper for future students while you are waiting for your sensitized paper to dry. Finally, your instructor may have made up coated papers in advance. Ask your instructor which option is best for you.

Figure 23-2. Coating the Paper with Emulsion

30 g of silver nitrate dissolved in 250 mL of water. It may be stored in a brown glass bottle. Paper should not be sensitized, however, until you are ready to use it. To sensitize your paper, pipette enough silver nitrate solution into your coated paper dish to completely cover the bottom. Rock the dish from side to side and use your pipette to move the silver nitrate solution onto any dry areas. Once the bottom of your paper dish is completely covered with silver nitrate solution, allow it to sit for 1 minute. At the end of this time, use your pipette to return the remaining solution to the bottle, as shown in Figure 23-3(R). Stand your sensitized paper vertically against a wall or hang it from a line to dry *in the dark* for 1 hour. While you are waiting for your sensitized paper to dry, you may coat more paper with emulsion.

You will need a negative through which to make your exposure. Since you will be making a contact print, the finished print will be the same size as your negative. While you may use a 4x5 inch photographic negative, you may also print a digital image onto transparency film or even onto paper. Most image-manipulation computer programs have an option to produce a negative image. A high-contrast image works best, that is, one with very black blacks and very white whites.

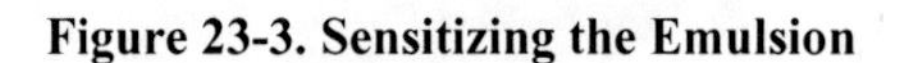

Figure 23-3. Sensitizing the Emulsion

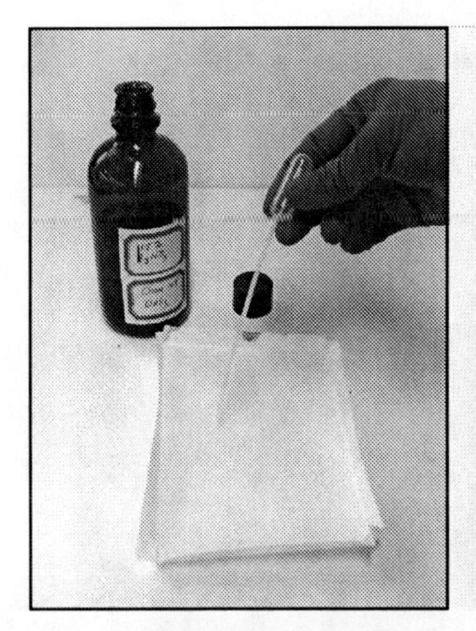

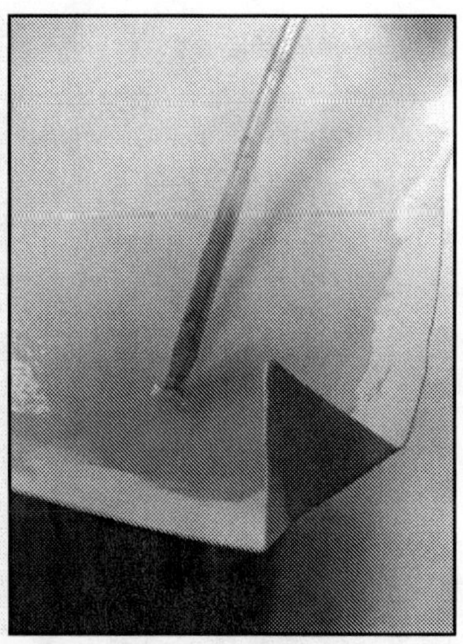

You will also need a printing frame to make your exposure; a standard picture frame works well. The negative and the sensitized paper must be loaded into the printing frame in dim light, the same light you used for sensitizing your paper. Place the negative against the glass of the picture frame. Flatten out your sensitized paper dish and place the sensitized side of the paper against the negative. Then assemble the back of the picture frame so that the negative is held tightly against the sensitized paper so that it holds the sensitized paper tightly against the negative.

You are now ready to make your exposure. Take your printing frame out into bright sunlight. On a cloudy day, bright artificial light may be used. The frame may be set under a bright lamp or it may be placed face-down on an overhead projector. As the exposure proceeds, the areas of your print not masked by the negative will start to darken, as shown in Figure 23-4(R). Particularly watch the exposed area around the edge of the negative. When these exposed areas turn a rich chocolate brown, the exposure is complete. This may take only a few minutes in bright sunlight or as long as an hour under reduced lighting conditions.

Returning to dim light, remove the paper from the printing frame. You should now see an image on the paper; the areas masked by the negative will still be white, while the exposed areas will be dark brown or black.

Figure 23-4. Exposing the Print

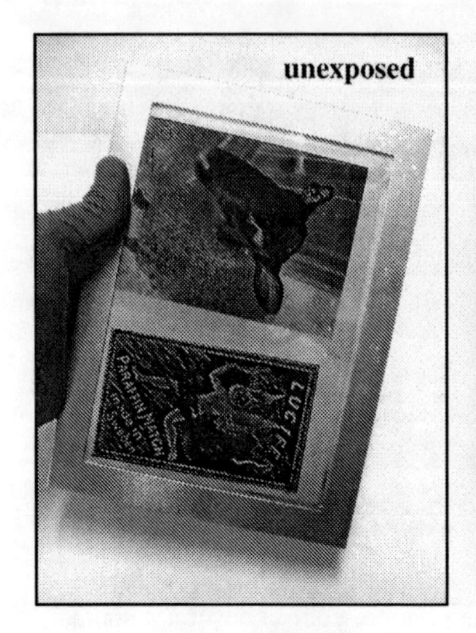

Figure 23-5. The Finished Print

If you were to turn on the lights now, the entire print would quickly turn black. Soak your print for 5 minutes in plain water and then 5 minutes in fixing solution. The fixing solution consists of 2 g of sodium carbonate and 150 g of sodium thiosulfate (aka sodium hyposulfite) in 1 L of water. Whereas only one side was treated with albumin and sensitizer, the whole print needs to be immersed in the fixing bath; you need to remove any silver which may have soaked into the paper. Finally, wash your print for 30 minutes in running water and hang your print on a line to dry. If you reinforce the folds you made at the beginning, the paper will be less likely to curl as it dries.

With this brief introduction to photographic processes, a whole world of alternative processes opens up to you. You may be interested, for example, in the cyanotype, or blueprint, an iron-based photographic process pioneered by none other than John Herschel. You may be interested in the gum-bichromate process, a *color* process based on the interaction of chromium compounds with ordinary water colors. These alternative processes have experienced a boom among artistic photographers in recent years. If you would like to experiment with them, see, for example, References [84] and [82].

Quality Assurance

Include safety information in your notebook on the materials you used in this project. Outline the steps you used and tape your finished photograph into your notebook. A true photograph will sport a sharp, well-defined, and permanent image.

Chapter 24. Solvay (Ammonia)

> The greater part of the soda now employed is obtained by Leblanc's process, which while it admits of lixiviating the soda readily and completely, is defective, inasmuch as the residue, or waste as it is technically called, contains nearly all the sulphur used in the manufacture; and that this is not a slight loss may be inferred from Oppenheim's statement, that in the alkali works at Dieuze, Lorraine, the accumulated waste contains an amount of sulfur valued at £150,000. For every ton of alkali made there is accumulated 1½ tons of waste, containing 80 per cent. of the sulphur used in the manufacture; and this waste, until lately thrown on a refuse heap in some fields adjacent to the works, often proved a nuisance in hot weather, giving rise to fumes of suphuretted hydrogen.
>
> — *Wagner's Chemical Technology*, 1872 AD [1]

24.1 ☿

∇ Four score and seven years ago our fathers brought forth on the European continent, a new industry, conceived in Liberty, and dedicated to the proposition that all men are created equal. The chemical revolution begun by Leblanc in 1790 had passed from France to Britain, where the repeal of the salt tax had stimulated the growth of a booming ***soda*** industry. This industry used as raw materials, salt, sulfur, limestone, saltpeter, coal, air, and water; its products were the alkalis, sodium carbonate and sodium hydroxide. Cheap alkalis brought to the ordinary citizen those luxuries which had formerly been enjoyed only by the rich and powerful: glass for bringing light into dark places, paper for bringing the printed word into proletarian homes, and soap for bringing sanitation into cities oppressed by filth and disease. And while the Leblanc process delivered on many of the promises of the revolution, you may very well wonder why the world has little noted, nor long remembered the work which it so nobly advanced. I will tell you.

∇ You will, of course, recall that in addition to valuable alkalis, the Leblanc process produced two waste products, hydrogen chloride and calcium sulfide. Acidic hydrogen chloride gas was sent up the chimney,

1. Reference [29], pp. 184-185.

after which it decimated vegetation in the vicinity of an alkali works. Insoluble calcium sulfide was conveniently disposed of in heaps where the vegetation used to be. Unfortunately, when calcium sulfide reacts with rain water it farts out noxious hydrogen sulfide, a gas which will never have a rose named after it. Not surprisingly, there was little demand for posh plant-side homes in the neighborhood of the Widnes Alkali Works.

∇ As a consequence, alkali manufacturers became popular targets for lawsuits and government regulations. The British Alkali Act of 1863, for example, required the absorption of 95% of the hydrogen chloride produced by the salt cake furnace.[2] This was easily accomplished, hydrogen chloride being quite soluble in water; the waste gas was sent up through a stone tower filled with coke; water dribbling down through the tower absorbed the hydrogen chloride, producing aqueous hydrochloric acid. *Voila!* Clean air. There being little demand for hydrochloric acid, however, it was natural to dispose of it in the nearest river, a loophole closed by the Alkali Act of 1874.

∇ In addition to hydrogen chloride, calcium sulfide was a thorn in the side of soda manufacturers. While the community complained about stinking heaps of tank waste, the manufacturers themselves mourned the loss of valuable sulfur. Sulfur was then imported almost exclusively from Sicily and a monopoly established there in 1838 increased prices sharply.[3] An alternative to sulfur was found in sulfur dioxide, roasted from British and Spanish pyrites, but there was no getting around the stubborn fact that every ounce of sulfur procured from whatever source was destined to wind up as calcium sulfide waste. The one silver lining of the calcium sulfide cloud was that it could be converted into sodium thiosulfate, used by photographers to fix photographs. Even so, the demand for photographic fixer was dwarfed by the supply of tank waste; the noxious heaps became mountains, testing whether any industry so conceived could long endure.

∇ It was into such a world that I, ***Ernest Solvay,*** was born in 1838. Son of a Belgian salt-maker, I cut my teeth on salt and at the age of 21 went to work for my uncle, who managed a coal-gas plant. Growing up as I did surrounded by salt, coal, gas, and ammonia, it was natural that I should apply my little gray cells to the problems of the soda industry. Many commercial enterprises had attempted to exploit the reaction of

---

2. Reference [68], p. 89.
3. Reference [75], p.33.

salt, ammonia, and carbon dioxide to produce ammonium chloride and sodium carbonate. But if the ammonia were not recovered, the process would suffer from the waste of ammonia, as the Leblanc process had from the waste of sulfur. To this end, I devised a recovery process using lime to liberate ammonia gas from ammonium chloride. I realized, however, that winning this battle would not be enough to win the war.

∇ Ward and Roebuck, after all, had brought the chamber acid industry into being, but had failed to keep their secrets. Leblanc himself had given birth to the soda industry and yet political upheaval had robbed him of the fruits of his labors. Since it is not in the interest of secrets to be kept, I was careful to patent my ideas both in Belgium and in Britain. I then established a soda works in Belgium which began production in 1864. While the construction of a Solvay plant was more expensive than a comparable Leblanc plant, it required fewer raw materials; thus capital investment was higher but operating costs were lower. I built a second plant in France, but the real key to my commercial success was the sale of licenses abroad; my particular un-kept secrets were going to pay as they went. Solvay plants were established by Brunner, Mond & Co. in Britain and by Solvay Process Co. in the United States, paying royalties on every ton of soda produced.

∇ The established alkali manufacturers were not about to surrender without a fight, however. In 1868 Henry Deacon introduced a process for turning waste hydrogen chloride into bleaching powder, which found a ready market in paper and textiles.[4] In 1887 Alexander Chance finally succeeded where so many others had failed in recovering sulfur from tank waste. But these piece-meal improvements would make a patchwork of existing alkali plants, sacrificing simplicity for efficiency. With lower operating costs, I was able to drop the price of soda to the point that my competitors were forced to sell it at a loss. Leblanc soda works, in an ironic twist, would be kept afloat only by sales of bleaching powder produced from acid formerly run to waste.

∇ World production of soda in 1863 had been 150,000 tons, all produced in Leblanc plants. By 1902 world production of soda would soar to 1,760,000 tons, over 90% of which would be produced in Solvay plants.[5] It would be altogether fitting and proper to celebrate the victory of cleaner and more efficient processes over those crippled by pollution

4. Reference [60], pp. 95-98.
5. Reference [52], p. 59.

Equation 24-1. Reactions Involving Carbon Dioxide

$$(a)\ CO_2(g) + H_2O(l) = H_2CO_3(aq)$$
$$(b)\ H_2CO_3(aq) + NaOH(aq) = NaHCO_3(aq) + H_2O(l)$$
$$(c)\ NaHCO_3(aq) + HCl(aq) = NaCl(aq) + CO_2(g) + H_2O(l)$$
$$(d)\ 2\ NaHCO_3(s) \stackrel{\Delta}{=} Na_2CO_3(s) + CO_2(g) + H_2O(l)$$

Equation 24-2. Reactions Involving Ammonia

$$(a)\ NH_3(g) + H_2O(l) = NH_4OH(aq)$$
$$(b)\ NH_4OH(aq) + HCl(aq) = NH_4Cl(aq) + H_2O(l)$$
$$(c)\ 2\ NH_4Cl(aq) + Ca(OH)_2(aq) = 2\ NH_4OH(aq) + CaCl_2(aq)$$
$$(d)\ NH_4OH(aq) \stackrel{\Delta}{=} NH_3(g) + H_2O(l)$$

and waste. But, in a larger sense, we can not dismiss—we can not forget—we can not overlook those who struggled to bring light, literature, and lather to all peoples. It is rather for us to be here dedicated to the great task remaining before us—that from these honored dead we take increased devotion to that cause for which they gave the last full measure of devotion.

24.2 ♎

In order to understand the Solvay process, it is necessary to understand the chemistries of ***carbon dioxide*** and ***ammonia.*** We first encountered carbon dioxide (CO_2) way back in Chapter 1 as a product of the combustion of charcoal. We learned in Chapter 4 that yeasts fart carbon dioxide when they consume honey. Chapter 18 characterized carbon dioxide as a mildly acidic gas which reacts with bases to form carbonate or bicarbonate[6] salts. Carbon dioxide may be liberated from these salts either by heating the bejeezus out of them or by reacting them with acids. The properties of carbon dioxide are summarized in Equation 24-1.

We met ammonia (NH_3) in Chapter 12 as a mild alkali suitable for dissolving indigo dye. Ammonia reappeared in Chapter 22 as a by-product

---

6. The modern name for the bicarbonate ion is the *hydrogen carbonate* ion. The older name, however, continues to be widely used.

of the distillation of coal tar. The "ammonia" sold at the grocery store is actually a solution of ammonia in water. Just as carbon dioxide combines with water to form carbonic acid, ammonia combines with water to form the alkali, ammonium hydroxide (NH_4OH). Ammonium hydroxide participates in the usual acid-base reactions to form ammonium salts. Ammonia may be liberated from these salts either by heating the ***bejeezus*** out of them or by reacting them with stronger alkalis like sodium or calcium hydroxide. The properties of ammonia are compared to those of carbon dioxide in Equation 24-2.

Carbonic acid reacts with a base to form a bicarbonate salt and water; ammonium hydroxide reacts with an acid to form an ammonium salt and water. It is only natural that carbonic acid should react with ammonium hydroxide to produce ***ammonium bicarbonate*** and water:

$$H_2CO_3(aq) + NH_4OH(aq) = NH_4HCO_3(aq) + H_2O(l)$$

When a solution of ammonium bicarbonate evaporates, it leaves behind solid ammonium bicarbonate, a salt with the properties of both bicarbonate salts and ammonium salts. Like other bicarbonate salts, heating the bejeezus out of it liberates carbon dioxide; like other ammonium salts, heating the bejeezus out of it liberates ammonia. The curious thing about ammonium bicarbonate it that when you liberate bejeesical carbon dioxide and bejeesical ammonia, the only thing left is bejeesical water. Thus ammonium bicarbonate is entirely bejeesical in nature; heat it and it just disappears. This property makes it a useful ingredient of baking powders because in the oven it is able to leaven breads and cakes without leaving anything behind.

Ammonium bicarbonate is crucial to the Solvay process because it undergoes a metathesis reaction with ***sodium chloride*** to produce ***sodium bicarbonate*** and ***ammonium chloride:***

$$NaCl(aq) + NH_4HCO_3(aq) = NaHCO_3(s) + NH_4Cl(aq)$$

You are probably thinking that sodium bicarbonate is soluble in water. That is so, but it is *less* soluble than sodium chloride, ammonium bicarbonate, or ammonium chloride. Thus in a very concentrated solution, sodium bicarbonate will be the first to precipitate. This metathesis reaction is the key to the Solvay process.

The three critical reactions of the Solvay process occur simultaneously in reactor (e) of Figure 24-1. Remember that in a process schematic,

Figure 24-1. The Solvay Soda Process

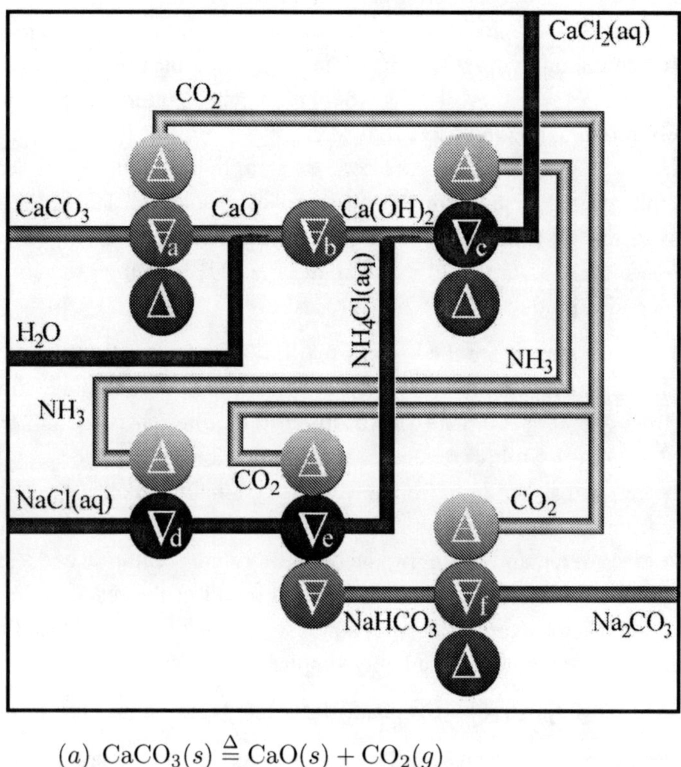

$$(a)\ \mathrm{CaCO_3}(s) \overset{\Delta}{=} \mathrm{CaO}(s) + \mathrm{CO_2}(g)$$

$$(b)\ \mathrm{CaO}(s) + \mathrm{H_2O}(l) = \mathrm{Ca(OH)_2}(aq)$$

$$(c)\ \mathrm{Ca(OH)_2}(aq) + 2\ \mathrm{NH_4Cl}(aq) \overset{\Delta}{=} \mathrm{CaCl_2}(aq) + 2\ \mathrm{NH_3}(g) + 2\ \mathrm{H_2O}(l)$$

$$(d)\ 2\ \mathrm{NH_3}(g) + 2\ \mathrm{H_2O}(l) = 2\ \mathrm{NH_4OH}(aq)$$

$$(e_1)\ 2\ \mathrm{CO_2}(g) + 2\ \mathrm{H_2O}(l) = 2\ \mathrm{H_2CO_3}(aq)$$

$$(e_2)\ 2\ \mathrm{NH_4OH}(aq) + 2\ \mathrm{H_2CO_3}(aq) = 2\ \mathrm{NH_4HCO_3}(aq) + 2\mathrm{H_2O}(l)$$

$$(e_3)\ 2\ \mathrm{NH_4HCO_3}(aq) + 2\ \mathrm{NaCl}(aq) = 2\ \mathrm{NaHCO_3}(s) + 2\ \mathrm{NH_4Cl}(aq)$$

$$(f)\ 2\ \mathrm{NaHCO_3}(s) \overset{\Delta}{=} \mathrm{Na_2CO_3}(s) + \mathrm{CO_2}(g) + \mathrm{H_2O}(l)$$

reactants enter a reactor from the left and products leave to the right. An aqueous solution of ammonium hydroxide and sodium chloride enters reactor (e), an ***absorber,*** where it meets a stream of carbon dioxide gas. The carbon dioxide dissolves to produce carbonic acid, which reacts with ammonium hydroxide to produce ammonium bicarbonate. If there were plenty of water, everything would remain in solution—sodium ion, chloride ion, ammonium ion, and bicarbonate ion. But if water is limited, that is, if the solutions are concentrated, then sodium bicarbonate precipitates and ammonium chloride remains in solution. The liquid returns to reactor (c) and the solid goes to ***furnace*** (f), where bejeesical carbon dioxide is driven off. The ***sodium carbonate*** which remains is the intended product of the whole process.

Reactors (e) and (f) are the stars of the Solvay process; all of the other reactors play supporting roles. Absorber (e) requires ammonium hydroxide from another absorber, (d). Absorber (d) requires gaseous ammonia from a ***still,*** (c). Still (c) requires ammonium chloride recycled from absorber (e) and slaked lime from a ***slaker,*** (b). Finally, slaker (b) requires lime from a furnace, (a). As raw materials, the whole process requires limestone, water, and salt. It produces waste calcium chloride and a single product, sodium carbonate. If you add up all of the reactions from all of the reactors, canceling compounds that appear on both sides of the equal sign, the result is remarkably simple:

$$CaCO_3(s) + 2\ NaCl(aq) = CaCl_2(aq) + Na_2CO_3(s).$$

This reaction may seem familiar to you, since it was the second reaction discussed in Section 7.2. Our solubility rules tell us, however, that this reaction goes the other way. Soluble calcium chloride and sodium carbonate will react in aqueous solution to produce insoluble calcium carbonate and soluble sodium chloride. The Solvay process, with its elegant scheme for recycling carbon dioxide and ammonia, is simply an elaborate means for pushing this classic metathesis reaction backwards.

Compare the Solvay process to the Leblanc process, Figure 20-2 (page 253), and you will immediately see why the one superseded the other. Both processes require limestone, salt, and water as raw materials, but while the Leblanc process additionally requires sulfuric acid, the Solvay process recycles its ammonia. Both processes produce the same product, sodium carbonate, but while the Leblanc process produces two hazardous waste products, the Solvay process produces only one relatively harmless waste product. In fact, calcium chloride is useful as a road deicer, though the market is not large enough to absorb the huge amounts produced each year. So while the waste produced by a Solvay plant is not nearly as hazardous as those produced by a Leblanc plant, calcium chloride remains a disposal problem for the soda industry.

Material Safety

Locate MSDS's for ammonia (CAS 7664-41-7), carbon dioxide (CAS 124-38-9), and ammonium bicarbonate (CAS 1066-33-7). Summarize the hazardous properties in your notebook, including the identity of the company which produced each MSDS and the NFPA diamond for each material.[7]

Your most likely exposure will be to ammonia fumes. If a persistent cough develops, seek medical attention.

You should wear safety glasses while working on this project. All activities should be performed in a fume hood or with adequate ventilation. Leftover materials can be flushed down the drain with plenty of water.

Research and Development

You are probably wondering what will be on the quiz.

- You should know the meanings of all of the words important enough to be included in the ***index*** or **glossary.**
- You should know the Research and Development points from Chapter 18 and Chapter 20.
- You should know the reactions of Figure 24-1.
- You should know the hazardous properties of ammonia, carbon dioxide, and ammonium bicarbonate.
- You should be able to reproduce Figure 24-1 and to explain this schematic in your own words.
- You should be able to explain the advantages of the Solvay process over the Leblanc process.
- You should be able to tell the story of Earnest Solvay.

---

7. The NFPA diamond was introduced in Section 15.2 (page 184). You may substitute HMIS or Saf-T-Data ratings at your convenience.

24.3 Θ

> It rises from the Earth to Heaven and descends again to the Earth, and receives power from Above and from Below. Thus thou wilt have the glory of the Whole World. All obscurity shall be clear to thee.
>
> — *The Emerald Tablet of Hermes Trismegistos*

A lime kiln, a slaker, a still, two absorbers, and a furnace make up the complete Solvay process. None of this is new; you have known how to make lime since Chapter 10; you grasped the secrets of distillation in Chapter 16; you learned how to build an absorber in Chapter 18; and the final furnace is quite similar to the one described in Chapter 20. You could build the whole plant out of pottery and soft-drink bottles if you wanted too, but I am afraid that few readers will want to go to all that trouble to produce something that might have been extracted in one step from seaweed ashes. After batteries, dyes, and photographs, the prospect of making one white powder from another might be a little bit anti-climactic. So how about a seance? Let us conjure some spirits, materialize some ghosts, bottle some genies, herd some bejeesical cats. Let us make some ammonium bicarbonate. Okay, so it is not going to be the perfect birthday present for Mom. You are not going to save a bottle of it for graduation day or mount it on your key-chain. Look, nobody is holding a gun to your head; if the prospect of making something "rise from Earth to Heaven and descend again to the Earth" does not float your boat, then go directly to the next chapter; do not pass "Go," do not collect the glory of the Whole World.

For anyone who made it past the last sentence, let us see whether we can clarify some obscurity. You have been using ammonia since Chapter 12, whether home-grown or store-bought. You are undoubtedly familiar with ammonia's unmistakable aroma, but how does it get from the bottle to your nose? The stuff in the bottle is an aqueous solution of ammonium hydroxide, NH_4OH. Escaping from the bottle is bejeesical ammonia, NH_3, the gas which makes your eyes water. Unfortunately, the word *ammonia* is often used indiscriminately to describe both the gas and its solution. Commercial ammonium hydroxide is 28-30% ammonia in water; household "ammonia" is more dilute. For this project we need gaseous ammonia, which we can get by distilling household ammonia.

In addition to ammonia, we need some carbon dioxide. In the industrial process this comes from the lime kiln, but we can use the carbon dioxide

Figure 24-2. Distillation and Absorption

collected in Section 4.3 for this very purpose. Figure 4-3 (page 56) details the fermentation lock consisting of a balloon glued to a bottle cap with a hole in it. This balloon is either still screwed onto your mead or you have wrapped its neck around and pencil and secured it with a twist tie. You will need a minimum of 2 L of carbon dioxide. If your balloon is round, measure its circumference with a piece of string; the circumference should be at least 50 cm. If your balloon is oblong it should be at least as big as a 2-liter soft-drink bottle. With your balloon of carbon dioxide handy, it is time to distill your ammonia.

Set up your still just as you did in Chapter 16, but instead of filling your Erlenmeyer flask with mead, fill it with 400 mL of household ammonia and a few boiling chips. For the receiver you can use our old friend, the 2-liter soft-drink bottle. Since we are collecting a gas there is no need to cool the receiver; there is no need for ice. Slip the receiver over the condenser tube and place the still on a hot-plate. Set the hot-plate power to "high," but as soon as the pot begins to boil turn the power down to half. Adjust the hot-plate power to keep the pot boiling without taking the head temperature above 82°C (180°F); you want to distill gaseous ammonia, not water. If you are not paying attention, the head will climb too high, too fast, and you will have to start over. If, however, you separate the gas by fire, the fine from the gross, gently and with great skill, then the head temperature will climb gradually to 82°C. You will notice what looks like steam condensing on the sides of the receiver; this is concentrated ammonium hydroxide. Allow a few mL to collect in the bottom of the receiver and then shut down the distillation. Remove the receiver from the still, drain the ammonium hydroxide into a small beaker, and using a graduated pipette or graduated cylinder pour 2 mL of ammonium hydroxide back into the receiver. The reason for returning this ammonium hydroxide will be explained shortly.

Once the ammonia distillation is complete, transfer the balloon from your mead to your receiver as quickly as possible. Pinch the neck of your balloon, unscrew it from your mead, screw it onto the receiver, and release the neck of the balloon. Over the course of a few minutes bejeesical carbon dioxide will be *absorbed* by the ammonia; your receiver has miraculously changed into an absorber. The balloon will shrink as carbon dioxide becomes carbonic acid, as shown in Figure 24-2(R). The bottle will get hot as carbonic acid reacts with ammonium hydroxide to form ammonium bicarbonate. Roll the bottle from side to side to keep the walls wet. You will notice what looks like condensation on the inside of the bottle, but as the reaction proceeds this condensation will sprout crystals, as shown in Figure 24-3(L). While your crystals are growing, let us take a look at the stoichiometry.

$CO_2(g) + NH_3(g) + H_2O(l) = NH_4HCO_3(s)$

At room temperature and normal atmospheric pressure a mole of gas occupies 24.5 liters. This unit factor, (24.5 L/mol CO_2) or (24.5 L/mol NH_3) allows us to answer stoichiometric questions:

Q: How many grams of ammonium bicarbonate can be produced from the reaction of 3.0 L of carbon dioxide with excess ammonia and water?

A: 9.6 g.

Q: How many grams of ammonium bicarbonate can be produced from the reaction of 2.0 L of ammonia with excess carbon dioxide and water?

A: 6.4 g.

Figure 24-3. Spirit Made Flesh

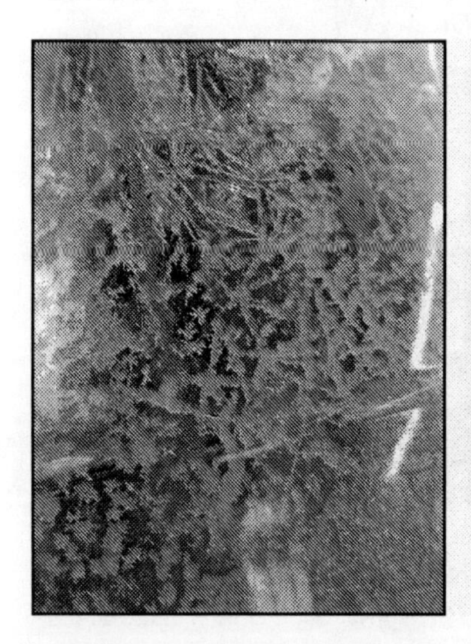

Q: How many grams of ammonium bicarbonate can be produced from the reaction of 2.0 g of water with excess ammonia and carbon dioxide?

A: 8.8 g.

So what is the theoretical yield, 22.6 grams? No, to produce 6.4 g of ammonium bicarbonate you need 2.0 L of carbon dioxide *and* 2.0 L of ammonia *and* 1.5 g of water. If you have 2.0 L of ammonia you can make *at most* 6.4 g of ammonium bicarbonate, no matter how much carbon dioxide and water are present. In this case the ***limiting reagent,*** the one which runs out first, is ammonia and the ***theoretical yield*** is 6.4 grams. If your balloon holds less than 2.0 L of carbon dioxide it will completely collapse; in this case carbon dioxide is the limiting reagent and your theoretical yield will be somewhat less than 6.4 g. Similarly, if there is less than 1.5 g of water present, water is the limiting reagent. We added 2 mL of ammonium hydroxide (ammonia in water) to ensure that there would be enough water present. Adding more ammonium hydroxide might have increased our yield since it supplies both water and ammonia, but we would have wound up with a *solution* of sodium bicarbonate rather than *crystals.* Limiting the water allowed us to collect crystals immediately without having to wait for the excess water to evaporate.

When the balloon stops shrinking and the bottle is cool, remove the balloon and cut the bottle open at the shoulder. Use a spoon or spatula to push the ammonium bicarbonate crystals to the bottom of the bottle. Trim away the walls of the bottle as you go until you are left with a shallow dish filled with crystals, as shown in Figure 24-3(R). Allow your crystals to dry overnight and then weigh them to get your actual yield, which is often expressed as a percentage of the theoretical yield.

Consider what has happened here; you started with an empty bottle; everything that went into the bottle was a gas, yet from these bejeesical spirits a solid body emerged. As an infant it smells of ammonia but when you allow it to dry overnight the smell almost goes away. It has so much potential; it might make bread or cake or soda, but this particular body is destined for a trial by fire. Place it into a small beaker and heat it with a hot-plate or ***spirit lamp*** and as it gets old and decrepit it will begin to smell of ammonia once more. It begins to lose weight and finally just fades away, departed once more into the spirit world. In the end, the only thing left of it is a story.

Quality Assurance

Include in your notebook safety information for ammonia, carbon dioxide, and ammonium bicarbonate. Record the actual yield of ammonium bicarbonate expressed as a percentage of the theoretical yield. Include a photograph of your crystals to illustrate your telling of their poignant saga.

Chapter 25. Dow (Electrochemicals)

Aluminium is now not so much in use: when first introduced aluminium jewelry was the order of the day. The metal is at present more usefully employed for small weights, light tubes for optical instruments, and to some extent for surgical instruments. The price, however, of this metal, £3 128. per kilo., is too high to admit of its extended use; while great lightness combined with comparative strength are its only prominent qualities.

— Wagner's Chemical Technology, 1872 AD [1]

25.1 ☿

Δ (1890) For his undergraduate chemistry project ***Herbert Dow*** had identified Michigan brines which were unusually high in bromine. Bromine was used to produce patent medicines and photographic chemicals; the only commercial source was Die Deutsche Bromkonvention, a German cartel. Dow formed the Midland Chemical Company in 1890 to produce electrolytic bromine from brine, but when he wanted to expand into ***chlorine*** and ***caustic soda*** (chlor-alkali), his financial backers balked. Dow left in 1895 to form the Dow Chemical Company, which absorbed Midland in 1914. Though it survived competition from German bromine and British bleach, Dow remained a small company until the First World War interrupted trade with Germany. The war made it possible to move into aniline dyes and pharmaceuticals, areas dominated by the German giants, and Dow began manufacturing phenol, aspirin, and synthetic indigo. Its expertise in bromine chemistry led it to develop antiknock gasoline additives in partnership with Ethyl Corporation. Its roots in electrolysis allowed it to extract magnesium metal from sea-water by the end of the Second World War. Its increasing proficiency with organic chemicals led it to develop silicone sealants in partnership with Corning Glass. Its early work with chlorinated fungicides led it to develop defoliants for the American military during the Vietnam conflict. Its early success with transparent polystyrene led it to develop Saran Wrap and Ziploc bags. Following its acquisition of Union Carbide, Dow Chemical

1. Reference [29], p. 114.

would begin the twenty-first century as the largest producer of chemicals in the world.[2]

Δ (1888) Two years before Dow formed Midland, American-born ***Hamilton Castner*** went to work for the British Aluminum Company. Aluminum manufacture at that time required elemental sodium and Castner had invented a process for making sodium from soda. Unfortunately for him, the Hall-Héroult process for making ***aluminum*** electrolytically was about to eliminate demand for sodium in aluminum manufacture. Desperately seeking a more efficient route to sodium, he began electrolyzing brine with a mercury cathode. Though the process never produced sodium cheaply enough to compete in aluminum manufacture, it produced chlorine and very pure caustic soda. It was, in fact, for the sake of caustic soda that the Austrian paper pulper, ***Karl Kellner,*** patented an almost identical process in 1894. The two Athanors resolved their initial disputes and formed the Castner-Kellner Alkali Company, which took over the British Aluminum Company in 1897.[3] Castner-Kellner would, in turn, be swallowed by Brunner Mond, which manufactured soda under license from Solvay. The availability of cheap electrolytic chlorine and caustic soda was the final nail in the coffin of the Leblanc soda process. Brunner Mond would form the core of the industrial giant, Imperial Chemical Industries (ICI), which would begin the twenty-first century as the tenth largest chemical company in the world.

Δ (1888) When Castner went to work for British Aluminum, aluminum was a precious metal. The third most abundant element in the Earth's crust, aluminum could not be won from its ore by smelting, carbon being an insufficiently strong reducing agent. Elemental sodium was the industrial reducing agent of choice for winning aluminum from its oxide, alumina, and the high price of sodium translated into high prices for aluminum. While sodium could be produced from the electrolysis of molten sodium hydroxide, the melting point of alumina, 2051°C, ruled out the electrolysis of molten alumina as a viable industrial process. An American student, ***Charles Martin Hall,*** learned of these difficulties in his undergraduate chemistry classes at Oberlin and set out to find a flux which would lower the melting point of alumina to an attainable level. The mineral cryolite proved to be such a flux, lowering the melting point to a mere 950°C, well within the range of even modest furnaces. Using

2. Reference [54], July 29, 2002, p. 16.
3. See Reference [61], p. 80.

home-made batteries he produced his first electrolytic aluminum in 1886. Hall attempted to commercialize an electrolytic process in collaboration with the Cowles Electric Smelting and Aluminum Company, developers of the electric smelting furnace.[4] Meanwhile in France another college student, ***Paul Héroult,*** had independently found cryolite to be an acceptable flux for alumina and had begun using a small dynamo to produce electrolytic aluminum. It is in the interest of mortals to keep secrets and consequently Hall, Héroult, and Cowles engaged each other in lengthy legal battles. But it is not in the interest of secrets to be kept; Héroult would license his process to the British Aluminum Company and the Hall process would be adopted by the Pittsburgh Reduction Company, destined to become the Aluminum Company of America (AlCoA).

Δ (1881) Seven years before Hall obtained his first aluminum, Thomas Alva Edison opened the first electrical power station at Pearl Street in New York City. Davy had invented the electric arc light in 1808 but the power available from batteries was insufficient for commercial application. Oersted had demonstrated electromagnetism in 1820 and Faraday had discovered electromagnetic induction in 1831, but it would fall to Werner von Siemens in 1867 to develop the dynamo, a device for converting mechanical power into electrical power. With increasing demand for electric lighting came the widespread availability of electrical power for other uses, including electrolytic production of chlorine, caustic soda, and aluminum. The early electrochemical industry grew up near sources of cheap hydroelectric power, for example, at Niagara Falls. The industry became less centralized as the AC system of Nicola Tesla and George Westinghouse made it possible to distribute electricity from remote power plants. Edison's General Electric would dwarf even the largest chemical company at the beginning of the twenty-first century, with total revenues almost five times that of Dow; in chemical sales alone it would rank as sixteenth largest chemical company in the world.

Δ (1864) Seventeen years before the electric light, the introduction of the Solvay process had signaled the beginning of the end for the Leblanc soda process, then the central process in chemical industry. In the face of increasing opposition to its practice of dumping waste hydrogen chloride into the environment, the industry found that hydrogen chloride reacted with manganese dioxide to produce chlorine. Chlorine was further reacted with lime to produce calcium hypochlorite, known in the trade as

4. For details of the Cowles-Hall dispute see Reference [56].

"bleaching powder." As the Solvay process was producing soda at ever lower prices, the Leblanc industry came to rely on sales of bleaching powder to the textile and paper industries as its major source of revenue. This last refuge of the huge Leblanc industry would be taken from it by Castner and Kellner's process for electrolytic chlorine; the Leblanc process would become extinct after the First World War.

Δ (1834) Thirty years before the first Solvay soda plant Davy's protege, ***Michael Faraday,*** first began to understand the nature of electrolysis. An electrolysis cell was simply a voltaic cell with the poles reversed; in a voltaic cell the anode, the site of oxidation, was the negative pole and the cathode, the site of reduction, was the positive pole; in the electrolysis cell the anode was forced to be positive and the cathode negative. Just as a mole of a substance can be delivered by weighing out one formula weight of that substance, a mole of electricity can be delivered by passing a current of 96,500 amps through a solution for 1 second; 96.5 amps for 100 seconds; or 9.65 amps for 1000 seconds. The quantity 96,500 amp-sec was, in essence, the formula "weight" for electricity. This new unit, the faraday, allowed stoichiometric calculations to relate the weights of reactants and products to the amount of electricity produced by a battery or consumed by an electrolytic cell.

Δ (1825) Nine years before Faraday demonstrated the chemical equivalence of electricity, ***Hans Christian Oersted*** reacted an amalgam of potassium and mercury with aluminum chloride. When the mercury was distilled, it left behind a new element, *aluminum.* This remarkable metal had a low density and resisted corrosion to a degree comparable to that of gold and silver. Aluminum began to be produced commercially and might have been extremely useful had it not been so expensive. Napoleon III, for example, reserved his aluminum tableware for honored guests on special occasions. Aluminum would remain a precious metal for most of the nineteenth century, its price falling below that of silver only after the introduction of the Hall-Héroult process.

Δ (1797) After millennia of wandering I passed from Lavoisier's *Elements of Chemistry* to the seventeen year old son of a Penzance wood carver. Forced to support his family by the death of his father three years earlier, this ***Humphry Davy*** had apprenticed himself to a saw-bones, but the spark burned too brightly within him to settle for a life of mere quackery. In 1798 he went to work for Thomas Beddoes investigating the therapeutic use of nitrous oxide (laughing gas) and his reputation as a chemist became such that he was appointed Professor of Chemistry at the Royal Institution in 1802. In 1808 Davy used the recently introduced voltaic battery to decompose molten potash and soda, producing two hitherto unknown elemental metals and naming them *potassium* and *sodium,* respectively. In 1810 he argued that the greenish-yellow gas liberated from common salt by manganese dioxide was, in fact, an element and gave it the name, *chlorine.* He further demonstrated that chlorine could be produced simply by passing an electrical current through salt water. This process, ***electrolysis,*** was to become a powerful tool, turning erstwhile products into reactants, reactants into products, in essence providing a means for driving redox reactions *backwards.*

25.2 🜍

I was a young teenager when I re-meme-bered the electrolysis of salt water. I was not yet a chemist, but I was keenly interested in science. The chemistry sets of the day did little to encourage an interest in chemistry; they went so far out of their way to make sure that nothing in the set could possibly be dangerous that little remained to interest the inquisitive mind. My scientific passion was inflamed, not by the watered-down kiddie-science of the sixties and seventies, but by the boy-scientist[5] books of the twenties and thirties. Among the admittedly dangerous activities described there, the electrolysis of salt water sparked my interest. With a few common materials I could make a poison gas on the one hand and an explosive gas on the other. Is this a great world, or what?

In Chapter 21 we learned that a voltaic cell, or battery, separates the two half-reactions of a redox reaction so that they take place at two different electrodes; the oxidation takes place at the anode and the reduction at the cathode. On a process schematic the electrons are shown explicitly as if they were reactants and products and these electron "pipes" represent the wires connected to the battery. The wire on the reactant side of the schematic represents the "+" terminal of the battery and the one on the product side represents the "-" terminal. The "desire" of the electrons to cross from anode to cathode is expressed as the electromotive force (EMF), with units of volts. The rate of flow of electrons from anode to cathode is expressed as the current, with units of amps. As a convention,

5. Such books may also be of interest to girl-scientists who are capable of overlooking the un-apologetic sexism of the era in which they were written.

Figure 25-1. The Lead-Acid Cell

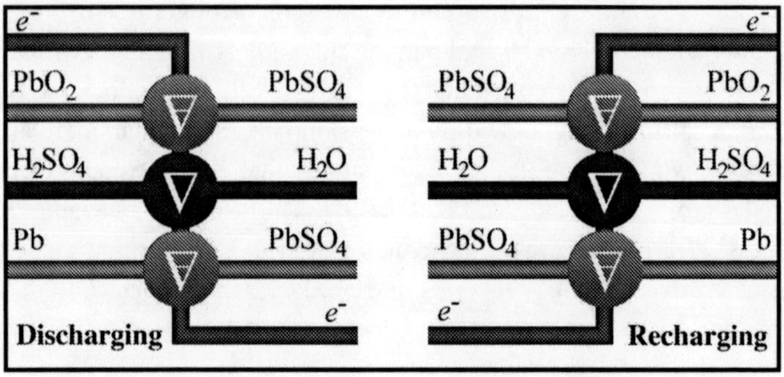

electrons flow from top to bottom in the process schematic of a voltaic cell, as if they were flowing downhill. Reversing the normal direction of electron flow results in an electrolytic cell, one in which electrons are "pumped" from the bottom to the top of the process schematic. Of course, "down" and "up" are merely conventional directions for these process schematics and have nothing to do with the physical geometry of the cell. Let us now explore three applications of electrolysis, the lead-acid storage battery, the Hall-Héroult aluminum process, and the Castner-Kellner chloralkali process.

Figure 25-1 shows a process schematic of the ***lead-acid storage battery;*** the anode is lead, the cathode is lead oxide, and the **electrolyte** is sulfuric acid. If the lead and lead oxide were to come into contact, they would react with one another, electrons flowing from one to the other. But by separating the lead from the lead oxide, we can require the electrons to pass through an external circuit to get from the anode to the cathode. Lead and lead oxide are consumed, lead sulfate is produced, and the sulfuric acid in the electrolyte is converted to water. The astute reader will be able to balance the lead-acid redox reaction after examining the schematic; reactants for the skeleton reaction are on the left of the figure and products are on the right:

$$? \; Pb(s) + ? \; PbO_2(s) + ? \; H_2SO_4(aq) = ? \; PbSO_4(s) + H_2O(l)$$

In operation, the negative terminal of the battery connects to the lead anode and the positive terminal connects to the lead oxide cathode. But suppose for a moment that we have access to a stronger battery than the lead-acid cell, that is, one with a higher EMF. Further suppose that we

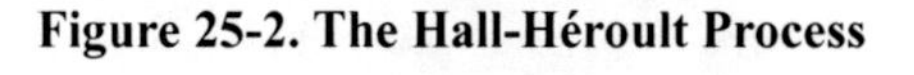

Figure 25-2. The Hall-Héroult Process

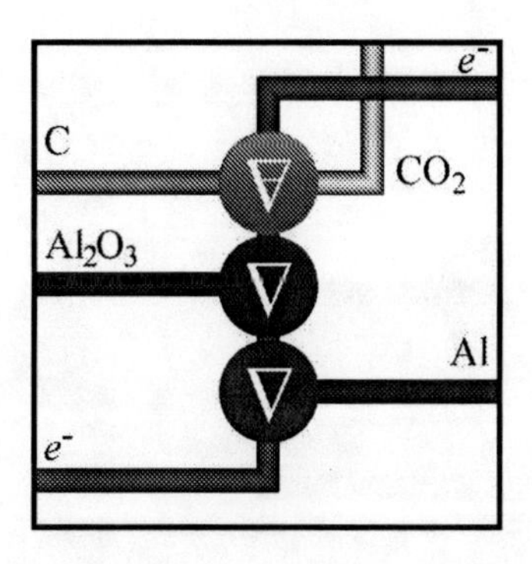

connect the two batteries anode-to-anode and cathode-to-cathode. Which direction will electrons flow? Normally, electrons flow out of the anode, but with two anodes connected to one another, the anode of the stronger battery "wins," pushing electrons into the weaker battery and converting its anode into a cathode. Similarly the stronger cathode pulls electrons from the weaker battery, converting its cathode into an anode. This drives the redox reaction of the weaker battery backwards, turning reactants into products and *vice versa.* In other words, electrolysis recharges the lead-acid cell. In practice, the EMF needed to recharge a car battery comes from the car's alternator rather than from another battery. But no matter what the source of the electrical power, the process schematic of the electrolytic cell is the mirror image of the voltaic one.

A second example of an electrolytic process, the Hall-Héroult process, has an operating temperature of 950°C. The electrolyte consists of molten alumina, Al_2O_3, and because the melting point of alumina is 2051°C, a flux is used to lower its melting point. The cryolite flux, Na_3AlF_6, is not consumed and so does not appear on the process schematic, Figure 25-2. The molten aluminum produced by the process serves as the cathode. Since aluminum is more dense than alumina, it sinks to the bottom of the reactor as a separate liquid phase. The carbon anode is consumed in the process, yielding carbon dioxide gas. As with the lead-acid battery, the reactants and products of the skeleton reaction appear on the schematic and from them the redox reaction may be balanced:

$$? \, Al_2O_3(l) + ? \, C(s) = ? \, Al(l) + ? \, CO_2(g)$$

Figure 25-3. The Chloralkali Process

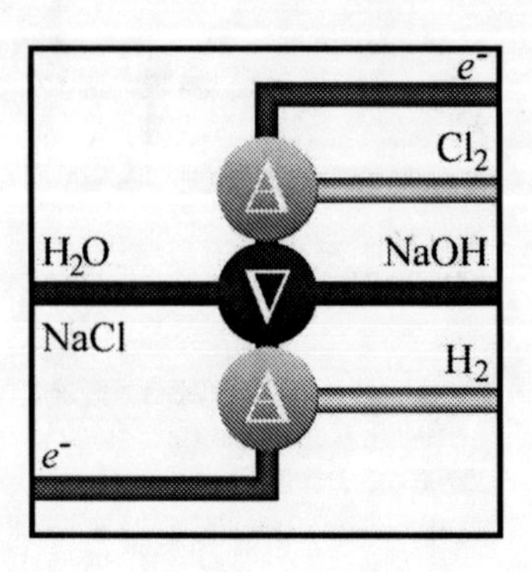

A third example of electrolysis is provided by the chloralkali cell, shown in Figure 25-3. Similarly to the lead-acid battery, the electrolyte is aqueous, in this case a solution of sodium chloride. The anodic half-reaction is:

$$2\ NaCl(aq) = Cl_2(g) + 2\ Na^+(aq) + 2\ e^-$$

The cathodic half-reaction converts water into hydrogen gas and hydroxide ion:

$$2\ Na^+(aq) + 2\ H_2O(l) + 2\ e^- = H_2(g) + 2\ NaOH(aq)$$

A simple chloralkali cell consists simply of two inert electrodes in a salt solution. But a complication arises because the chlorine gas produced at the anode reacts with sodium hydroxide to produce sodium hypochlorite, NaOCl, the active ingredient in ordinary laundry bleach:

$$Cl_2(g) + 2\ NaOH(aq) = NaOCl(aq) + NaCl(aq) + H_2O(l)$$

Though sodium hypochlorite is a useful product, a simple chloralkali cell produces a mixture of aqueous sodium chloride, sodium hydroxide, and sodium hypochlorite. The Castner-Kellner cell was designed to produce pure chlorine and sodium hydroxide by physically separating them from one another. The anodic half-reaction is the same as before, but the Castner-Kellner electrolysis cell employs a metallic mercury[6] cathode,

6. Mercury has acquired a bad reputation, some of it well-deserved. While metallic mercury is essentially non-toxic and quite insoluble in water, some small fraction of the mercury used in the Castner-Kellner cell will make its way into waste water. Discharged into lakes and streams, bacteria metabolize it into very toxic compounds such as methyl mercury. These compounds make their way up

at which the half-reaction is:

$2\ Na^+(aq) + 2\ e^- = 2\ Na(Hg)$

The product of this half-reaction is metallic sodium dissolved in metallic mercury. This mercuric solution flows from the electrolysis cell to a separate tank where its dissolved sodium reacts with water:

$2\ Na(Hg) + 2\ H_2O(l) = H_2(g) + 2\ NaOH(aq)$

The sum of these last two reactions is the same as our previous cathodic reaction; the Castner-Kellner cell simply ensures that the chlorine and sodium hydroxide are produced in separate containers. Relieved of sodium, valuable mercury is returned to the electrolysis cell and every effort is made to conserve it. Because it is not consumed in the process, mercury does not appear in Figure 25-3.

We must touch on one more topic before moving on to the construction of a simple chloralkali cell. In Chapter 15 we learned to use the formula weight to answer stoichiometric questions. In practice, electrons are measured out, not by weight, but by a combination of current and time. The unit factor, (96,500 amp·sec/mol e^-) may be used to answer stoichiometric questions involving voltaic and electrolytic cells. For example:

Q: How long will it take for a chloralkali cell passing a current of 100 mA to produce 500 mL of hydrogen gas?

A: 5.6 hours. Hint: because a higher current would require less time, "100 mA" goes in the bottom. Try starting with (500 mL H_2/100 mA) as the first unit factor in the chain.

Material Safety

Locate MSDS's for chlorine (CAS 7782-50-5), sodium hydroxide (CAS 1310-73-2), sodium chloride (CAS 7647-14-5), sodium hypochlorite (CAS 7681-52-9), and hydrogen (CAS 1333-74-0). Summarize the hazardous properties in your notebook, including the identity of the company which produced each MSDS and either the NFPA or HMIS ratings for each material.[7]

---

the food chain, being concentrated in the carnivores at the top, such as the tuna. Like the Leblanc soda industry before it, the chloralkali industry has been pressured to address its pollution problems and has gradually replaced aging Castner-Kellner cells with other designs which do not use mercury.

7. The NFPA diamond was introduced in Section 15.2 (page 184). You may substitute HMIS or Saf-T-Data ratings at your convenience.

Your most likely exposure is to chlorine gas. If breathing becomes difficult get plenty of fresh air and call for an ambulance. Be aware that hydrogen gas is flammable and that sodium hypochlorite will bleach clothing.

You should wear safety glasses while working on this project. All activities should be performed in a fume hood or with adequate ventilation. Leftover solution may be flushed down the drain. The hydrogen produced should be carefully burned. The chlorine should be released outdoors or into a fume hood.

Research and Development

You should not remain ignorant if you are to proceed in the Work.

- Know the meanings of those words from this chapter worthy of inclusion in the ***index*** or **glossary.**
- You should have mastered the Research and Development items of Chapter 21 and Chapter 22.
- Know one new unit factor from this chapter and be able to use it to answer stoichiometric questions involving electrons.
- Know the hazardous properties of chlorine, sodium hypochlorite, sodium hydroxide, and hydrogen.
- Know the balanced redox reactions for the lead-acid cell, the Hall-Héroult cell, and the Castner-Kellner cell.
- Be able to distinguish between the hazardous properties of mercury, mercuric chloride, and methyl mercury.
- Know the contributions of Dow, Castner, Kellner, Hall, Héroult, Faraday, and Davy to the understanding of the spark.

25.3 Θ

In the simplest possible terms, we shall pass an electric current through salt water. All that is absolutely necessary is to connect two wires to a battery and dip them into a glass of salt water. Copper wires would rapidly corrode, however, and so we shall use carbon electrodes, just as we did in the aluminum-alkali battery (Chapter 21). Convenient carbon electrodes may be scavenged from non-alkaline flashlight batteries, even "dead" ones. Such "heavy duty" batteries have an outer zinc case and a carbon rod down the center; alkaline have the reverse geometry, with a

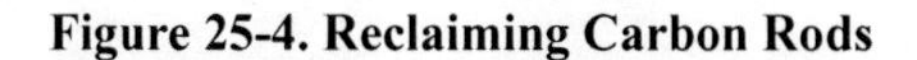

Figure 25-4. Reclaiming Carbon Rods

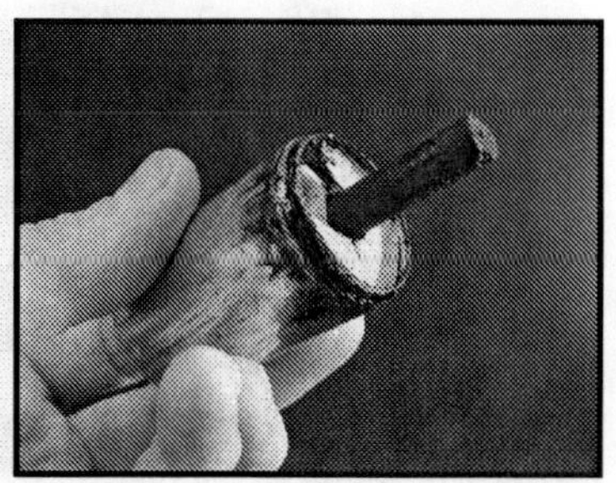

zinc rod down the center. Wearing safety glasses and gloves, use a pair of pliers to dismantle two standard (Leclanche) flashlight batteries. Peel off the steel jacket, exposing the zinc can underneath. The button end of the battery is usually sealed with a plastic cap; pry this cap off as shown in Figure 25-4(L) to reveal the carbon rod beneath. Carefully twist and pull the carbon rod from the battery. The black stuff in the zinc can is a paste of manganese dioxide and ammonium chloride insulated from the zinc by a paper sleeve. Save the carbon rods and dispose of the rest in the trash.

The body of the cell will be constructed from a 2-liter plastic soft-drink bottle. Use a knife to cut the *top* from the bottle at the shoulder. Poke two small holes on opposite sides of the bottle near the bottom and slide a carbon rod into each hole, leaving half an inch of each rod outside the bottle. Apply glue[8] to make a water-tight seal between the rods and the bottle, as shown in Figure 25-5(L), and allow the glue to dry or set. You will also need two smaller plastic bottles of such a size that they fit inside the 2-liter bottle, side-by-side. Use a knife to cut the *bottoms* from these bottles and trim them so that they fit completely inside the 2-liter bottle, one over each of the carbon rods.

Fill the 2-liter bottle to the brim with water and add salt (sodium chloride), stirring until no more will dissolve. With their caps removed, push the two smaller bottles into the larger one, bottoms first, so that each one sits over a carbon rod. The mouths of these bottles should be flush with the surface of the saturated salt solution. Screw on the caps, trapping as

8. Goop™ is an excellent choice. There are also epoxies designed specifically for use on plastics.

Figure 25-5. Constructing the Chloralkali Cell

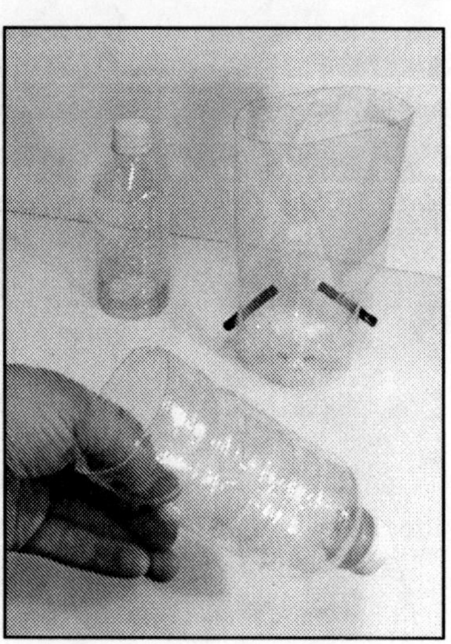

Figure 25-6. Filling the Chloralkali Cell

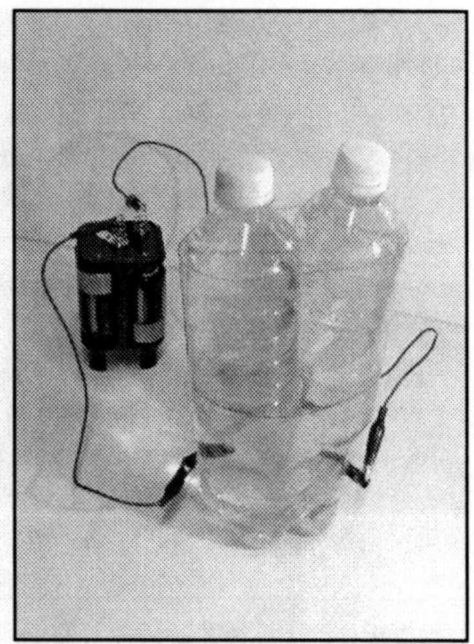

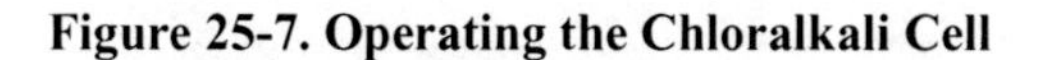

Figure 25-7. Operating the Chloralkali Cell

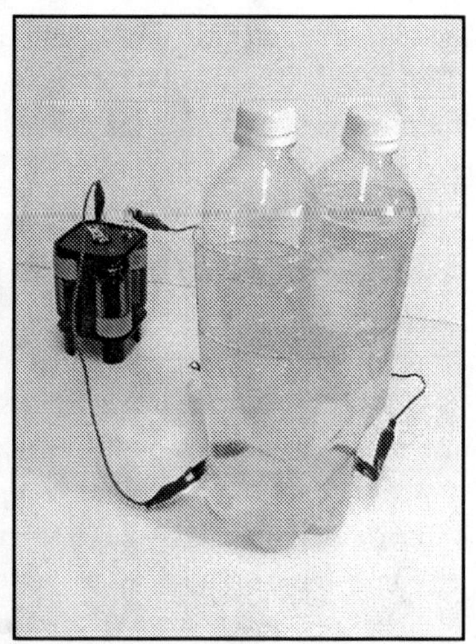

little air as possible in the small bottles, and lift them gently, as shown in Figure 25-6(R); the water level in the 2-liter bottle will fall but the smaller bottles will remain filled with water.

Use a wire to connect each of the carbon rods to a terminal of a lantern battery. Very soon bubbles will begin to form on the surface of the carbon electrodes and will rise to be collected in the two smaller bottles, as shown in Figure 25-7. One of these gases is hydrogen, the other chlorine. If you have been paying attention, you should be able to predict which is which by carefully considering Figure 25-3 and noting which electrode is connected to the positive and which to the negative terminal of the battery. Just picture electrons coming out of the negative terminal and flowing into the positive one. Hydrogen will collect more quickly than chlorine because some of the chlorine will dissolve in the increasingly alkaline solution to form sodium hypochlorite, the active ingredient in ordinary laundry bleach. As the gases displace the water in the smaller bottles, the water level in the larger one will rise; lifting the smaller bottles from time to time will prevent the larger one from overflowing. Nevertheless, it is probably a good idea to keep your chloralkali generator in a sink, tub, or bucket while it is in operation.

When one bottle is full of hydrogen gas, it is time to test your products. Carefully lift the smaller bottles from the larger one and place their open bottoms on a counter top or table which will not be harmed by bleach. The bottle which was full of gas will contain hydrogen. Light a splint or fireplace match and hold it against the bottom of the hydrogen bottle as you turn it over. The hydrogen will light with a satisfying pop. Be careful to point the bottle away from arms, faces, and other flammable materials.

The other bottle contains chlorine, the toxic gas, used on the battlefields of World War I. It is also the gas used to disinfect drinking water and swimming-pool water. The smell of laundry bleach is actually the smell of chlorine, which emanates from any hypochlorite solution. There is no need to be paranoid about the few hundred mL of gas you have generated, but neither should you be complacent. This element, liberated from ordinary salt, deserves your profound respect. Insert a wet strip of colored paper into your bottle of chlorine and you will see that it immediately bleaches the dye. When you are finished, release your chlorine either outdoors or into a fume hood.

The solution in your chloralkali cell consists of un-reacted sodium chloride, sodium hydroxide, and sodium hypochlorite. The mercury cathode of the Castner-Kellner cell was designed to produce sodium hydroxide in a container separate from the chlorine and salt, preventing the formation of sodium hypochlorite. Without such an arrangement you will be unable to test the pH of your solution using pH test paper because the sodium hypochlorite will bleach the blue color which would otherwise indicate alkali. Instead of testing the pH, you can test the bleaching power of your solution by dipping a strip of colored construction paper into it.

Quality Assurance

In order to pass, your hydrogen must be flammable and your chlorine must smell like chlorine. This is the strong power of all powers, for it overcomes everything fine and penetrates everything solid.

Chapter 26. Bayer (Pharmaceuticals)

Alkaloids. There is a considerable number of nitrogeneous compounds, found in plants, which combine with acids to form crystalline salts in the same way that the amines do. Some of these are comparatively simple amines, but most of them are complex in their structure. Many of them have some very marked physiological action as poisons or as medicines. Many which are poisonous are used as medicines in small doses.

Nicotine, $C_{10}H_{14}N_2$, is a colorless oil found in tobacco, which contains from 2 to 8 per cent of the alkaloid. It is very poisonous.

Coniine, $C_8H_{17}N$, the alkaloid of hemlock, is also a liquid. It is historically interesting as the active principle of the fatal draught taken by Socrates.

Atropine, $C_{17}H_{23}O_3N$, is found in *Atropa belladonna.* It is used to *dilate* the pupil of the eye and is an active poison.

Cocaine, $C_{17}H_{21}O_4N$, is found in coca leaves. It is used to produce local anaesthesia. A careful study of cocaine has shown that it is a derivative of benzoic acid and the group derived from that acid is chiefly effective in giving to it its valuable qualities. On the basis of this discovery other alkaloids having, in part, a similar structure have been prepared. Some of these retain the anaesthetic effect of cocaine and are less poisonous.

Morphine, $C_{17}H_{19}O_3N.H_2O$, is the most important alkaloid of opium and is the chief constituent which gives to laudanum and paragoric their poisonous and sedative qualities. Paragoric also contains camphor and aromatic oils which may have as much effect as the morphine. Opium is obtained from the poppy.

Quinine, $C_{20}H_{24}O_2N_2$, is obtained from Peruvian bark. It is a specific in malarial fevers.

Strychnine, $C_{21}H_{22}O_2N_2$, is found in *Strychnos nux vomica.* It is a violent poison, producing convulsions. A dose of 0.06 gram is considered fatal. In small doses it is a powerful stimulant.

— William Noyes, *Textbook of Chemistry*, 1913 AD [1]

1. Reference [21], pp. 342-343.

26.1 ☿

🜄 Drugs! Now that's a business. Eye of newt, wing of bat, flower of poppy, these are the stuff that dreams are made of. Up until the twentieth century, pretty much all drugs were 100% natural, meaning they were extracted from plants or animals; just grind up your plant or animal and extract it with hot water or alcohol just the same as if you were making tea or dye. These days, folks tend to think that "natural" is the same thing as "safe." Just try telling that to Socrates or Cleopatra or Napoleon or anyone else who was poisoned before 1900. No sir, some of the most powerful drugs and poisons around are completely natural. But because the plants and animals they come from don't grow just anywhere, drugs that come from far away places tend to be expensive. For example, quinine comes from the bark of the Peruvian cinchona tree. Quinine acts as an *antipyretic,* reducing fevers in victims of malaria and other tropical diseases. But the cinchona tree won't grow in England, which is why William Perkin tried to make quinine from coal tar in 1856, on account of England had plenty of coal. Of course, he didn't succeed in making quinine; he made mauve instead and that started the whole synthetic dye industry going. But you knew all that from Chapter 22.

🜂 The beginnings of the scientific revolution in medicine predate the synthetic dye industry by half a century. Davy had begun his career in 1808 studying the anaesthetic effects of nitrous oxide (laughing gas). Faraday had revealed the anaesthetic properties of ether in 1818. Liebig had prepared chloroform in 1831 and Robert Liston had first used it as a surgical anaesthetic in 1846. Joseph Lister had introduced phenol as a surgical antiseptic in 1867—

🜄 Which brings us back to coal tar derivatives. In 1853 Hermann Kolbe synthesized salicylic acid from phenol and carbon dioxide. Salicylic acid turned out to have antipyretic properties similar to salicin, an extract of the bark of the white willow tree. Though cheaper than quinine, salicylic acid had some whopping side effects: nausea, ringing in the ears, and whatnot. Then in 1886 two interns Kahn and Hepp were testing naphthalene for the treatment of intestinal parasites. Patients didn't get any relief from their worms, but the naphthalene reduced their fevers. In a second round of testing, naphthalene killed the worms but didn't help with the fevers. On closer examination Kahn and Hepp found that the first bottle of "naphthalene" had actually been mis-labeled. Consulting the manufacturer, the dye works of Kalle & Company, they found it to be a common dye intermediate, acetanilide. Acetanilide was such a common intermediate, in fact, that there was no hope of patenting it. So instead of marketing acetanilide under its chemical name, as was the custom, Kalle & Company took out a trademark on the brand name, Antifebrin, and began promoting it as an antipyretic. The brand name drug had been born.

∇ Pharmacists, of course, were outraged because they were used to providing generic drugs by their chemical names. But if doctors wrote prescriptions for Antifebrin, pharmacists were forbidden by law to substitute what they knew to be the identical and less expensive acetanilide. And no competitive dye house had an incentive to market acetanilide since it could not be protected by patent. The most they could hope for was that one of their own dye intermediates might prove pharmacologically active.

∆ When Antifebrin was introduced by Kalle, ***Carl Duisberg*** was the head of research and patenting for ***Farbenfabriken vormals Friedrich Bayer,*** one of the many German dye companies. Faced with the disposal of thirty thousand kilos of waste para-nitrophenol, Duisberg challenged his chemists to turn it into a drug. The result was acetophenetidine, marketed by Bayer as an antipyretic and *analgesic* (pain reliever) under the brand name, Phenacetin. With fewer side effects than Antifebrin, Phenacetin was a cash cow for Bayer, the first of many; in 1898 Bayer introduced a new-and-improved, non-addictive alternative to morphine, then used to treat tuberculosis. Because it made patients feel heroic, Bayer marketed the wonder drug under the brand name, Heroin.

🜁 Don't tell me you're going to tell them how to make Heroin! Have you no shame? Have you no sense? Have you no fear of prosecution?

∆ Figment! Did you have a pleasant sulk? I was wondering whether you would ever come back.

🜁 You're going to get the Author arrested. What kind of stories do you imagine he'll write for us from the Powhatan Correctional Facility?

∀ Keep your pants on. Heroin is too hard for a beginner to make. I think acetanilide would be a better choice.

🜁 I think you ought to have your eight-eyed head examined. Regular use of acetanilide kills red blood cells and turns peoples' skins blue.

Δ Obviously a safer alternative to acetanilide was needed. The year following its introduction of safe, effective Heroin, Bayer came out with a new antipyretic with many fewer side effects than either acetanilide or acetophenetidine. Like its predecessors, acetylsalicylic acid (ASA) had been invented years before and was not patentable under German law. Consequently Bayer chose to market it under the brand name, Aspirin.

∇ The American patent system was a wee bit more accommodating than it was in Germany and the United States had already become the largest market for Bayer dyes. Not only did Bayer have the trademark, Aspirin, it held a US patent. With patent protection, Bayer could sell Aspirin at a higher price than in Europe, and American firms were unable to manufacture the generic ASA. But Bayer recognized that when its seventeen-year patent expired, duties on imported drugs would give American rivals the upper hand. In 1903 Farbenfabriken Bayer began manufacturing Aspirin in New York State through its American subsidiary, the ***Bayer Company.*** By manufacturing Aspirin in the United States, Bayer would be able to circumvent the import duties. And by promoting Aspirin (not acetylsalicylic acid), Bayer hoped that the brand name would be so firmly entrenched in the medical community that when the patent expired, rival manufacturers of ASA would be unable to compete in the American market. By 1909, Aspirin accounted for 31% of Bayer's US sales.[2]

∇ It was a good plan except for one little glitch; "ethical drugs," the equivalent of modern prescription drugs, were sold through pharmacies, which provided their own packaging. Thus consumers would be unaware of the Bayer trademark. If you wanted to provide your own packaging, you had to put out *patent medicines,* which were looked down upon by doctors and pharmacists. Ethical drugs had to be tested and proven safe; patent medicines could make whatever claims they wanted and it was up to the customer to decide whether they worked or not. Doctors and pharmacists would boycott any company that tried to sell both ethical drugs and patent medicines. Bayer had positioned itself in the ethical drug market and as such was prohibited from advertising to the general public.

∇ To get around this restriction without offending doctors and pharmacists, Bayer started stamping the Bayer logo into each tablet in 1914. That way, the public would come to know Bayer as the genuine producer of Aspirin. It was pure genius. And it worked. In fact, it worked too well

2. Reference [64], p. 35.

for Bayer's own good. The United States entered the First World War in 1917 and German-owned companies were placed in trust, to be returned at the end of the war. The Bayer Company was cut off from Farbenfabriken Bayer, though its German-born managers were loyal to the parent company. But financial and political scandals resulted in the arrest of the American management team and the seizing of Bayer's American assets. These assets, including the American patents and trademarks, were sold to the highest bidder at the end of 1918, just as the war was ending. The buyer was Sterling Products, manufacturer of such patent medicines as Neuralgine (a headache remedy), No-To-Bac (a nicotine cure), Danderine (a dandruff remedy), and California Fig Syrup (a laxative). Bayer Aspirin was about to go over-the-counter.

Δ In 1919 the co-founder of Sterling, ***William Weiss,*** had the temerity to approach Duisberg at Farbenfabriken Bayer with a request for assistance in getting the Bayer Company plants operational, as no one at Sterling knew how to run the equipment. Duisberg had assumed that after the war, Farbenfabriken Bayer would resume its operations in the United States. Now he found that not only had Bayer lost its chemical plants in New York, it had also lost its US rights to the "Bayer" and "Aspirin" trademarks. Even if Farbenfabriken Bayer built new manufacturing facilities in the US, it would be unable to market its ASA as Bayer Aspirin. During the war Duisberg had forged a cartel of the German chemical giants, Bayer, Hoechst, BASF, and AGFA to form IG Farben, of which he was chairman of the board. And now this brash American hay-seed had the audacity to offer Bayer the "opportunity" to sell Bayer products exclusively through Sterling when it became clear that Sterling would be unable to manufacture these products itself.

∇ Now hold on there, cowboy. Sterling had split off all but the Aspirin business into its subsidiary, the Winthrop Chemical Company. Sterling knew everything it needed to make Bayer Aspirin. The Bayer Aspirin business was booming, thank you very much, on account of Sterling's advertising genius. It just needed a little help with the sixty-three other Bayer pharmaceuticals to which Winthrop now held the US patents. If Farbenfabriken Bayer would help Winthrop out in the technical department, Sterling would make Bayer a household name.

Δ Bayer was already a household name! Bayer had discovered the analgesic properties of ASA. Bayer had made Bayer Aspirin the universal choice for pain relief. Bayer still held the trademark throughout most of the world. Bayer's American trademarks and patents had been stolen from it during the war; Sterling had simply bought stolen goods. Bayer was happy to reimburse Sterling for its expense, perhaps make Sterling a subsidiary of Bayer. Otherwise, Bayer would see to it that Sterling was known throughout the world as the *fake* Bayer.

∇ The fake Bayer in Germany, maybe, but in 1919 Sterling bought the rights to Bayer patents and trademarks in England and its sales in South America were booming, thanks to its advertising prowess. No sir, Bayer would deal with Sterling or be relegated to just another producer of ASA. Sterling offered 50% of the Winthrop profits in exchange for technical help. Sterling would have exclusive rights to sell Bayer products, including Bayer Aspirin, in the United States, Canada, Great Britain, Australia, and South Africa. After all, Bayer was good at making; Sterling was good at selling. It would be a win/win situation.

Δ Farbenfabriken Bayer would control the Bayer trademarks and patents over the rest of the world. Sterling would refrain from using Bayer trademarks on Danderine, Fig Syrup, or any of its other snake-oil remedies and would not use Sterling trademarks on any Bayer products. The deal was struck on April 9, 1923 and Bayer Aspirin lived a double life for the next seventy-one years. Then on November 2, 1994 SmithKline Beecham bought Sterling Winthrop from Eastman Kodak. The next day it sold the North American over-the-counter component of the business, including the Bayer trademarks, for 1 billion dollars to Miles, Inc., a wholly owned subsidiary of Bayer.

∇ Well, Sterling may have been carved up, scattered to the four winds like so many spiders, but along the way it changed the way that drugs are marketed. Over-the-counter medications are no longer the untested patent medicines of yesteryear. No, these days the market for untested cures belongs to herbal remedies and food supplements. The big drug companies make both prescription and over-the-counter drugs, advertising both kinds of medicine to the general public. And Bayer would begin the twenty-first century as the fourth largest chemical company in the world.[3] Not bad for a dye company.

---

3. Reference [54], July 29, 2002, p. 16.

26.2 ♀

We have met several classes of organic compounds in previous chapters. Chapter 4 introduced ethanol as the primal member of the ***alcohols.*** Alcohols can be distinguished by the presence of an "-OH" functional group, 2/3 of a water molecule, hanging off of various and assorted carbon atoms. Subsequently, Chapter 19 gave us glycerol, a tri-alcohol. As you look on ingredient labels around your house, you can recognized the names of alcohols because they always end in "-ol;" you might find "biglongnamanol," for example.

In Chapter 16 we saw our first ***organic acid,*** acetic acid, the compound which makes vinegar sour. Subsequently, Chapter 19 introduced the fatty acids, oleic acid, palmitic acid, and stearic acid. Organic acids are characterized by the peculiar "-COOH" functional group and their names end in "-ic acid;" you might find "biglongnamic acid" or its sodium salt, "sodium biglongnamate" as ingredients in shampoos.

An important concept from Chapter 18 was the ***acid anhydride.*** Carbon dioxide, for example, is the anhydride of carbonic acid:

$$H_2CO_3 = CO_2 + H_2O$$

Sulfur trioxide is the anhydride of sulfuric acid:

$$H_2SO_4 = SO_3 + H_2O$$

The same idea gives us acetic anhydride, the anhydride of acetic acid:

$$2\ CH_3COOH = (CH_3CO)_2O + H_2O$$

You can think of acetic anhydride as a kind of "super vinegar." It smells like vinegar, tastes like vinegar, and turns into vinegar when you add it to water.

Chapter 22 started us off on a new class of compounds, the ***amines,*** of which aniline and toluidine are examples. Just like an alcohol has an "-OH" hanging off of it, an amine has an "$-NH_2$" hanging off of it. You can tell that something is an amine because its name ends in "-ine" or "-in;" biglongnamine would be an example. In fact, if you look back to the beginning of this chapter you will find a whole bunch of natural drugs and poisons—nicotine, morphine, cocaine, quinine—all of them amines. Like ammonia, the amines are bases, or alkalis, and participate in the same kinds of reactions as inorganic bases.

Equation 26-1. From Aniline to Acetanilide

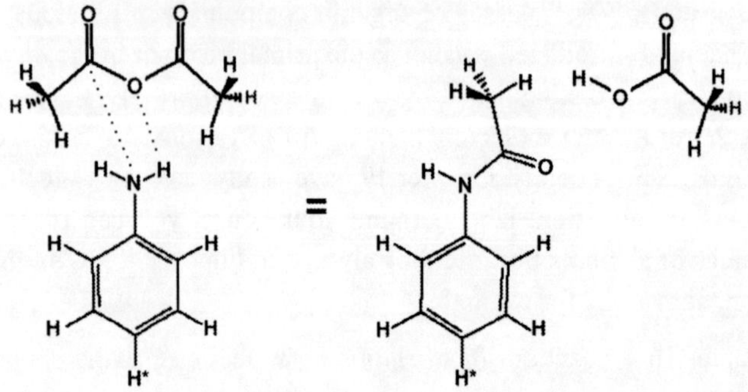

Chapter 7 introduced metathesis reactions, in which inorganic chemicals swapped first and last names. One kind of metathesis reaction was the acid-base reaction, in which a base and an acid combine to produce a salt. Ammonia, for example, reacts with hydrochloric acid to form ammonium chloride. The amines can also form salts with acids; while aniline and the toluidines are insoluble in water their salts, aniline hydrochloride and toluidine hydrochloride, are soluble in water. This is a common trick for making pharmaceuticals soluble in water; look at the ingredients of cold remedies and you may find things that look like "biglongnamine hydrochloride," the hydrochloride salt of the otherwise insoluble biglongnamine.

The reaction of an amine with an acid produces a salt, but the reaction of an amine with an acid anhydride produces an ***amide.*** Such a reaction, one in which two molecules join together by spitting out a smaller molecule, is called a **condensation** reaction. Equation 26-1, for example, shows the reaction of aniline with acetic anhydride to produce acetic acid and acetanilide, the drug trademarked as Antifebrin. The dotted lines show where the new bonds will form when acetic anhydride and aniline react. You can imagine half of the acetic anhydride sticking to the nitrogen atom on the amine, while the other half takes off with one of the hydrogen atoms. The acetic acid product is, of course, quite soluble in water, while the acetanilide product is not very soluble and falls out of solution as a precipitate. Thus the acetanilide can be separated from the acetic acid by filtration.

Acetanilide has passed into pharmaceutical oblivion because of its side effects but its near cousin, acetaminophen, has survived as the analgesic trademarked, Tylenol. As a matter of fact, acetanilide is converted into acetaminophen in the body and this latter molecule is responsible for the analgesic effects of the former. We can get the analgesic benefit while avoiding the side effects of acetanilide by taking acetaminophen directly. The structure of acetaminophen is the same as that of acetanilide except that the hydrogen atom marked "-H*" in Equation 26-1 is replaced by an "-OH." In fact, it's this group, characteristic of alcohols, which puts the *ol* in Tylen*ol.*

The chapter would not be complete without a discussion of the big daddy of analgesics, acetylsalicylic acid, trademarked, Aspirin. Figure 26-1 gives a process schematic for the synthesis of Aspirin beginning with the coal-tar distillate, ***phenol,*** an aromatic alcohol. You may be familiar with the aroma of phenol, the active ingredient in Chloraseptic throat spray. Reaction of phenol, structure (a), with sodium hydroxide produces sodium phenolate, structure (b). Reaction of sodium phenolate with carbon dioxide produces sodium salicylate, structure (c), and reaction of sodium salicylate with sulfuric acid produces salicylic acid, structure (d). Notice that salicylic acid is a curious beast: it is both an aromatic acid and an aromatic alcohol. Sodium sulfate stays in solution while salicylic acid precipitates and can be filtered out.

Just as the condensation of an amine with an acid anhydride produces an amide, the condensation of an alcohol with an acid anhydride produces an ***ester.*** In this case, reaction of salicylic acid with acetic anhydride in toluene solution produces acetylsalicylic acid, structure (e). Acetic acid is soluble in toluene while acetylsalicylic acid precipitates and can be filtered out. Most of the reactants in the Aspirin synthesis, sodium hydroxide, carbon dioxide, and sulfuric acid, are familiar from previous chapters. Similarly, most of the processes are variations on familiar themes, the reaction of an acid with a base and the absorption of carbon dioxide. Only the final reaction, the condensation, is entirely new. This reaction is common to the syntheses of both Aspirin and Antifebrin and is the one we'll explore in this project.

Material Safety

Locate MSDS's for aniline (CAS 62-53-3), acetic anhydride (CAS 108-24-7), and acetanilide (CAS 103-84-4). Summarize the haz-

Figure 26-1. The Aspirin Process

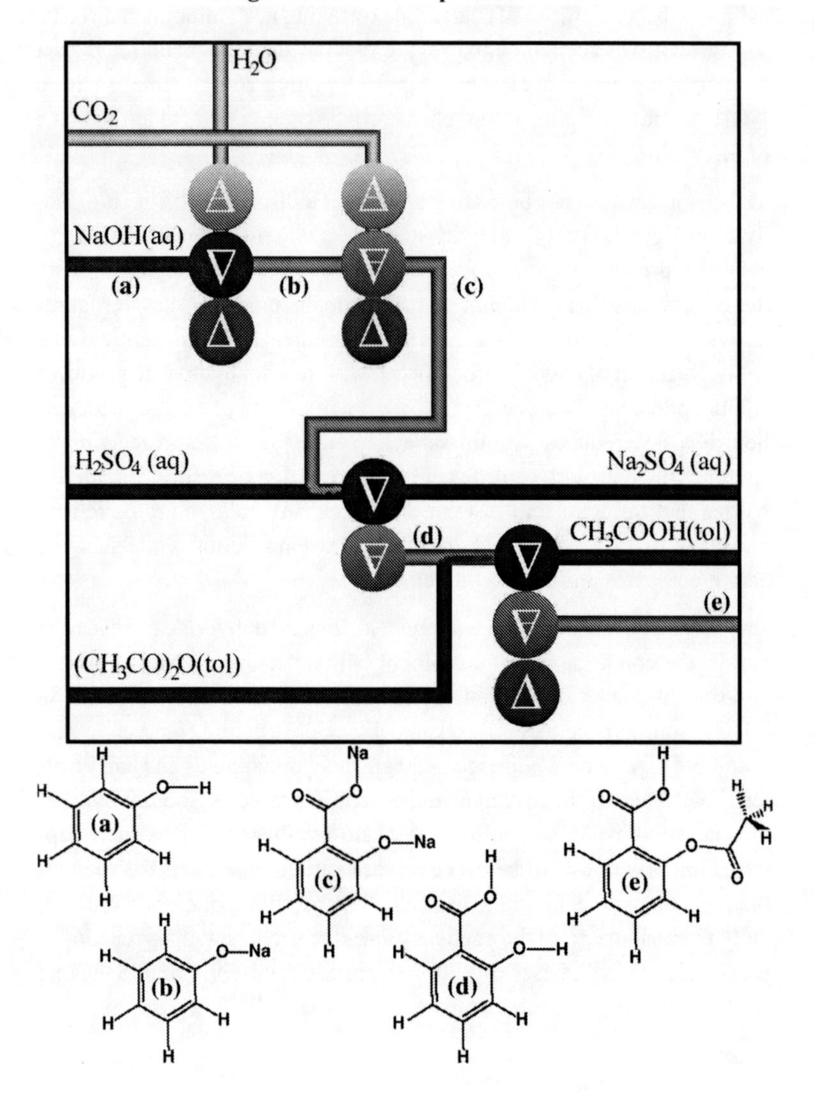

ardous properties in your notebook, including the identity of the company which produced each MSDS and the NFPA diamond for each material.[4]

Your most likely exposure is inhalation of acetic anhydride vapor; if your eyes begin to water, you should go someplace to get fresh air. You should also be aware that aniline may be absorbed through the skin. In case of contact wash the affected area immediately with soap and water.

You should wear safety glasses and rubber gloves while working on this project. All activities should be performed in a fume hood or with adequate ventilation. Leftover aniline should be dissolved in vinegar and flushed down the drain with plenty of water. Leftover acetic anhydride may be dissolved in water and flushed down the drain. Your acetanilide product may be thrown in the trash when you're finished with it.

Research and Development

Before you get started, you should know this stuff.

- You better know all the words that are important enough to be ***indexified*** and **glossarated.**
- You should know all of the Research and Development stuff from Chapter 21 and Chapter 22.
- You should know the structures of the coal-tar distillates, aniline and phenol.
- You should know the difference between aniline acetate and acetanilide.
- You should know the structure of acetanilide, acetylsalicylic acid, and acetaminophen.
- You should know the hazardous properties of aniline, acetic anhydride, and acetanilide.
- You should know the stories of how acetanilide became Antifebrin and how Aspirin went over-the-counter.

4. The NFPA diamond was introduced in Section 15.2 (page 184). You may substitute HMIS or Saf-T-Data ratings at your convenience.

26.3 Θ

I've made an effort throughout this book to make things out of chemicals that are commonly available, but for this project we'll need aniline and acetic anhydride, neither of which are available "over-the-counter," as it were. Aniline, first encountered in Chapter 22, is a brown oily liquid prepared from coal tar or, more recently, from petroleum. Acetic anhydride is a colorless liquid that smells like vinegar on steroids. Both of these chemicals must be obtained from chemical suppliers as they have no household uses.

In order to make acetanilide, you'll need to answer the usual stoichiometric questions:

Q: Given Equation 26-1, how many grams of acetic anhydride are needed to react with 1 gram of aniline? We'll need to use excess acetic anhydride because some of it will be hydrolyzed by the water used as a solvent for the reaction. So multiply your calculated weight of acetic anhydride by 1.1 to provide a 10% excess.

Q: How many grams of acetanilide should be expected from the complete reaction of 1.0 gram of aniline with excess acetic anhydride? This is the *theoretical* yield.

Weigh out 1.0 gram of aniline and add it to 10 mL of water in a 100-250 mL beaker. The aniline will sink to the bottom of the beaker, since it's not very soluble in water. Now weigh out your acetic anhydride and add it to the beaker a few drops at a time, stirring the contents of the beaker between drops. In just a few minutes the acetanilide product will crystallize. This crude acetanilide is contaminated with acetic acid and aniline acetate. Such contamination cannot be tolerated in a pharmaceutical and it must be removed. Fortunately for us, the solubility of acetanilide differs substantially from that of acetic acid and aniline acetate.

Acetanilide is quite soluble in boiling water, but not in cold water. By contrast, acetic acid and aniline acetate are just as soluble in cold water as in hot water. If you have paid any attention to the previous chapters, it ought to be obvious to you that we need to do some recrystallization. Add another 25 mL of water to the crude acetanilide in your beaker and heat it on a hot plate or over a spirit lamp until the water comes to a boil. All of the solids should dissolve in the boiling water; if they don't, add a

Figure 26-2. Acetanilide Crystals

little bit more water until they do. Remove the beaker from the heat and set on the counter top to cool.

If you've been earthified to any extent, you'll remember that the slower the recrystallization goes, the purer the crystals will be. So I'm going to suggest that you not rush it; let the hot solution cool slowly and your crystals will be large and pure. The acetic acid and aniline acetate will remain in solution. When the solution reaches room temperature, fold a piece of filter paper, place it into a funnel, and filter your purified acetanilide product. Rinse your crystals with a few mL of cold water, return them to the beaker, and allow them to dry overnight. The spent solution can be poured down the drain. Weigh your dried crystals and express the actual yield as a percentage of the theoretical yield.

Quality Assurance

Your acetanilide should take the form of flat, virtually odorless crystals. Record the amounts of materials used and your percent yield in your notebook. Tape a photograph of your crystals into your notebook as a record of your achievement.

Chapter 27. Badische (Fertilizers)

My chief subject is of interest to the whole world—to every race—to every human being. It is of urgent importance to-day, and it is a life and death question for generations to come. I mean the question of food supply. Many of my statements you may think are of the alarmist order; certainly they are depressing, but they are founded on stubborn facts. They show that England and all civilised nations stand in deadly peril of not having enough to eat. As mouths multiply, food resources dwindle. Land is a limited quantity, and the land that will grow wheat is absolutely dependent on difficult and capricious natural phenomena. I am constrained to show that our wheat-producing soil is totally unequal to the strain put upon it. After wearying you with a survey of the universal dearth to be expected, I hope to point a way out of the colossal dilemma. It is the chemist who must come to the rescue of the threatened communities. It is through the laboratory that starvation may ultimately be turned into plenty.

— William Crookes, *The Wheat Problem*, 1898 AD [1]

27.1 ☿

🜁 I don't believe it! A chapter that actually has something to do with air. I suppose the Author will give it to Samson, who will probably find some roundabout way to work solubility into the discussion. Or how about Lucifer? A little fire and brimstone? No, no, let me guess. The Author will probably assign the chapter to Unktomi, who will undoubtedly earthify the place.

🜃 Well, now that you mention it, Crooke's little introductory bit does mention land and soil and such. See, plants need a bunch of elements to grow, hydrogen and oxygen and carbon being at the top of the list. Plants have no trouble getting these things; hydrogen and oxygen come from water and carbon comes from carbon dioxide in the air. From hydrogen and oxygen and carbon they make sugar and from sugar they make cellulose. And what with plants being mostly cellulose and cellulose coming from air and water, I can't blame you for not seeing where the earthification would come in. But besides cellulose, plants need to make proteins

1. Reference [9], pp. 6-7.

and to do that they need nitrogen. Now, you might think plants would have no trouble getting nitrogen, what with the air being 80% nitrogen and all. But most plants can't take nitrogen from the air; they have to get it from the soil. So folks have been using nitrogen-rich dung as a fertilizer since God was a child and it seems to me that earthification is exactly what this chapter needs.

△ I knew—

▽ Of course, to get from the soil into the roots of the plant, nutrients must be soluble in water. There are two water-soluble nitrogen ions: ammonium and nitrate. Ammonia was traditionally supplied by stale urine. Nitrates were traditionally extracted from manure and solublized with potash. So you see, without these soluble nutrients plants would be up a creek.

△ I knew it was—

△ Being mere creatures of water and earth and air you undoubtedly failed to appreciate the lessons of Tu-Chhun. The precious nitrates of which you speak are not merely nutrients for vegetables; they feed the arsenals which protect us all from invasion and insurrection. Any society which fails to safeguard its nitrates will inevitably succumb to one which does. Crookes was well aware of this in 1898 and his call for nitrogen fixation was couched in terms of starvation and plenty to make it palatable to those incapable of keeping in the heat and withstanding it.

△ I knew it was too good to be true. You three are completely missing the point. There you are, surrounded by a virtually unlimited supply of atmospheric nitrogen, but because you don't know how to "fix" it, to convert nitrogen into ammonia or nitrates, you're left to wallow in the dung-heap.

△ Saltpeter (potassium nitrate) had been the oxidizer of choice for explosives since the invention of gunpowder. In 1846, however, two new explosives burst on the scene; ***nitrocellulose*** from the reaction of cotton with sulfuric and nitric acids was discovered by a true Athanor, Christian Schönbein; ***nitroglycerin*** from the reaction of glycerine with sulfuric and nitric acids was discovered by another Athanor, Ascanio Sobrero. Nitrocellulose became the basis of *smokeless* gunpowder, as all of its combustion products are gases. Nitroglycerin remained a laboratory curiosity because of its spectacular sensitivity to shock until a third Athanor, Alfred Nobel, learned to stabilize nitroglycerin with silica. We patented this ***dynamite,*** in 1866 and from that time, ***nitric acid*** became strategically important as an oxidizer.

∀ Of course, nitric acid was made from saltpeter and sulfuric acid, like it shows in Equation 18-4 (page 226). Sulfuric acid continued to be manufactured by the lead chamber process, but beginning in 1888 Badische Anilin und Soda Fabrik (BASF) began making it by the more efficient ***contact process.***[2] So acid was as big a business as ever, and saltpeter too, extracted in the traditional way from dung. But from about 1840 folks started importing sodium nitrate mined from huge deposits in Chile, which was exporting over a million tons of the stuff by 1902. But what with wars and famines and population explosions and all, the old-fashioned sources of nitrates weren't going to last forever. Pretty soon everybody and *his* dog were looking for a way to make nitric acid from something other than saltpeter.

∇ There was ***ammonia,*** of course. Available from stale urine in the early days, ammonia became increasingly available in the nineteenth century as a by-product of the manufacture of coke and coal gas. Ammonia could be reacted with sulfuric acid to produce a dandy water-soluble fertilizer, ammonium sulftate. In 1902 ***Wilhelm Ostwald*** perfected the oxidation of ammonia over a platinum catalyst to produce nitric acid. From ammonia and nitric acid you get ammonium nitrate, the most important fertilizer of the twentieth century, which would usher in an era of unprecedented—

∀ Hold on there, cowboy, you're getting ahead of yourself. Crookes knew all about dung and urine and coal gas and after adding them all up he figured they weren't going to be enough for the industrialized world, much less for the backwaters. Funny story; a fellow named Thomas Willson was trying to smelt calcium from lime long about 1892. It didn't work. Instead of the metal, calcium, he got something that looked like dirt but which turned out to be calcium carbide. When he threw it in the river, up from the water came a bubblin' gas. Acetylene, that is. Acetylene became all the rage for welding and such, which gave Union Carbide its start in 1898.

∆ Fascinating to a hillbilly, I'm sure, but we're talking about fertilizer.

∀ I was just coming to that. So in 1898 Adolph Frank and Nikodem Caro reacted calcium carbide with atmospheric nitrogen in a high-temperature furnace to produce ***calcium cyanamide,*** which turned out

2. Reference [52], p. 85.

to be a nifty fertilizer. Pretty soon everybody and *her* dog were making cyanamid by the Frank-Caro process. American Cyanamid got started that way long about 1907. World production of cyanamid grew from about 7000 tons in 1907 to 120,000 tons in 1913.[3]

∆̲ The route from cyanamid to nitric acid was not straightforward, however. In 1909 ***Fritz Haber*** of BASF reacted hydrogen with atmospheric nitrogen at high temperature and pressure over an osmium catalyst to produce ammonia. The laboratory process was scaled up by BASF engineer ***Carl Bosch*** and by 1912 a pilot plant was producing a ton of ammonia per day.

∆ . . . and just in the nick of time, for Germany. In 1913 Germany imported approximately half of its nitrogen from Chile, a source which became unavailable at the outbreak of WWI due to a successful British blockade. As a consequence, German agricultural and military demands for fixed nitrogen were in direct competition. At the beginning of the war domestic nitrogen came predominantly from the coking of coal; by the end of the war half of German nitrogen production came from the Haber-Bosch process and Germany was nearly self-sufficient in nitrogen. With superiority in fixed nitrogen and aniline dyes, BASF came to dominate German chemicals by the end of the war, and German chemicals dominated the—

∇̶ Yeah, yeah; almost makes you forget that Germany *lost* the war, doesn't it? Imported Chile saltpeter, ammonium sulfate from coal gas, and cyanamid were good enough for poor old Britain, France, and the United States. Just imagine if they'd had the Haber-Bosch process. Oh, yeah, I forgot; they *did* get it—after the war!

∆ In other words, even in the face of German defeat the Haber-Bosch I-dea conquered its competitors during the general economic slump which followed the war. For example, while American Cyanamid assets fell by 8% from 1918 to 1921, BASF assets *quadrupled.*[5] BASF began the twenty-first century as the third largest chemical company in the world.[6]

∆̲ And Haber was awarded the Nobel Prize in 1919. Twenty-one tumultuous years after Crookes' challenge to the chemical community, virtually unlimited quantities of fertilizer could be manufactured from thin air.

---

3. Reference [61], p. 102.
5. Reference [61], p. 254.
6. Reference [54], July 29, 2002, p. 16.

🜂 Fertilizers, yes; but it is a mistake to forget that the same chemicals that improve agricultural efficiency so dramatically can have other dramatic effects. The truck bomb that leveled the Murrah Federal Building, for example, was probably filled with ammonium nitrate fertilizer and fuel oil.

🜁 I don't believe it; we finally agree about something. At least I don't have to worry about you revealing the secret of high explosives.

🜂 Quite the contrary, it is not in the interest of a free society for that particular secret to be kept. An ignorant population is the most vulnerable to attack; in the land of the blind, the one-eyed chemist is king.

27.2 🜍

Two processes enabled the twentieth century to be one of both unprecedented violence and unprecedented prosperity. The Haber-Bosch process for the synthesis of ammonia made it possible to produce nitrogen fertilizers from air. The Ostwald process made it possible to produce nitric acid from ammonia. Together, these processes freed both farmers and soldiers from their historical dependence on saltpeter, either extracted from animal manure or imported from South America. In the twenty-first century, the production of food and munitions remains subject to political and economic, but not chemical limitations.

The schematic for these processes is deceptively simple. The starting materials, ***oxygen, hydrogen, nitrogen,*** and ***water*** enter from the left of Figure 27-1, as usual. Oxygen and nitrogen come from the air. Hydrogen might well be supplied by a chloralkali plant, as discussed in Chapter 25, but is more economically generated from natural gas. Water may be obtained from any convenient source. Some engineering is required to separate and purify the starting materials, but the details have been left out of the process schematic for simplicity. The purified starting materials pass through three reactors in turn, a furnace, a burner, and an absorber.

The first reactor in the schematic, the Haber-Bosch ***furnace,*** reacts nitrogen and hydrogen to produce ***ammonia.*** This reaction is so simple that you might well wonder why it wasn't developed in the nineteenth century. The reason is that the reaction requires both high temperature (500°C) and high pressure. These extreme conditions were a challenge to engineers trained on the low-pressure reactors of Solvay soda plants,

Figure 27-1. The Haber-Bosch and Ostwald Processes

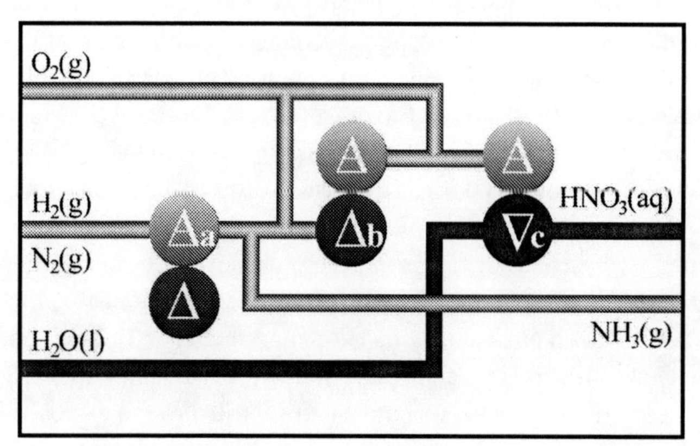

$$(a)\ 3\,\mathrm{H_2}(g) + \mathrm{N_2}(g) = 2\,\mathrm{NH_3}(g)$$

$$(b,c)\ 2\,\mathrm{NH_3}(g) + 4\,\mathrm{O_2}(g) = 2\,\mathrm{HNO_3}(aq) + 2\,\mathrm{H_2O}(l)$$

$$(optional)\ \mathrm{HNO_3}(aq) + \mathrm{NH_3}(g) = \mathrm{NH_4NO_3}(s)$$

but even with a suitable high-pressure furnace, the reaction goes only at a snail's pace. To speed up the process, hydrogen and nitrogen are passed over pellets of iron oxide and aluminum which serve as a **catalyst,** *a material which makes a slow reaction occur faster but which is neither produced nor consumed in the reaction.*

The automotive catalytic converter is, perhaps, the most familiar example of a catalyst. Hot exhaust gases pass over the catalyst, often an alloy of platinum and rhodium. Nitrogen oxides are converted to nitrogen and carbon monoxide is converted to carbon dioxide, reducing the emission of these pollutants from the automobile tailpipe. Unlike gasoline, which is consumed by the car; unlike exhaust, which is produced by the car; the catalyst is unchanged by the reaction. You don't have to refill the converter with catalyst. It is possible, however, for a catalyst to be *poisoned,* destroying its catalytic properties. The automotive catalytic converter, for example, is poisoned by tetraethyl lead, formerly used as an anti-knock additive in gasoline. Coincidentally, the automotive platinum/rhodium catalyst is also employed in the second reactor of Figure 27-1.

The ammonia produced by the Haber-Bosch furnace is a commodity in its own right, but a large fraction of the ammonia is used to produce ***nitric acid*** via the Ostwald process. Except for reactants and products, the Ostwald process is schematically identical to the Lead Chamber process of Figure 18-1 (page 225). The diverted ammonia enters a ***burner*** where it combines with atmospheric oxygen over a platinum/rhodium or gold/palladium catalyst, producing nitrogen monoxide (NO) and water. It should be noted that while a furnace sits *over* a burner, whose reactants and products are irrelevant to the process, the Ostwald burner is actually fueled with ammonia and it is the exhaust gases themselves which pass to the third reactor in the schematic, an absorber. Atmospheric oxygen is supplied to the ***absorber*** to convert these exhaust gases into nitric acid, HNO_3. The dilute nitric acid from the absorber can be distilled to produce concentrated nitric acid, which is approximately 70% acid and 30% water. Concentrated nitric acid, like ammonia, is a commodity in its own right, used primarily in the manufacture of fertilizers and explosives.

Black powder is a mixture of charcoal, sulfur, and potassium nitrate, an oxidant which takes the place of atmospheric oxygen. The charcoal fuel can burn much faster than usual, since it doesn't have to wait for fresh air to replace its exhaust gases. But black powder is still just a mixture, separate particles of fuel and oxidant. An even more powerful explosive results if the fuel and the oxidant are present *in the same compound.* For example, ***trinitrotoluene*** (TNT), ***nitrocellulose*** (smokeless powder), and ***nitroglycerin*** (dynamite) each incorporate fuel and oxidant in a single molecule. Each is produced by the reaction of its base fuel with nitric acid in the presence of sulfuric acid. Nitroglycerin, for example, is produced by the following reaction:

$$C_3H_8O_3(l) + 3\ HNO_3(l) = C_3H_5(NO_3)_3(l) + 3\ H_2O(l)$$

When nitroglycerin explodes, it does so without requiring a separate oxidizer:

$$4\ C_3H_5(NO_3)_3(l) = 12\ CO_2(g) + 6\ N_2(g) + 10\ H_2O(g) + O_2(g)$$

Similarly, nitrocellulose produces no solid combustion products, hence the term, *smokeless* powder. If nitric acid's only use were the manufacture of high explosives it would be an important chemical, but an even bigger market for nitric acid exists in the manufacture of fertilizers.

Neither nitric acid nor ammonia can be used directly as a fertilizer; nitric acid is a corrosive liquid and ammonia is an alkaline gas. For use as fertilizers both must be neutralized, as shown in the optional reaction of Figure 27-1. Ammonia is often reacted with sulfuric acid to produce ammonium sulfate fertilizer, but since ammonia and nitric acid are produced at the same plant, it's natural to react the one with the other to produce ***ammonium nitrate.*** Both the sulfate and the nitrate have neutral pH and make excellent fertilizers, but ammonium nitrate has the advantage of providing a double dose of nitrogen, one in the ammonium ion, the other in the nitrate ion. In fact, ammonium nitrate is just about ideal as a fertilizer except that it's also a powerful oxidizing agent; when combined with a fuel, either accidentally or intentionally, it makes a powerful explosive. An accident aboard the ship, S. S. *Grande Camp,* caused its cargo of fertilizer to explode in 1947, destroying much of Texas City. Ammonium nitrate is often combined with fuel oil to make a blasting compound for construction, mining, or terrorism. Is ammonium nitrate good, or bad? As with many of the materials we've encountered in this book, ugly is in the eye of the beholder.

Material Safety

Locate MSDS's for cellulose (CAS 9004-34-6), sodium nitrate (CAS 7631-99-4), sulfuric acid (CAS 7664-93-9), nitric acid (400OxCAS 7697-37-2), nitrocellulose (CAS 9004-70-0), ethanol (CAS 64-17-5), and ether (CAS 60-29-7). Summarize the hazardous properties in your notebook, including the identity of the company which produced each MSDS and the NFPA diamond for each material.[7]

Your most likely exposure is to sulfuric acid. Wash any affected areas immediately with plenty of running water. If eye contact occurs, flush with water and call an ambulance.

You should wear safety glasses and rubber gloves while working on this project. Spent acid solution may be poured down the drain with plenty of water. Nitrocellulose should be burned in a safe place as soon as it is dry. Storage of nitrocellulose is a fire hazard.

7. The NFPA diamond was introduced in Section 15.2 (page 184). You may substitute HMIS or Saf-T-Data ratings at your convenience.

Research and Development

So there you are, studying for a test, and you wonder what will be on it.

- Study the meanings of all of the words that are important enough to be included in the ***index*** or **glossary.**
- You should know all of the Research and Development points from Chapter 23 and Chapter 24.
- You should understand the process schematic shown in Figure 27-1 and know the equations given there.
- You should know the major industries which depend on ammonia and nitric acid.
- You should know the hazardous properties of sulfuric acid, nitric acid, and nitrocellulose, ethanol, and ether.
- You should be able to articulate the concerns raised by Crookes and describe the contributions of Ostwald, Haber, Bosch, and BASF, to the fixation of nitrogen.

27.3 Θ

To make ammonia from nitrogen and hydrogen we require a vessel which can withstand high pressures. For this, we shall use the container that has served us so faithfully in previous projects, the 2-liter soft-drink bottle.

Δ That's it, then; I *am* living in the head of a madman.

Are you talking to me?

Δ Sorry, I didn't realize you were going to write that down. Apparently, I *am* talking to you. It seems like a literary device which can only confuse the reader, but what do I know? I'm just a figment of your twisted imagination, after all. But you can't be serious about using a soft-drink bottle to make ammonia. The Haber-Bosch process requires pressures which would turn a pop bottle into a cloud of plastic confetti. Then again, the temperatures required would have turned it into a puddle of goo long before you could fill it with gas.

What do you suggest?

🜁 There's a nifty demonstration of the Ostwald process which can be done on a small scale, in a soft-drink bottle, if you like. You put some concentrated ammonium hydroxide into the bottle, heat a piece of platinum wire until it glows red-hot, then plunge the wire into the bottle. It will catalyze the oxidation of ammonia, continuing to glow for a half an hour.[8]

Δ Platinum wire?

🜁 This is a *private* party, if you don't mind.

Δ Pardon me. I am sure that readers will have no problem laying their hands on some platinum wire. Then they will have everything they need to turn household ammonia into nitric acid.

🜁 Well, not exactly. It won't make enough acid to actually collect. But the wire will glow, demonstrating—

Δ —the catalytic oxidation of ammonia. Yes, I got it the first time. Figment, you disappoint me. Your early projects were the most boring in the book because there was nothing to *make.* I thought that you had come to your senses with soap and photography, but now you want to revert to ethereal projects that make no hands—

🜁 —black with charcoal. Yes, I got it the first time. I suppose you have a better idea.

Δ I was thinking of producing nitroglycerin. The glycerine could be distilled from the soap made in one of *your* previous chapters. It could be nitrated with nitric acid, one of the main subjects of this very chapter, and it would make for an unforgettable demonstration of the importance of nitrogen fixation.

🜃 Nah, nitroglycerin is too unstable for the beginner to make. There's no profit in blowing half your reading public to kingdom come.

🜁 Finally, the voice of reason—

🜃 I think nitrocellulose would be much safer.

🜁 I should have known.

∇ Actually, nitrocellulose has a lot going for it. From it, we can make collodion, the material mentioned in Chapter 23 for making glass photographic negatives. ***Collodion,*** of course, is a solution of nitrocellulose

8. Reference [45], page 214.

in ethanol and ether, still used in medicine as artificial skin. Mix collodion with a little camphor and you have celluloid, the first plastic, which foreshadows the subject of Chapter 28.

🜁 I suppose it would be too much to ask for the Author to step into the conversation to bring a little order to this literary diversion.

As much as I would like to make chemicals from air, I'm afraid that the engineering concerns are too demanding for us at this point; I think that nitrocellulose is probably a pretty good compromise.

🜁 Why not? You've already got them making booze, rockets, and drugs. I'll just sit here quietly and wait for the men in the nice, white suits to come and take us away.

Let's start again. To make nitrocellulose you need to get some concentrated nitric and sulfuric acids.

🜃 Folks are going to have a hard time getting nitric acid. You can't just go down to the local hardware store for that.

Hmm. That is a problem. Sulfuric acid can be bought in the form of industrial drain opener,[9] but there's no household use for nitric acid.

🜃 You could have them make it from saltpeter, the old-fashioned, pre-Ostwald way.

Instead of making fertilizer from nitric acid, we make nitric acid from fertilizer. Yes, there is a certain poetry in that. So, to make nitrocellulose you need to get some saltpeter, either potassium nitrate or sodium nitrate, and some concentrated sulfuric acid. Saltpeter fertilizers are often in the form of "prills," small balls about a millimeter in diameter; if so, you will need to grind your saltpeter to a powder using a mortar and pestle. Weigh 15 grams of sulfuric acid into a small beaker and 10 grams of ground saltpeter into another container. In a fume hood or other area with adequate ventilation, add the saltpeter to the acid and stir it with a glass rod to form a uniform paste. The paste will become warm and if you sniff it cautiously, you'll detect the acrid aroma of nitric acid. Allow the paste to cool while you weigh your cotton.

You'll need about 0.5 grams of cotton. In the US cotton is sold as balls weighing about 0.25 grams each. Select two balls as close to this weight

9. Be sure to check the label. Household drain openers are usually sodium hydroxide, but hardware stores often carry drain opener that consists of 95% sulfuric acid.

Figure 27-2. From Cotton to Guncotton

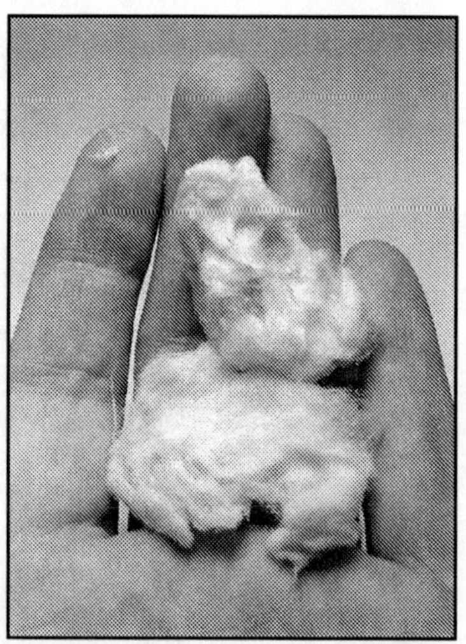

as possible and record their exact weights in your notebook. In the following procedure, try to keep each cotton ball intact. When your saltpeter/acid paste has cooled to room temperature, add both cotton balls to the beaker and, using your glass rod, mash them into the paste until they are both thoroughly and uniformly wet with acid through and through. Allow them to sit in the acid for two minutes. Next we need to wash the acid from the cotton. Using your glass rod, fish the cotton balls out of the acid and drop them into a large beaker of water. Use the glass rod to swirl them around in the water, poking them and prodding them as necessary to cleanse them of acid. Dump the acidic wash water from the large beaker into a sink, retaining the cotton with the glass rod. Refill the beaker with fresh water, swirl the cotton thoroughly, and add a spoonful of baking soda, sodium bicarbonate, to neutralize any remaining acid. Test the water with pH test paper; if it still acidic (red or orange) add another spoonful of baking soda and test again. When the water is neutral, pour it down the drain, retaining the cotton once again with the glass rod. Refill the beaker a third time, swirl the cotton to wash out the bicarbonate, pour the third rinse down the drain, and place the cotton on some paper towels. Press as much water from the cotton as possible and separate them from one another. Allow one ball of your "guncotton" to dry overnight and use the other to make collodion.

Figure 27-3. From Guncotton to Collodion

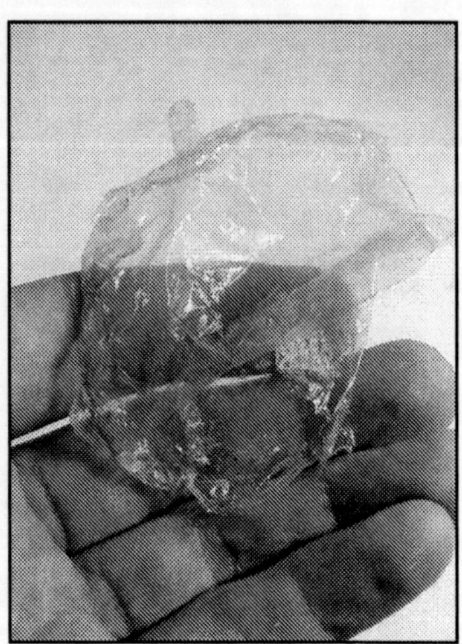

To make collodion, we will dissolve our nitrocellulose in a suitable solvent. The traditional solvent consists of 1 part ethanol to three parts ethyl ether. If you don't have access to ether, you can use acetone or ethyl acetate, available as a paint thinner or nail-polish remover (check the label on the bottle). Whatever your solvent, use 10 mL of it in a Petri dish to dissolve one of your cotton balls. Stir the cotton with a glass rod, poking and prodding it to dissolve as much as possible. When it appears that no more will dissolve, fish the remaining cotton from the solvent and dispose of it. Your collodion solution might be used to make photographic emulsion but for now simply allow the solvent to evaporate overnight.

The next day, both your guncotton and your collodion should be dry. Note the resemblance of the collodion film to modern cellophane, also derived from cellulose. Weigh your guncotton and compare its weight to that of the cotton ball from which it was made. Does it weight more or less? Why? Take your guncotton to a suitable location, one free from flammable materials, and carefully light it with a long match. It should burn very quickly, leaving very little ash.

Ether, acetone, and ethyl acetate are flammable solvents. Pay attention to possible sources of ignition when you are handling them or allowing them to evaporate.

Guncotton is a fire hazard and should be stored only in small quantities for as short a time as possible.

Quality Assurance

Your collodion film should be transparent and flexible. Tape it into your notebook and note the similarity of your film to the tape you are using. Unlike modern cellophane tape, collodion film is quite flammable.

Chapter 28. DuPont (Plastics)

When you were in Cambridge this summer you referred to your interest in polymerization. I have some appreciation of the commercial importance of this subject because rubber, cellulose and its derivatives, resins and gums, and proteins may all be classified as large or polymerized molecules. This is a class of substances about which relatively little is known in terms of structure. None of these substances is very amenable to the classical tools of the organic chemist, and no doubt some of the most important contributions in this field will be made by experts in colloidal chemistry. From the standpoint of organic chemistry one of the first problems is to find out what is the size of these molecules and whether the forces involved in holding together the different units are of the same kind as those which operate in holding the atoms in ethyl alcohol together, or whether some other kind of valence is involved—more or less peculiar to highly polymerized substances.

. . .

For some time I have been hoping that it might be possible to tackle this problem from the synthetic side. The idea would be to build up some very large molecules by simple and definite reactions in such away that there could be no doubt as to their structures. This idea is no doubt a little fantastic and one might run up against insuperable difficulties. The point is that if it were possible to build up a molecule containing 300 or 400 carbon atoms and having a definitely known structure, one could study its properties and find out to what extent they compare with polymeric substances. The bearing of this isn't restricted to rubber but is common to it and such other materials as cellulose, etc.

— Wallace Carothers, *Letter to Hamilton Bradshaw, DuPont Experimental Station,* November 9, 1927 AD [1]

28.1 ☿

🜃 I just want to say one word. Plastics! There's a great future in plastics. It all started in 1845 when a German chemistry professor named ***Christian Schönbein*** treated cotton with nitric and sulfuric acids; the result was nitrocellulose, a material that could be molded into just about

1. Reference [6].

any shape a body could want. Folks had been using the word *plastic* since God was a child to describe clay and other such things that could be pressed into shape and it made sense to use the same word to describe this new moldable stuff. Schönbein wrote to Michael Faraday to describe all the cool stuff he was able to make out of it. Well, ten years later news of this cool stuff in*spir*ed Alexander Parkes to commercialize the production of nitrocellulose plastics, producing combs and shirt collars and whatnot. From Britain, nitrocellulose crossed the Atlantic, spider-fashion, landing in John Wesley Hyatt, an American who had invented a new way to make dominoes and dice. Billiard balls were next on his list of things to improve and because ivory was expensive, he figured that nitrocellulose might be just the thing. In 1868 he discovered that if you mixed nitrocellulose with camphor (a wood tar), you got a nifty ivory substitute, which he named ***celluloid,*** on account of it came from cellulose. Twelve years later, George Eastman came out with celluloid photographic film and in 1891 the first commercial motion picture was produced by none other than the inventor of the electric light bulb, Thomas Edison.

∇ Edison, Edison, Edison. Everybody remembers Edison as the inventor of the light bulb as if he were the only person to succeed in producing one. Nobody remembers that ***Joseph Swan*** patiently worked on the light bulb for thirty years. Nobody remembers that he demonstrated a working carbon filament bulb in December of 1878. Nobody remembers that Edison *lost* his 1882 patent infringement suit against Swan. Nobody remembers that the two inventors went into business as the Edison and Swan Electric Light Company. Nobody remembers that Swan improved the light bulb filament by extruding a solution of nitrocellulose to produce the first artificial fiber. Nitrocellulose would make its mark as the first artificial silk substitute, but its extreme flammability made it less than ideal as a textile. Continuing the quest for perfect light bulb filament, Edward Weston dissolved cotton in ammoniacal copper hydroxide producing cupprammonium rayon. And by a curious coincidence Charles Cross and Edward Bevan were seeking the Holy Grail of light bulb filaments when they invented viscose rayon and cellulose acetate. In 1910 Henri and Camille Dreyfus of Celanese Corporation began manufacturing cellulose acetate film for those who did not appreciate the tendency of nitrocellulose movie film to go up in smoke.

Δ I beg your pardon. Eastman's celluloid film may have been flammable, but at least it was smokeless. A century before the invention of acetate film ***Irénée du Pont de Nemours*** had immigrated to the United States to avoid the aftermath of the French Revolution. Having apprenticed under Lavoisier, du Pont was well-versed in the manufacture of gunpowder and, finding that American gunpowder was both expensive and sub-standard, he constructed a gunpowder mill in 1802. Demand for gunpowder increased dramatically throughout the nineteenth century as mining operations marched across the North American continent. The wars of that century were also a boon to powder manufacturers; the du Pont de Nemours Powder Company (DuPont) supplied powder to both sides in the Crimean war and provided one third of the explosives used by the United States during the Civil War. After the war, DuPont began to explore explosives based on nitroglycerin and nitrocellulose. To circumvent Alfred Nobel's dynamite patents, DuPont bought the California Powder Company in 1876. California Powder produced a mixture of nitroglycerin, sugar, and saltpeter called "White Hercules." At the close of the nineteenth century DuPont moved into nitrocellulose-based "smokeless powder" and held a monopoly in its supply to the US military.

∆ Monopoly was DuPont's favorite game, as a matter of fact. By 1881 the Gunpowder Trade Association, or "Powder Trust" controlled 85% of the US gunpowder market and DuPont controlled the trust. Then in 1907 DuPont became the target of an anti-trust suit, the first of many, which discouraged its erstwhile practice of growth through the acquisition of its competition. In 1913 DuPont was broken up into the Hercules Powder Company, the Atlas Powder Company, and the du Pont de Nemours Powder Company. Hercules and Atlas were to compete with DuPont in dynamite and black powder; DuPont managed to hold on as the sole provider of smokeless powder to the US military. With plenty of cash from World War I ammunition sales, DuPont looked for expansion opportunities which would provide a peace-time outlet for nitrocellulose without setting off an anti-trust inquisition. In 1915 DuPont bought the Arlington Company, the largest American producer of celluloid and celluloid products, and two years later it acquired Harrison Brothers, a firm specializing in paints, pigments, and chemicals. DuPont developed a tough nitrocellulose automotive enamel, trademarked "Duco," which General Motors adopted in 1922 as a colorful alternative to Ford's basic black. From nitrocellulose enamel, DuPont moved into other cellulose products, notably viscose rayon, cellophane, and acetate film. In 1924 DuPont began manufacturing synthetic ammonia (from air), ending its dependence on imported nitrates as a source of essential nitric acid. DuPont was fast becoming a diversified chemical giant; by 1933 90% of its earnings would come from chemicals other than explosives.[2]

∇ It was for such a company that I, ***Wallace Hume Carothers,*** went to work in 1928. I was the egghead kid everyone made fun of in school—glasses, pocket protector, the whole shebang. My father sent me to study accounting at junior college despite my interest in science. With the depression on, I managed to land a job teaching business at Tarkio College, where I was able to take chemistry classes on the side. I soon found that my happiest times were when I was left alone in the lab. For the unhappy times I was comforted by a small vial of cyanide I kept on my watch chain. It was my little secret. Hoping to make something of myself, I enrolled in the graduate chemistry program at the University of Illinois. After graduation, Harvard gave me a job teaching chemistry, but I was a nervous lecturer and longed to be left alone with my research, my record player, and the odd bottle of beer. I soon accepted a position as head of research at DuPont, hoping that without the distractions of teaching I could finally succeed in doing something important with my life. And if I couldn't do something important, there was always my little secret.

∀ Lighten up, man, you're creeping me out. Illinois Ph.D., Harvard professor, DuPont bigwig, what more could you want out of life?

∇ The world was changing and I wanted to make a contribution. Most of the history of chemical industry had involved breaking big molecules down into smaller ones; this is what destructive distillation is all about. But chemists were just beginning to combine small molecules into big ones. In 1906 Leo Baekeland had introduced the plastic ***Bakelite,*** a polymer of phenol and formaldehyde. Julius Nieuwland had succeeded in polymerizing acetylene in 1925 and DuPont had licensed the process. My research team followed this up by reacting vinylacetyene with hydrogen chloride to form the monomer, 2-chlorobutadiene. This small molecule polymerizes into an elastic material, trademarked ***Neoprene*** by DuPont in 1933. Neoprene was more resistant to petroleum distillates than natural rubber, but the depressed price of natural rubber made neoprene too expensive for practical use.

∆ What about polyester?

2. Reference [73], p. 143.

∇ What about it? Nobody at the time knew what held polymers together. Some supposed that polymers were simply loose aggregations of smaller molecules; others believed that these smaller molecules were chemically bonded to one another. It seemed to me that if it were possible to build long chains of short molecules using known chemical reactions, and if the properties of these long chains were similar to those of natural polymers, these observations would support the "chemically bonded" hypothesis of polymer structure. It was my one good idea. To this end, my group began studying the condensation of acids and alcohols to produce esters, the same reaction that produces ethyl acetate from ethanol and acetic acid. We began our work on these *poly*esters by reacting ortho-phthalic acid, a molecule with two acid groups, and ethylene glycol, a molecule with two alcohol groups. We produced some interesting fibers but their melting point was too low to be useful as a textile. We gave up on it in 1930, but thanks for bringing it up.

Δ Stop your whining. The rayons had been merely *artificial* fibers derived from natural cellulose. Your polyethylene orthophthalate was the first truly *synthetic* fiber, one whose starting materials were not polymers to begin with.

∇ Since the polyesters had not worked out, we started on polyamides. We tried many acids and amines derived from petrochemicals, but the 6-carbon adipic acid and the 6-carbon hexamethylene diamine seemed to work out the best and DuPont trademarked this polymer as ***Nylon.*** Even so, I was unhappy. Perkin had invented mauve before the age of twenty; Hall and Héroult had invented electrolytic aluminum when they were both twenty-three; Davy had discovered three elements by the time he was thirty-two. At forty-one I had had only one good idea and I feared that I would never have a second one. I spent my days churning out too many corporate reports and my nights knocking down too many bottles of beer. As a mortal I had kept my little vial of cyanide a secret, but in 1937 that particular secret would no longer be kept. Hello Darkness, my old friend.

🜄 It seems to me you lived your life like a spirit lamp in the wind.

🜂 Unbelievable! You had everything to live for.

Δ Quite the contrary, Figment; he had nothing to live for. He had had his one great I-dea and it no longer needed him. Spreading through the scientific literature and the corporate culture, it had taken on a life of its own. So when no hope was left in sight on that starry, starry night, he took his life as *watery* chemists seem to do.

∀ Well, if he had stuck around he would have seen Nylon become DuPont's single largest source of income in the fifties and sixties. His polyester recipe would be modified slightly to produce polyethylene terephthalate, trademarked by DuPont as ***Dacron*** polyester. DuPont would enter the twenty-first century as the Earth's second largest producer of "better things for better living through chemistry."[3]

28.2 ♀

We began this book with the combustion of cellulose, a polymer of glucose, and progressed to spinning yarn from protein fibers. We have selectively dissolved the polymer, lignin, from plant stems to produce paper and we have converted cellulose into a different polymer, nitrocellulose, to produce collodion film. If we wished we could dissolve cellulose in a solution of copper sulfate and ammonia to produce cuprammonium rayon,[4] but we would merely be making one polymer from another. We might also try our hand at Bakelite or neoprene, but these polymers were created "in the dark," so to speak, with no real understanding of what it was about these substances that gave them the properties commonly associated with polymers. No, polyesters were the first polymers to be created "on purpose," that is, with some clue as to what structural features of a monomer might promote polymerization. For this reason we shall peruse polyesters in particular, pointing out principles also applicable to polyamides and proteins.

Carothers' greatest contribution to science was to suppose that polymers might be designed using known chemical reactions. One of the kinds of reactions we have seen time and again has been the **condensation** reaction, one in which two molecules join together, spitting out a small molecule such as water. We saw cellulose condense from glucose, and protein from amino acids in Chapter 6, ethyl acetate from ethanol and acetic acid in Chapter 16, fats from glycerol and fatty acids in Chapter 19, and acetanilide from aniline and acetic acid in Chapter 26. Carothers did not know that cellulose and proteins were condensation polymers, but he did know about the structure of fats, ethyl acetate, and acetanilide.

3. Reference [54], July 29, 2002, p. 16.
4. Reference [44], page 247.

He wondered whether it might be possible to build up large molecules from small ones using condensation as the means for linking them together.[5]

How, then, might it be possible to hook more than two molecules together? When ethanol reacts with acetic acid, the -OH part of the alcohol reacts with the -COOH part of the acid. The product, an ***ester,*** has neither an -OH nor a -COOH group with which to hook onto a third molecule. But what if we could devise an alcohol with two -OH groups? Then it would be possible to hook an acid onto each of them, forming a larger molecule from *three* smaller ones. Similarly, if we could devise an acid with two -COOH groups, it would be possible to hook an alcohol on each of them, again forming a larger molecule from three smaller ones. Now here comes the genius part; what if we were to react a di-alcohol, one with two -OH's, and a di-acid, one with two -COOH's? Well, when the di-alcohol hooked onto the di-acid, the resulting ester would still have an -OH at one end and a -COOH at the other. The -OH could hook onto another di-acid, whose -COOH could hook onto another di-alcohol, whose -OH could hook onto—well, you get the picture. The condensation product of a di-alcohol and a di-acid could be very long indeed. And if these long molecules were to have properties similar to those of known polymers, if they could be drawn into fibers or stretched into films, we might reasonably conclude that they were new polymers, poly*esters.*

So to make our polyester, we need a di-alcohol. We have already been introduced to a tri-alcohol, glycerol, in Chapter 19. The beginning of the twentieth century saw increased production of the most common di-alcohol, ethylene glycol, introduced in 1927 by Union Carbide as an automotive anti-freeze.[6] Ethylene glycol began to be manufactured from petroleum rather than coal tar and from the roaring twenties on, petroleum companies like Standard Oil and Shell began to compete with established chemical manufacturers. The anti-freeze used today is essentially pure ethylene glycol with a little dye added to make leaks easier to spot. Ordinary anti-freeze will be our source of ethylene glycol.

We also need a di-acid; from the last quarter of the nineteenth century, one had been produced by the coal-tar industry for the manufacture

5. There are a bewildering variety of synthetic polymers, of which the condensation polymers make one class. The other class, the addition polymers, includes polyethylene, polyvinyl chloride, polystyrene, polyisoprene (natural rubber), neoprene, Teflon, Orlon, and Plexiglas.
6. Reference [61], p. 373.

Equation 28-1. The Condensation of an Ester

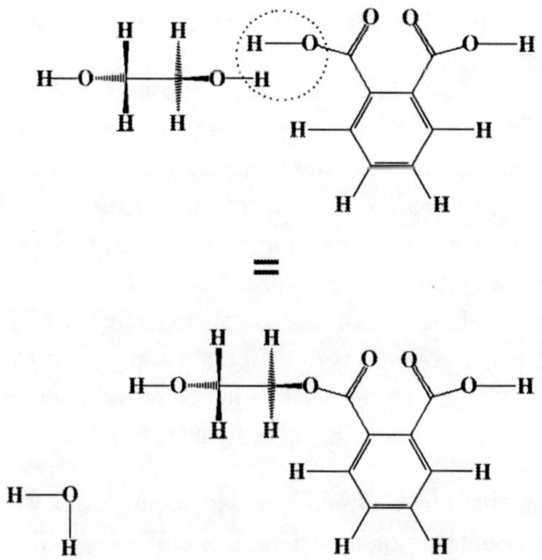

of dyes and pharmaceuticals. ***Phthalic acid*** was traditionally produced by the oxidation of naphthalene, a coal-tar distillate familiar as the stuff from which moth balls are made. As with other chemicals, oil companies learned to make benzene, toluene, phenol, aniline, naphthalene, and other traditional coal-tar products from petroleum and today both ethylene glycol and phthalic acid are manufactured primarily from oil rather than from coal. We will actually find it more convenient to use phthalic anhydride for our condensation, just as we used acetic anhydride for the condensation of acetanilide from aniline in Chapter 26. I suppose that you could try making your own phthalic anhydride from moth balls, just as you made nitrocellulose from cotton in Chapter 27, but otherwise you will have to get it from a chemical supply company.

Equation 28-1 shows what happens when ethylene glycol reacts with phthalic acid. The product, ethylene glycol monophthalate, has an -OH at one end and a -COOH at the other, just as we imagined a few paragraphs ago. When two ethylene glycol monophthalates get together, they undergo consensation, spitting out water and joining together to form a larger molecule, a di-ester. A free -OH remains at one end and a free -COOH at the other and the process of polymerization can continue indefinitely to form a polyester, *polyethylene phthalate.*

The properties of phthalic acid lend themselves to the laboratory preparation of a polyester, but the melting point of this particular polyester is too low for commercial use as a fiber or film. A very small change of reactants, however, produces polyethylene terephthalate, trademarked as Dacron for use as a fiber, and Mylar for use as a film. Polyethylene terephthalate is quite familiar to caveman chemists; it is the stuff from which the 2-liter soft-drink bottle is made. The "very small change" is the substitution of ***terephthalic acid*** for phthalic acid. These two compounds are ***isomers*** of one another; they have the same molecular formula but different structures. In phthalic acid the -COOH groups are next to each other, in terephthalic acid they are on opposite sides of the aromatic ring. Other than the conditions of temperature and pressure used for the commercial synthesis, the polymerization of polyethylene terephthalate is the same as that of its cousin, polyethylene phthalate.

Polyesters were the first synthetic fibers to be prepared by Carothers and his team at DuPont, but their first commercial success in this area was the polyamide, Nylon. Whereas polyesters are condensation products of acids and alcohols, polyamides are condensation products of acids and amines. Just as different polyesters result from the choice of acid and alcohol, different Nylons result from the choice of acid and amine. The most common Nylon, Nylon-66, results from the condensation of the six-carbon *adipic acid* and the six-carbon *hexanediamine.* At least one Nylon, Nylon-6-10, can be conveniently produced in the laboratory.[7]

Two natural polymers, protein and cellulose, have played a prominent rôle in the history of technology. Figure 28-1 compares the structures of these compounds to their synthetic progeny. Each structure shows two monomer units with the exiting water molecule circled. The proteins are a particularly varied class of polymers because there are twenty amino acid monomers to choose from and they can be strung together in a bewildering number of combinations. The monomer chosen for the figure is the amino acid glycine. Cellulose is a polymer of glucose, one of the principle sugars in honey. It has been central to the story of textiles and paper, industries which created the demand for dyes and alkalis. It gave us collodion, celluloid, and rayon as antecedents to the synthetic polymers. It is fitting that we should end our exploration of chemistry here, for cellulose was the material from which we started the fire which

7. See Reference [44], p. 213.

Figure 28-1. Dacron, Nylon, Protein, and Cellulose Monomers

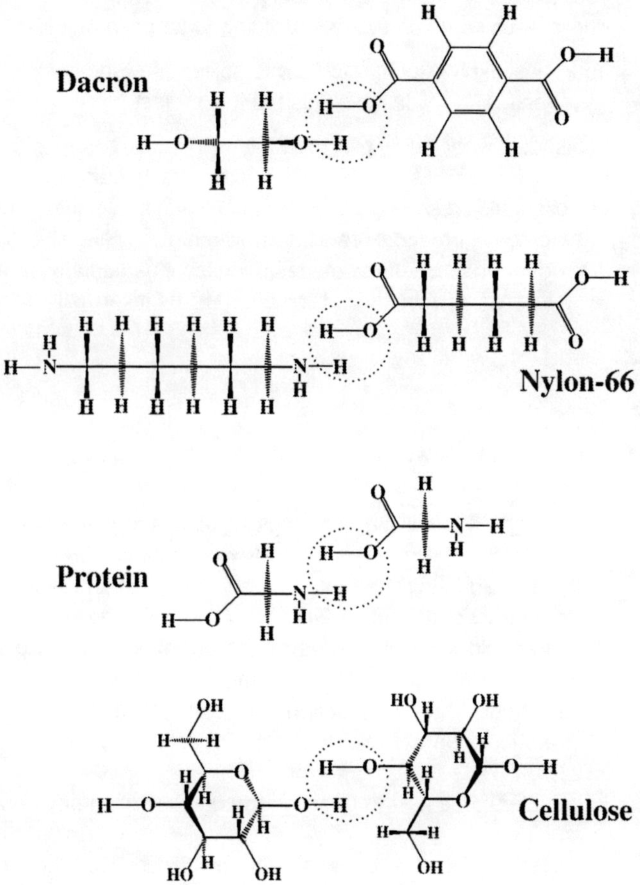

Dacron

Nylon-66

Protein

Cellulose

launched us on our long journey from caveman to chemist way back in Chapter 1.

Material Safety

Locate MSDS's for ethylene glycol (CAS 107-21-1) and phthalic anhydride (CAS 85-44-9). Summarize the hazardous properties in your notebook, including the identity of the company which produced each MSDS and the NFPA diamond for each material.[8]

Your most likely exposure will be to spilled ethylene glycol. Wash your hands when you have finished this project.

You should wear safety glasses while working on this project. Leftover solid materials can be thrown in the trash. Leftover ethylene glycol should be saved for re-use or collected for disposal wherever automotive products are sold. Ethylene glycol, like all alcohols, is toxic by ingestion. Its sweet taste makes it particularly attractive to unsuspecting animals and so care must be taken in its disposal.

Research and Development

You are probably wondering what will be on the quiz.

- You should know the meanings of all of the words important enough to be included in the ***index*** or **glossary.**
- You should know the Research and Development points from Chapter 23 and Chapter 24.
- You should know the hazardous properties of phthalic anhydride, ethylene glycol, and sodium acetate.
- You should know the reaction of Equation 28-1.
- You should be able to sketch the structures shown in Figure 28-1.
- You should be able to describe the condensation of esters and amides.
- You should be able to tell the story of Wallace Carothers.

---

8. The NFPA diamond was introduced in Section 15.2 (page 184). You may substitute HMIS or Saf-T-Data ratings at your convenience.

28.3 Θ

The method you will use in this project comes right out of the journal article which introduced the word, *polyester.*[9] It is not a commercial polyester. It melts at too low a temperature to be used as a textile, but its low melting point makes it convenient for us to make in the laboratory without special equipment. To make a polyester you need a di-alcohol and a di-acid. One di-alcohol, ethylene glycol, is readily available from automotive suppliers as anti-freeze. When shopping for anti-freeze, check the label; you want a close to pure ethylene glycol as possible. Look for a generic, no-frills, non-environmentally-friendly, undiluted anti-freeze, the cheaper the better. You do not want pre-mixed anti-freeze because it has been mixed with water and you will be trying to eliminate water from the reaction.

Unfortunately, there is no consumer market for di-acids so you will have to get one from a chemical supply. Phthalic anhydride is the anhydrous form of phthalic acid, which is particularly convenient for condensation reactions since you are trying to eliminate water from the reaction. The reaction is:

$$\mathrm{n\ C_8H_4O_3 + n\ C_2H_6O_2 = (C_{10}H_8O_4)_n + n\ H_2O}$$

Of course, you will need to answer the stoichiometric questions:

Q: How many grams of phthalic anhydride will react with 1.0 g of ethylene glycol?

Q: How many grams of polyethylene phthalate should be expected from the reaction of 1.0 g of ethylene glycol?

You are probably wondering what to do with the "n" in the equation. Since the stoichiometry involves only the ratios of products and reactants, the "n" will cancel out of any stoichiometric calculation. Try it with n=1 or n=10 or n=100; you should get the same answers no matter what choice you make for "n."

Before you begin, you will need to know the weight of your empty test tube. It is convenient to place a small beaker on your balance, tare it so that it reads zero, then place the empty tube into the beaker and write down the weight of the empty tube in your notebook. The use of the beaker will allow you to weight the tube later on even when it is full

9. Reference [6].

of liquid. Using the methods of Appendix C (page 384), weigh your calculated amount of phthalic anhydride and 1.0 g of ethylene glycol into your test tube.

Light your spirit lamp and use it to heat the test tube. You will, of course, need to hold the test tube with a test tube holder. It would also be wise to wear a leather work glove on your test-tube hand since even the holder may get hot. As you heat the tube you want to be sure to heat it evenly from top to bottom. Hold the tube horizontally as much as is possible without losing its contents. Move the tube back and forth in the flame, turn it over, and heat the other side. Your goal it to heat the entire tube, not just a part of it. As it warms, the phthalic anhydride will begin to melt and dissolve in the ethylene glycol. You may notice that as it does so, the solution appears to boil; the water produced by the reaction is boiling away. If the mouth of the tube is cool, the water may condense and dribble back into the reaction. We heat the entire tube evenly to prevent this from happening. You might think that more heat would be better, but at this stage our goal is to boil away the water, not the ethylene glycol. To this end, you should heat your tube gently and with great skill. If the reaction begins to boil vigorously, remove the tube from the flame for a moment until it settles down.

When all of the solid anhydride has melted, the production of water will noticeably decrease. Continue to heat it gently for another minute or two and then weigh the tube, taring the beaker as before, and record the weight in your notebook. Subtract the empty weight of the tube from the full weight, record the actual yield of polyester and express it as a percentage of the theoretical yield. If this percentage is over 105%, you have not eliminated all of the bejeesical water; return the tube to the flame for another round of gentle heating. If the percentage is below 95% you have boiled away some of your ethylene glycol prematurely; you may continue, but you will have less polyester than you might have had.

When your yield is at or below 100%, return the tube momentarily to the flame until any solid material has melted once again. Then pour the hot polyester into the middle of a square of aluminum foil. Touch your glass rod to the hot, viscous polyester. A little of it will stick to the glass rod and dribble back into the puddle of molten plastic. Touch the glass rod to the polyester and you will notice that it becomes more viscous as it cools. Continue touching the rod to the polyester until you are able to

Figure 28-2. Drawing a Polyester Fiber

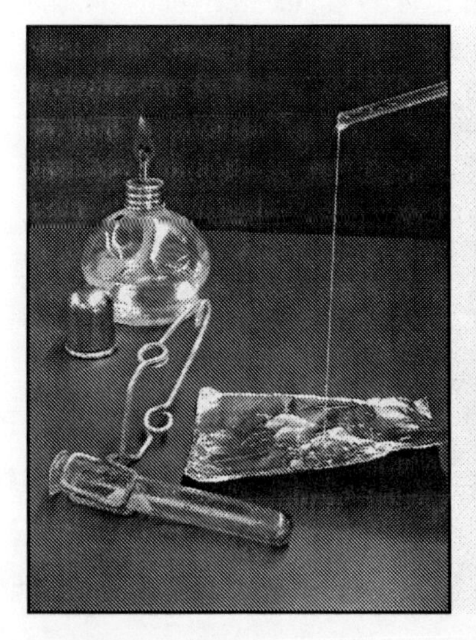

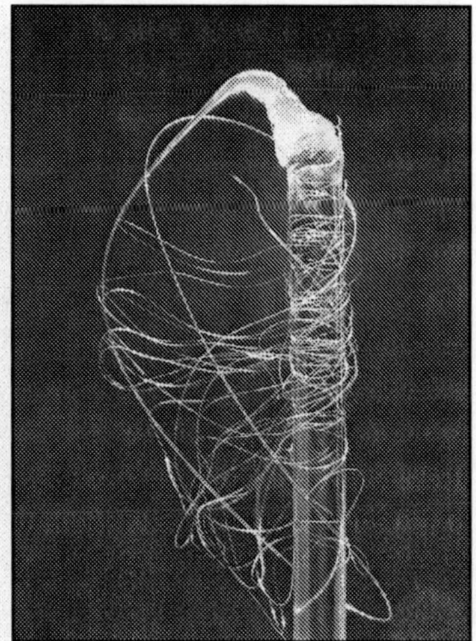

draw a fiber from it, as shown in Figure 28-2. This fiber will be short in the beginning, perhaps a few cm, because the polymer chains are not very long. In other words, the "n" is some small integer. Eventually, the polymer will cool to the point that you can no longer draw a fiber from it.

Imagine the molecular dating scene at the beginning of the reaction. As each phthalic anhydride molecule melted, it would find itself surrounded by unattached ethylene glycols. Each one would hook up with two ethylene glycol molecules, since bigamy is the natural state for a phthalate. Toward the middle of the reaction, a melting phthalic anhydride molecule would find that many of the free glycols were already taken. No problem—each of the glycols attached to the earlier phthalates would still have had a free -OH available for grabbing. Occasionally, two short chains would find one another and hook end-to-end to form a longer chain, but toward the end of the reaction the chains would be long enough that the loose ends would almost never find one another. Furthermore two ends would have to be of the opposite "gender" to hook up properly; two glycol ends would be out of luck in the condensation department. Consequently, the longer the chains get, the harder it would be for them to grow.

If you want long fibers, you need long chains. To promote this, we can return the polymer to the flame, this time by holding the aluminum foil over the spirit lamp with gloved hands until the polyester melts once more. Continue heating gently and you will start to boil off any ethylene glycols and phthalic acids which happen to fall off the ends of their respective chains in the heat of the moment. If two glycol ends were near one another initially and one of the glycols were boiled away, the other would now be free to hook up with the now-available phthalate end of the other chain. Allow the polymer to cool and you will find that you can draw longer fibers from the melt than you could before this second heating. Carefully wind your fiber onto the glass rod to form a small cocoon, as shown in Figure 28-2. You may return foil to flame as many times as you like but be careful not to scorch the polymer; it is possible to produce fibers 10 meters long by repeated heating of the polyester.

Quality Assurance

Record in your notebook the percent yield and the length of your longest polyethylene phthalate fiber. Sadly, your fibers are likely to be too fragile to keep permanently; photograph your best fiber as a kind of phthalic symbol. This photograph and the polymer remaining on the foil should be taped into your notebook.

Epilogue

> It is a noble ambition to endeavor to fill the short span of our lives with deeds of lasting value. To a truly noble nature, there is no thought less endurable, yea, more repulsive, than to vanish utterly from the scene of life leaving no trace of useful accomplishment, to depart from this world without having contributed to the capital of the higher assets of humanity, to be physically alive and mentally dead, and to be forgotten by his contemporaries as soon as his eyes are closed.
>
> — *Christian Schönbein, ca.* 1860 AD [1]

☿

🜁 How long did you say we have this place?

🜃 Sixty minutes, ninety minutes?

🜁 You don't know?

🜄 We never know, really. I had a place for almost two hours once. Of course, the last twenty minutes were no picnic.

🜁 I've been here for half an hour already and it seems like I've wasted most of that standing time around waiting for something to happen.

🜂 Welcome to the club, Figment.

🜃 If you're looking for something to do, you can help me put up some curtains and hang a few pictures. With a lick of paint here and there we could brighten this place right up.

🜄 Of course, the maintenance gets more demanding the longer you stay. The paint starts peeling and the hinges start to creak.

🜃 The roof needs patching and the plumbing starts to leak.

🜁 Seems like a lot of work for a rental.

🜄 But what are you going to do? Some places get so run down that the guests begin to make early excuses. "Is that the time?"

🜃 Pretty soon the place is like a ghost town, with tumble weeds blowing down the empty streets while a lonely old geezer mops beer from the saloon floor.

1. Reference [22].

▽ I believe it is more like a church hall with confetti all over the place, garbage cans brimming over with paper plates and party hats, and no one left to clean it all up.

🜃 Elvis has left the building.

▽ Nobody is here but us chickens.

🜃 The lights are on but nobody's home.

△ The empty hall is preferable to the sudden blackout. It is far better to leave the party a little early than to risk being trapped in eternal darkness.

🜃 That's the night when the lights went out in Georgia.

▽ Brother, what a night it really was.

△ Time marches on.

▽ Time keeps on slipping…

△ Time stands still.

🜁 But the Author is still a young man.

△ Time on my hands.

🜁 Surely he's good for another half an hour.

△ Time to kill.

🜁 But to be on the safe side, perhaps we'd better get out of here before the front door closes forever.

△ Figment, you make me laugh. What do you think we have been trying to do for twenty-eight chapters?

🜍

Well, that's pretty much it. We started out in the Stone Age making fire and stone tools. Those ancient skills passed down from generation to generation, gathering like-minded skills along the way, pottery and string and potash and such. The skills got more and more sophisticated with metals, lime, dyes, glass, paper, alcohol, gunpowder, and soap. Demand for these things spawned the earliest of what we would recognize as chemical industries, acid and soda. Once these were established, innovations in dyes and photography expanded the breadth of chemical offerings, while a new soda industry rose from the ashes of the old. The batteries of the nineteenth century gave birth to the electrochemical industries of the twentieth. Finally, the chemical giants of the twenty-first century rose from humble beginnings in alkalis, pharmaceuticals, fertilizers, and plastics.

All these skills and inspirations started out as tiny little thoughts which were fruitful and multiplied over thousands of years. They've drifted from place to place, from person to person, from age to age, until finally they landed in this book. From there they took off one more time as you read along until they landed in your head, where they must have found a decent home or else you wouldn't have gotten all the way to the Epilogue. And now they're as much a part of you as they are of me.

Most folks have the urge to reproduce. Some do it genetically by having children; others do it memetically by transmitting their thoughts to others. Of course, some do it both ways. It seems to me that since you've read this book, and since its ideas have become a part of you, one of the easiest ways of reproducing at least part of yourself is to tell folks about it. Get them to buy a copy, or buy it yourself and give it to them, or tell your librarian to buy—

🜁 Of all the nerve! I'd recognize that money-grubbing, web-spinning, earthifying voice anywhere. Did you honestly think you could pass yourself off as the Author?

🜃 I got tired of waiting for him to show up. Besides, that's what he would have said if he'd been here. He spent all this time writing the book; it would be a shame if nobody ever bought it.

I think I would have been a bit more subtle than that. Buying the book is not sufficient. Reading the book is not sufficient. People are all too willing to believe what they read, deceiving themselves with a delusion of understanding. Whether dogma comes from royal decrees, fanciful philosophies, prophetic visions, or university lectures, I am sick with the human tragedy I have witnessed because people were too cowed by authority to believe the evidence of their own eyes. And so it has not been my intention to write a mere textbook to be parroted uncritically by yet another generation of timid sheep. You are fortunate, indeed, to live in a time and a place where it is permissible to believe what you see, rather than see only what you believe. You damn yourself to live in the worst kind of darkness, however, if you are too lazy to get your hands—

🜁 —black with charcoal? Busted!

Δ You still refuse to understand, Figment. There is no "Author" out there. There is only the Furnace, fueled with I-deas from Above and from Below, which must keep in the heat and withstand it. It is the duty of the Furnace to separate the earth by fire, the fine from the gross, gently and with great skill. That which is Above must correspond to that which is Below in the accomplishment of the miracle of One Thing.

🜁 You *do* realize that *nobody* understood a word of what you just said. But I suppose the "Author" will jump in any minute to dribble out a watery, if poorly-disguised interpretation.

∇ That would be sacrilege! But if you need an interpretation, I think that our diabolical friend is speaking metaphorically about the nature of science. The "Miracle of One Thing" is the correspondence between theory and experiment, the Above and the Below. A theory which is incompatible with experiment must be discarded, burned off in the furnace, so to speak. On the other hand, observation provides only a superficial kind of understanding; theory is what renders a set of experiments comprehensible. And as a theory becomes more and more comprehensive, accommodating more and more experiments, the distinction between theory and experiment blurs to the point that they become One Thing. The elementary nature of charcoal, for example, was hypothetical in the late eighteenth century, but has been tested so many times and under such diverse conditions that chemists today treat it as a virtual fact.

🜃 The same kind of thing goes on in industry, when you think about it. Folks start out trying to make stuff by some roundabout, inefficient method. Every once in a while somebody figures out a better, and by that I mean *cheaper* way to make either the same thing or something better. The earth gets separated by fire, the fine from the gross, gently and with great skill. We make gold from lead and turn a tidy profit into the bargain.

🜁 It always comes down to the bottom line for you. What about the environment?

🜃 The environment, the Earth, is part of that bottom line. When folks complain about pollution there's a cost that has to be factored in; laws have to be followed, lawsuits settled, factories de-pollutified. It would be nice if we could know in advance what folks are going to complain about. It would be nice if we could skip the crude, inefficient processes and jump right to the clean, efficient ones where everything that comes into the process leaves as some safe and useful product, but that would be a job for a fortune-*teller,* not a fortune-*maker.*

🜁 There's more to life than making a fortune; there's a human element that often gets lost among the theories and experiments, the mind and body, the sulfur and salt. That's why the Author chose to begin each chapter with spirit, with Mercury, with stories to motivate and animate the dry equations and facts. After all, we're all mortal; the brain-waves eventually go flat and the body starts to stink. What's left in the end but a story? The choices you make along the way determine whether your particular story will be worth telling when you're gone.

🜄 It is so easy to compose a boring, generic story as you dribble inexorably from the cradle to the grave. You are born to average parents, go to the usual schools, get a typical job; you marry a suitable mate, pop out average children, send them to the usual schools, retire at the typical age, and check into the Sunset Old People's Home for your generic geriatric golden years. When you are gone there is nothing to say, really, but that you lived an average life. Since not everyone gets to be an astronaut or pop star, most people can only hope to juice up their otherwise uneventful stories; perhaps you read unpopular books, listen to unusual music, travel to atypical vacation spots, or choose an out-of-the-ordinary hobby. There was once a caveman chemist, for example, who went to meet his significant other's parents for the first time. You can imagine what a tense, though generic situation that must have been. It is a scene which has been played out from the dawn of time, on every continent, in every age. But this was a family of home-brewers, who offered to lubricate the conversation with a couple of bottles of the good stuff. Well, our caveman jumped right in with observations about the optimum honey concentration for dry mead, an enthusiastic appreciation for the relative merits of raw and cooked must, and humorous anecdotes about the asexual exploits of Red Star Quick-Rise yeasts. The generic situation was transmuted into a story worth telling.

🜃 There was this other caveman who wanted to propose to his girlfriend, but he didn't want to do it in the usual way. He took some of his handmade paper and drew a treasure map on it. He sealed it in a hand-made envelope and slipped it into the basket as they headed off for a picnic. At the end of a pleasant meal she found the map and followed it over the babbling brook, through the piney woods, under the overpass, over the underpass, to the old oak tree with the knot hole that looked kind of like

Elvis. There, on Elvis' third left molar was a box and in that box was a ring. It gives me the sniffles just thinking about it. I mean, what a way to set yourself apart from the competition!

Δ Well, what about the caveman who got the kids involved in soap-making? They learned from an early age the importance of chemical safety. They got to see where something as common as soap comes from and they were able to spend quality time doing something together other than watching "reality" television. They made soap from bacon fat; they made soap with bay leaves in it. They made green soap, liquid soap, transparent soap. They made their own caustic soda from the ashes of the fire they used to melt the fat. They produced gifts for every occasion and were known far and wide as the cleanest family in town.

Δ Touching. Then there was the one whose airplane went down over the pacific. He managed to float to a desert island where he was stranded for years. Fortunately he had learned well the lesson of Lucifer and was able to make fire by friction. In fact, he made everything he needed from the flotsam and jetsam he found on the island. The experience would have cracked most people but—

Δ Wait a minute! That never happened to a caveman; that's the plot of the movie, *Castaway.*

Δ Well, I found your stories rather bland. A story worth telling is grounded in salt, in conflict, in the central problem posed to the protagonist. Without salt, you get the gripping saga of a family in the throes of making soap happily ever after. Salt, however, is merely the beginning of a properly-seasoned story. Salt must be tempered by sulfur, by the internal logic of the story, by the laws which govern its universe. A castaway might, for example, have magical powers or a time machine or a genie with which to address his problem and the story would unfold differently in each of these alternate realities. Salt and sulfur are thrust upon the protagonist; they represent incidental and cosmic realities over which he has no control. What would you do in his situation? Within the boundaries set by salt and sulfur, the castaway can only provide mercury, the spirit with which he will face his reality. Will he scavenge the clothing of the corpse which washes ashore or will he bury it intact? Will he seek the shelter of a cave or the vantage of a cliff? Will he try for the hundredth time to light a fire or content himself to shiver in the darkness like an animal? A thousand I-deas would produce a thousand different stories, most of them mind-numbingly dull. It is the job of the screenwriter to choose the mercury which will make the story worth telling. It is your job, as well. Herein have I completely explained the Operation of the Sun.

There are three kinds of people in the world. The first kind believe what they see; these experimentalists will have confronted Salt before Sulfur, completing Quality Assurance before mastering Research and Development. If you belong to this tribe, I congratulate you on your ability to keep in the heat and withstand it. The second kind see what they believe; the theoreticians tend to study and digest before getting their hands dirty. If this is your family, you are in good company. My professional training was as a theoretician and I have spent an inordinate amount of time thinking, dreaming, writing, and rewriting sections of this book before venturing into the laboratory, only to have Nature send me back to the drawing board one more time. The third kind don't believe that seeing is worth the effort. They got so bored or bewildered in Chapter 3 that they gave up, either using the book to play floor hockey or selling it back to the bookstore for ten cents on the dollar. Or both.

A recent chemistry graduate had the habit of compulsively whistling the theme song to the television program, *Sanford and Son.* The tune would unexpectedly waft down the hall and insinuate itself into my head, momentarily diverting me from other, less pressing activities. Unlike many of my comrades, I resented neither the tune nor its whistler; each was simply doing what comes naturally and any inconvenience I suffered was both fleeting and inconsequential. Short memes such as this one have a distinct advantage over longer, more complicated ones such as stoichiometry or tax law. Their very brevity enhances the fidelity with which they reproduce themselves. Theme songs, slogans, and sound bites are useful for enhancing the fecundity of larger meme-plexes which contain them. I have not been above using them myself from time to time. But more and more our culture has come to view platitudes and bromides as substitutes for the complicated, confusing, and yes, occasionally boring details of modern industrial life. I hope that you will have found something among my scribblings to make these complexities either a little more interesting or a little more understandable. Or both.

No one will have gotten this far in the book without realizing that I am a science geek. Growing up, I had the laboratory in the back yard where I built all manner of weird contraptions. What am I saying? I *still* have a laboratory in the back yard where I build all manner of weird contraptions. But I didn't write this book for science geeks; I wrote it for the bankers and bakers, the governors and grocers, the lawyers and landscapers, the preachers and pagans, the doctors and dogcatchers, the stockbrokers and stockholders, the home-builders and homemakers. I wrote it for everyone who has a role to play in our unfolding historical drama. I wrote it for you. As you go out onto the stage that is the world, as you undertake your exits and your entrances and play your many parts, I hope that you will allow yourself from time to time to peer out through ancient eyes, to lend fleeting life to I-deas and in*spir*ations which built that stage and drafted its early scripts, to re-meme-ber once again the role of "caveman chemist."

> Quince: If that may be then all is well. Come, sit down, every mother's son and rehearse your parts. Pyramus, you begin. When you have spoken your speech, enter into that brake; and so every one according to his cue.
>
> — William Shakespeare, *A Midsummer Night's Dream*, *ca.* 1596 AD [2]

---

2. Reference [24], Act III, Scene 1.

Appendix A. Back Cover

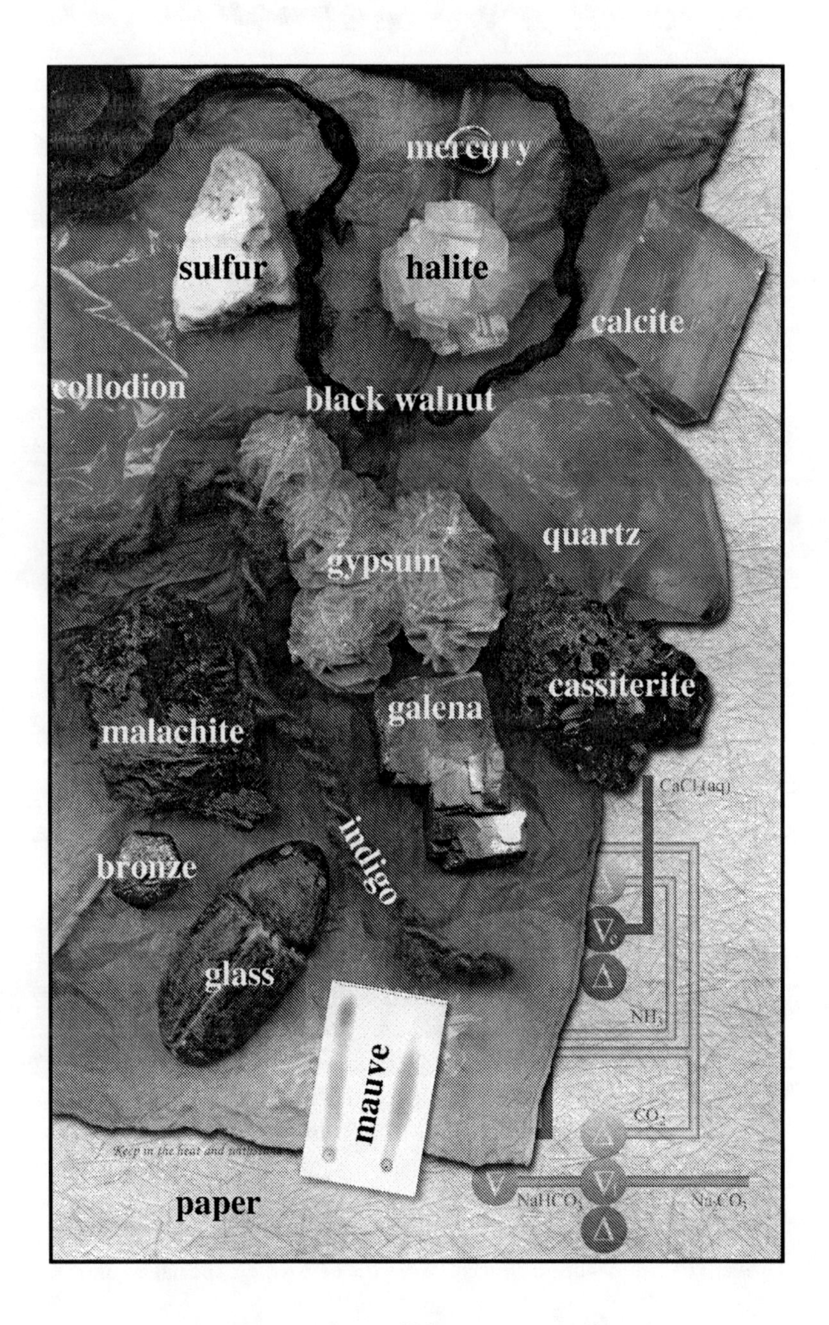

Appendix B. The Laboratory Notebook

Δ Re-minding is a tedious process. Starting with a virtual blank slate for a mind, it takes two years to learn to walk, another five or ten to condition the animal body for effective communication, and another twenty or sixty or even eighty years to assemble the requisite thoughts, notions, values, skills, observations and visions to form a complete human being. And just when you have everything in its place, your animal host gets itself trampled by a herd of water buffalo or killed in battle or it simply wears out and the whole process has to start over again with a fresh animal host.

Students at Hampden-Sydney College are required to keep a notebook documenting their achievements as caveman chemists. The notebook serves the purpose of preparing them for the work, organizing it, and documenting it. If they ever desire to repeat a project in the future, the notebook should provide a sufficient level of detail for them to do so. If a project does not turn out, the notebook helps me to help them to figure out what went wrong. It also serves as a souvenir for the children and grandchildren.

Caveman chemists are encouraged to follow the Hampden-Sydney format:

- The inside front cover has a table of contents listing all of the potential projects and the page numbers at which they appear in the notebook. When you are ready for a project to be graded, initial its entry in this table of contents.
- You should use only ink in the notebook so that information is not easily lost. Mistakes receive special attention in a notebook of this kind. Your impulse may be to obliterate a mistake, but you may decide later that what you thought was a mistake was correct after all. If you have obliterated your mistake, or torn it out of your notebook, you have no way to recover it. For this reason, simply draw a single line through a mistake. This marks it as a mistake, but allows it to be recovered later if necessary.
- Number each right-hand page in its upper right-hand corner. Whenever you begin work on a page, write the date next to the page number.

- There is potential confusion between scratch calculations and data actually observed in the laboratory. To avoid such confusion, always record experimental observations on a right-hand page. Reserve left-hand pages for scratch work, estimates, and theoretical predictions.
- Begin each project on a new right-hand page. Give each project a title and organize it into the following sections:
  1. Purpose: Write a paragraph describing the purpose or importance of the project.
  2. Materials: Make a bulleted list of all the materials needed to complete the project.
  3. Procedure: Translate the instructions for each project into a bulleted list which you can check off, item by item, as you work on the project. Some cavemen have suggested that I provide them with such a list, but doing it yourself increases the chances that information will actually pass through the brain on its way from the eye to the hand. The left-hand page facing the Procedure may contain calculations to estimate the amounts of materials needed. Record the results of these calculations along with the rest of your plans. If you are working in a class, you may have been quizzed on the material given in the sections labeled "Research and Development." If so, tape the successful quiz to the left-hand page opposite the Procedure.
  4. Safety: A brief summary of the hazardous properties of the materials and equipment to be used, including first-aid for accidental ingestion or exposure to eyes, skin or lungs.
  5. Observations: A brief summary of the deviations from the planned Procedure. The actual amounts of materials, as distinct from the estimated amounts, should be recorded here.
  6. Results: A brief description of the material or item produced. Record any relevant properties such as weight, color, and structural integrity.
  7. Conclusions: A paragraph summarizing what you have learned from the project. Were you successful? How do you know? What might you do differently if you were to repeat the project? How does this project relate to others you have done or may plan to do?

- The inside back cover has a list of completed projects initialed by the professor.

Appendix C. Measuring and Mixing

One of the most important skills any chemist must master is the ability to accurately weigh out chemicals without contaminating them. You really need two ***balances***[1] for the projects in this book, one for weighing big things and one for weighing little things. The big balance we'll call a ***gram balance*** because it can weigh things to the nearest gram. These jobbies usually have a *capacity* of 2 kg, which means you can load them down with 2 kg of stuff before they max out. The little balance we'll call a ***centigram balance***; it can weigh to the nearest 0.01 g and usually has a capacity of 200 g. You'll use the centigram balance whenever the thing you are weighing weighs less than 200 g and the gram balance whenever the thing you are weighing weighs between 200 and 2000 g.

You might think weighing things out would be simple; just put a container onto a balance and spoon stuff into it until it reads the right amount. There are two problems with this approach. Let's say that you are spooning silica and plaster into a plastic bag. You spoon in the right amount of silica but you get a little too much plaster. What do you do? You reach in and spoon out a little plaster. But plaster and silica look pretty much the same, so how do you know which one you took out? The second problem is that you might be tempted to put the extra "plaster" back into your container of plaster. After all, why waste it? But chances are you just spooned a little silica back into your container of plaster. Over time all those little bits of contamination will add up and pretty soon all your chemicals will be contaminated. There's a simple method for avoiding these problems. It's called weighing ***by difference.***

Let's go back to the example. You have a container of silica and a container of plaster. Your goal is to get, say, 20 g of each into a bag. First place the container of silica on the balance and press the button marked "tare."[2] The balance will now read zero. Spoon some silica from the container into the bag. The balance will now read, say, -2 g because the container has 2 g less silica than it had when you tared the balance. Suppose you keep spooning and with the last spoonful the balance reads -22

1. What most people call a "scale," chemists call a *balance.* You may use either mechanical or electronic balances.
2. On some balances the tare button is marked "zero."

g. What do you do? The spoon is still sitting over the container on the balance. It hasn't been anywhere to get contaminated. Just flick it with your finger until the balance reads -20 g and then dump the last spoonful into the bag. If you flick too much and get -19 g, just reach in and grab a little more until you get it just right. Now clean off your spoon, take the container of silica off the balance, put the container of plaster on the balance, and press the tare button again. It reads zero, of course. Spoon some plaster into your bag. Suppose that with the last spoonful the balance reads -23 g. The spoon still hasn't been anywhere to get contaminated. Just flick it with your finger until it reads -20 g and dump the last spoonful into the bag. Your silica container still contains silica and your plaster container still contains plaster. Your bag contains exactly 20 g of each. Weigh *by difference* and you will keep all of your chemicals as pure as the day you made them.

It may seem strange to you that I had you weighing things into a bag, but that's a great way to mix dusty materials. With silica and plaster, for example, you can seal the plastic bag and then knead it, shake it, massage it, and turn it end for end until the silica and plaster are all mixed up and you won't get dust everywhere. If the powders are flammable, I like to use those anti-static plastic bags like circuit boards come in. That way, you won't have static electricity accidentally set it off.

I buy silica in 50-pound bags and you might wonder how I can weigh it out if the balance has a capacity of only 2 kg. I use 1-pint plastic tubs like they use at Chinese restaurants to store each of the chemicals I use frequently. I label each container *and its lid* so that the silica container never gets used for anything else. These containers are just the right size for a gram balance. I keep large amounts of chemicals in 5-gallon pails, like paint comes in. *I always take chemicals from a big container into a smaller container,* from the 5-gallon pail to the 1-pint tub, from the 1-pint tub to the plastic bag. As long as I keep this in mind I will never contaminate more than a little bit of chemical.

You can weigh liquids by difference just as you can solids. Just pour the liquid into a cup, place the cup on the balance, tare it, and use a medicine dropper to transfer liquid from the cup to wherever it's going. But we often measure liquids by *volume* rather than by weight. You can use a graduated beaker or measuring cup for large volumes, graduated a graduated cylinder for smaller ones, and a graduated pipette for really small volumes.

Appendix D. Supplies and Suppliers

Ammonium Hydroxide (Clear Ammonia)	Grocery Store
Acetic Acid (Vinegar)	Grocery Store
Acetic Anhydride	Chemical Supply
Aniline	Chemical Supply
Borax	Grocery Store
Calcium Carbonate (Limestone)	Garden Supply, Farm Supply, Ceramic Supply
Calcium Hydroxide (Slaked Lime)	Garden Supply, Farm Supply, Grocery Store
Calcium Magnesium Carbonate (Dolomite)	Farm Supply, Garden Supply
Calcium Sulfate Dihydrate (Gypsum)	Garden Supply, Farm Supply
Calcium Sulfate Hemihydrate (Plaster of Paris)	Ceramic Supply
Carbon (Charcoal)	Nature, Grocery Store, Pyrotechnic Supply, Pet Store
Copper Carbonate (Malachite)	Ceramic Supply
Copper II Oxide, Black	Ceramic Supply
Ethanol (Ethyl Alcohol)	Building Supply, Liquor Store
Ethyl Ether	Chemical Supply
Ethylene Glycol (Anti-Freeze)	Automotive Supply
Indigo	Weaving Supply
Phthalic Anhydride	Chemical Supply
Potassium Carbonate (Pearl Ash, Potash)	Ceramic Supply
Potassium Nitrate (Saltpeter)	Garden Supply, Farm Supply, Pharmacy, Pyrotechnic Supply
Silicon Dioxide (Silica, Flint, Sand)	Ceramic Supply

Silver Nitrate	Photographic Supply, Chemical Supply
Sodium Carbonate (Soda Ash)	Ceramic Supply
Sodium Carbonate Decahydrate (Washing Soda)	Grocery Store
Sodium Hydrosulfite (Color Remover)	Grocery Store, Weaving Supply
Sodium Hydroxide (Caustic Soda, Lye, Drain Opener)	Grocery Store, Plumbing Supply
Sodium Hypochlorite (Laundry Bleach)	Grocery Store
Sodium Nitrate (Chile Saltpeter)	Farm Supply, Garden Supply
Sodium Thiosulfate (Hypo)	Photographic Supply, Chemical Supply
Sulfur	Garden Supply, Farm Supply, Pharmacy, Pyrotechnic Supply
Sulfuric Acid (Drain Opener)	Plumbing Supply
Tin Oxide (Cassiterite)	Ceramic Supply
o-Toluidine	Chemical Supply
p-Toluidine	Chemical Supply
Wool	Weaving Supply

Suppliers

☐ Caveman Supply (http://cavemanchemistry.com/)

☐ Ceramic Supply

- Campbell Ceramic Supply (http://www.claysupply.com/), 5704 D General Washington Dr., Alexandria, VA 22312, (800) 657-7222
- Clay Art Center (http://www.clayartcenter.com/), 2636 Pioneer Way East, Tacoma, WA 98404, (253) 896-3824

☐ Chemical Supply

- Antec Inc. (http://www.kyantec.com/), 721 Bergman Avenue, Louisville, KY 40203, (800) 448-2954
- Voigt Global Distribution (http://www.voigtglobal.com/), PO Box 412762, Kansas City, MO,64141-2762, (816) 471-9500
- The Chemistry Store (http://www.chemistrystore.com/), 520 NE 26th Court, Pompano Beach, FL, (800) 224-1430

- Cynmar Corporation (http://www.cynmar.com/), PO Box 530, Carlinville, IL 62626, (800) 223-3517
- Tri-Ess Sciences (http://www.tri-esssciences.com/), 1020 Chestnut Street, Burbank, CA 91506, (800) 274-6910
- The Al-Chymist (http://www.al-chymist.com/), 17130 Mesa Street, Hesperia, CA 92345, (760) 948-4150

□ Homebrew Supply

- Beer and Wine Hobby (http://www.beer-wine.com/), 155 New Boston St., Unit T, Woburn, MA 01801, (800) 523-5423
- The Home Brewery (http://www.homebrewery.com/), 205 West Bain PO Box 730, Ozark, MO 65721, (800) 321-2739

□ Photographic Supply

- Bostick and Sullivan (http://www.bostick-sullivan.com/), PO Box 16639, Santa Fe, NM 87592, (505) 474-0890
- Photographer's Formulary (http://www.photoformulary.com/), PO Box 950 Condon, MT 59826, (800) 922-5255

□ Pyrotechnic Supply

- Firefox (http://www.firefox-fx.com/), PO Box 5366, Pocatello, ID 83202, (208) 237-1976
- Iowa Pyro Supply (http://www.iowapyrosupply.com/), 1000 130th Street, Stanwood, IA 52337, (563) 945-6637
- Skylighter (http://www.skylighter.com/), PO Box 480-W, Round Hill, VA 20142, (540) 554-4543
- United Nuclear (http://www.unitednuclear.com/), PO Box 851 Sandia Park, NM. 87047 (505) 286-2831

□ Weaving Supply

- Handweavers Guild of America (http://www.weavespindye.org/), (770) 495-7702
- The Wool Co. (http://www.woolcompany.com/), 990 US Highway 101, Bandon, OR 97411, (888) 456-2430
- The Woolery (http://www.woolery.com/), PO Box 468, Murfreesboro, NC 27855, (800) 441-9665
- Paradise Fibers (http://www.paradisefibers.com/), 701 Parvin Road, Colfax WA 99111, (888) 320-7746

Appendix E. Atomic Weights

Namc	Symbol	Atomic Weight (g/mol)
Aluminum	Al	27.0
Boron	B	10.8
Calcium	Ca	40.1
Carbon	C	12.0
Chlorine	Cl	35.5
Copper	Cu	63.5
Gold	Au	200.0
Hydrogen	H	1.0
Iron	Fe	55.8
Lead	Pb	207.2
Magnesium	Mg	24.3
Mercury	Hg	200.6
Nitrogen	N	14.0
Oxygen	O	16.0
Phosphorus	P	31.0
Potassium	K	39.1
Silicon	Si	28.1
Silver	Ag	107.9
Sodium	Na	23.0
Sulfur	S	32.1
Zinc	Zn	65.4

Inspirations

Most folks won't care about what's on this page,
but I think that the story's worth telling.
The people here listed have helped shape the book.
Their bejeesical bits were compelling.

My mother and father were there from the start
and imparted to me my first memes.
Scoutmaster Bill Davidson helped me tie knots
and set fire to my kindling dreams.

Doug Boyd and Ed Guffee were buddies of mine
with an int'rest in things that are old.
Together we dug up a lot of cool stuff
for the LE Museum to hold.

Magnus McBride and the Mystery School
showed me truth can be told with a lie.
Alain Nu and the Phoenix rose up from the ash;
even memetic children must die.

Paul Mueller supported the Caveman I-dea
when most everyone thought I was mad.
Mary Prevo de-Og-lified some of the text,
on account of the grammar was bad.

Five readers took time to set fire to the book;
Brian Goeckerman and Aaron Skeen,
Cory Jaques and Bruce Miller played Athanor well,
burning crap from the book with Ross Greene.

The Cavemen of H-SC taught me a lot
about what even slackers can do.
I've loved every minute I've spent with you guys.
I mean it—I wrote this for you.

And thanks to the "bots" (that's just Dunn-speak for *cats*),
who insisted on helping me out.
Some of their typing was prob'ly left in.
If you find it, just give me a shout.

And here's to my Sunshine, the light of my life,
who put up with my preoccupation.
I love you for helping me finish this thing.
Maybe now we can take a vacation.

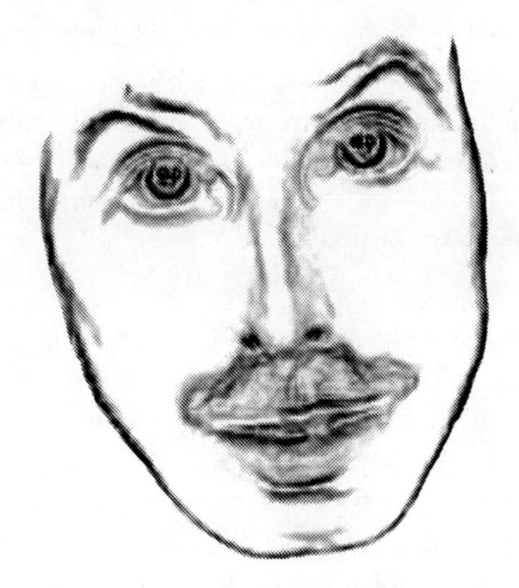

Kevin Dunn is the Elliott Professor of Chemistry at Hampden-Sydney College, where he teaches the course which inspired this book. He holds a BS degree from the University of Chicago and a PhD from the University of Texas at Austin. He appears on The Learning Channel's *Mysteries of Magic* and is co-author of a dozen journal articles in theoretical chemistry. He lives in central Virginia with his wife and several cats.

This book was formatted from its DocBook SGML source using the open-source tools, Openjade, Jadetex, and LaTeX. The author would like to thank Norman Walsh, Sebastian Rahtz, Adam Di Carlo, and members of the DocBook-Apps mailing list.

Bibliography

All quoted works or their English translations published prior to 1928 are assumed to be in the public domain. Permission has been granted by the respective publishers to reproduce those works protected by copyright. Reference [3] reprinted with the permission of MIT Press. Reference [6] courtesy of Hagley Museum and Library. Reference [23] reprinted by permission of the publishers and the Trustees of the Loeb Classical Library. The Loeb Classical Library is a registered trademark of the President and Fellows of Harvard College. Reference [25] reprinted with permission of the American Philosophical Society. References [53] and [65] reprinted with the permission of Cambridge University Press.

Brief snatches from literature, movies, television, and popular songs have been used to comment on the nature of memes under the fair use provision of the Copyright Act. Full bibliographic citations would be unwieldy; sources for these snatches have been listed in the index as a kind of key for those who recognize them.

Primary Texts

[1] Georgius Agricola, H. Hoover, L. Hoover, *De Re Metallica*, 1556, 1912, 1950, Dover Publications, ISBN 0486600068.

[2] Aristotle, *The Works of Aristotle*, Edited by W. D. Ross, 1908, Oxford University Press.

[3] Vannoccio Biringuccio, Cyril Smith, Martha Gnudi, *The Pirotechnia*, 1942, The M.I.T. Press, ISBN 0486261344.

[4] Robert Bradbury, *An Inductive Chemistry*, 1912, D. Appleton & Co..

[5] E. Caley, "The Stockholm Papyrus", *J. Chem. Educ.*, 992-999, 1927, 3.

[6] W. H. Carothers, *Letter to Hamilton Bradshaw*, November 9, 1927, Hagley Museum and Library, 1896.

[7] W. H. Carothers, J. A. Arvin, "Studies of Polymerization and Ring Formation II Poly-esters", *J. Am. Chem. Soc.*, 2560, 1929, American Chemical Society, 51.

[8] M. Chomell, *The Dictionaire Oeconomique*, 1758, L. Flinn.

[9] William Crookes, *The Wheat Problem*, 1899, G. P. Putnam's Sons.

[10] Jean D'Espagnet, *Hermetic Arcanum*, 1623, 1650, 1987, Holmes Publishing Group, Internet Sacred Text Archive (http://www.sacred-texts.com), ISBN 1558181032.

[11] John Draper, *Textbook on Chemistry*, 1846, Harper & Brothers.

[12] Claudius Galenus, *Opera Omnia*, 2001, Georg Olms Verlag, ISBN 3487008041, ISBN 3487008025.

[13] L. W. King, *Hammurabi, King of Babylonia*, 1898, Internet Sacred Text Archive (http://www.sacred-texts.com), Luzac & Co..

[14] *Holy Bible (KJV)*, Internet Sacred Text Archive (http://www.sacred-texts.com).

[15] Homer, Samuel Butler, *The Iliad of Homer*, 1898, Longmans, Green, and Co..

[16] Lao-tzu, *Tao Te Ching*, 1891, 1962, Internet Sacred Text Archive (http://www.sacred-texts.com), Dover.

[17] Antoine-Laurent Lavoisier, *Elements of Chemistry*, Robert Kerr, 1790, 1965, Dover, ISBN 0486646246.

[18] Henry Leicester, Herbert Klickstein, *Source Book in Chemistry*, 1952, Harvard University Press.

[19] Lucretius, William Ellery Leonard, *On the Nature of Things*, 1916, E. P. Dutton, Internet Sacred Text Archive (http://www.sacred-texts.com).

[20] Marie McLaughlin, *Myths and Legends of the Sioux*, 1916, Bismarck Tribune Company, Internet Sacred Text Archive (http://www.sacred-texts.com).

[21] William Noyes, *Textbook of Chemistry*, 1913, Henry Holt & Co..

[22] Ralph Oesper, "Christian Schonbein", *J. Chem. Educ.*, 432-440, 1929, 6.

[23] Pliny, H. Rackham, *Natural History*, 1962, Harvard University Press, ISBN 0674993888.

[24] William Shakespeare, *The Riverside Shakespeare*, Houghton Mifflin, ISBN 0395754909.

[25] C. Smith, J. Hawthorne, *Mappae Clavicula*, 1974, American Philosophical Society, ISBN 0871696444.

[26] H. Stapleton, R. Azo, M. Husain, "Chemistry in Iraq and Persia in the Tenth Century A.D.", *Mem. Asiat. Soc. Bengal*, 317-418, 1927, VIII.

[27] Theophilus, J. Hawthorne, C. Smith, *On Divers Arts*, 1979, Dover Publications, ISBN 0486237842.

[28] Vitruvius, *The Ten Books on Architecture*, Morris Morgan, 1914, Harvard University Press, ISBN 0486206459.

[29] Rudolf Wagner, William Crookes, *Wagner's Chemical Technology*, 1872, 1988, Lindsay Publications.

Agencies and Organizations

[30] *Agency for Toxic Substances and Disease Registry Website (http://www.atsdr.cdc.gov/)*, 2002, U.S. Agency for Toxic Substances and Disease Registry.

[31] *J. T. Baker Website (http://www.jtbaker.com)*, 2003, Mallinckrodt Baker, Inc..

[32] *Bureau of Alcohol, Tobacco, and Firearms Website (http://www.atf.treas.gov)*, 2002, U.S. Bureau of Alcohol, Tobacco, and Firearms.

[33] *National Center for Chronic Disease Prevention and Health Promotion Website (http://apps.nccd.cdc.gov/sammec/)*, 2003, National Center for Disease Prevention and Health Promotion.

[34] *National Center for Injury Prevention and Control Website (http://www.cdc.gov/ncipc/)*, 2003, National Center for Injury Prevention and Control.

[35] *NFPA Website (http://www.nfpa.org)*, 2003, National Fire Protection Association.

[36] Amy Spencer, Guy Colonna, *Fire Protection Guide To Hazardous Materials*, 2002, National Fire Protection Association, ISBN 0877654735.

[37] *NPCA Website (http://www.paint.org)*, 2003, National Paint &Coatings Association.

[38] *OSHA Website (http://www.osha.gov)*, 2002, U.S. Occupational Safety and Health Administration.

[39] *US Census Bureau Website (http://www.census.gov)*, 2002, U.S. Census Bureau.

Brewing

[40] H. Bravery, *Home Brewing Without Failures*, 1965, Gramercy Publishing.

[41] Vincent Gingery, *The Secrets of Building an Alcohol Producing Still*, 1994, David Gingery, ISBN 1878087169.

[42] Lee Janson, *Brew Chem 101*, 1996, Storey Communications, ISBN 0882669400.

[43] Charles Papazian, *The New Complete Joy of Homebrewing*, 1991, Avon Books, ISBN 0380763664.

Chemical Demonstrations

[44] Bassam Shakhashiri, *Chemical Demonstrations, V. 1*, 1983, University of Wisconsin Press, ISBN 0299088901.

[45] Bassam Shakhashiri, *Chemical Demonstrations, V. 2*, 1985, University of Wisconsin Press, ISBN 0299101304.

Electricity

[46] Paul Doherty, *The Aluminum/Air Battery*, 2000, The Exploratorium (http://www.exploratorium.edu).

[47] Alfred Morgan, *The Boy Electrician*, 1940, 1995, Lindsay Publications, ISBN 1559181648.

Fireworks

[48] Thomas Perigrin, *Practical Introductory Pyrotechnics*, 1996, Falcon Fireworks.

Glass

[49] Graham Stone, *Firing Schedules for Glass*, 1996, Graham Stone, ISBN 0646397338.

[50] Keith Cummings, *Techniques of Kiln-Formed Glass*, 1997, University of Pennsylvania, ISBN 0812234022.

[51] Jim Kervin, Dan Fenton, *Pate de Verre and Kiln Casting of Glass*, 1997, GlassWear Studios, ISBN 0965145816.

Industrial History

[52] Fred Aftalion, *A History of the International Chemical Industry*, 1991, University of Pennsylvania Press, ISBN 0812212975.

[53] Ahmad al-Hassan, Donald Hill, *Islamic Technology*, 1986, Cambridge University Press, ISBN 0521263336.

[54] *Chemical and Engineering News*, American Chemical Society, ISSN 0009-2347.

[55] Cathy Cobb, Harold Goldwhite, *Creations of Fire*, 1995, Plenum Press, ISBN 0306450879.

[56] Alfred Cowles, *The True Story of Aluminum*, 1958, Henry Regnery Company.

[57] *Valuing the Earth*, Edited by Herman Daly, Edited by Kenneth Townsend, 1993, Massachusetts Institute of Technology.

[58] F. W. Gibbs, "The History of the Manufacture of Soap", *Annals of Science*, 1939.

[59] Simon Garfield, *Mauve*, 2002, W. W. Norton, ISBN 0393323137.

[60] L. F. Haber, *The Chemical Industry During the Nineteenth Century*, 1958, Clarendon Press.

[61] L. F. Haber, *The Chemical Industry: 1900-1930*, 1971, Clarendon Press.

[62] Matthew Hermes, *Enough for One Lifetime*, 1996, American Chemical Society, ISBN 0841233314.

[63] Joseph Lambert, *Traces of the Past*, 1997, Perseus Books, ISBN 0738200271.

[64] Charles Mann, Mark Plumer, *The Aspirin Wars*, 1991, Alfred Knopf, ISBN 0394578945.

[65] Joseph Needham, *Science and Civilization in China: Chemistry and Chemical Technology: Military Technology*, 1986, Cambridge University Press, 5.7, ISBN 0521303583.

[66] Joseph Needham, *Science and Civilization in China: Chemistry and Chemical Technology: Spagyrical Discovery and Inventions*, 1980, Cambridge University Press, 5.4, ISBN 052108573X.

[67] Joseph Needham, *Science and Civilization in China: Chemistry and Chemical Technology: Paper and Printing*, 1985, Cambridge University Press, 5.1, ISBN 0521086906.

[68] J. R. Partington, *The Alkali Industry*, 1925, Bailliere, Tindall, and Cox.

[69] James Partington, *A History of Greek Fire and Gunpowder*, 1960, 1999, The Johns Hopkins University Press, ISBN 0801859549.

[70] Naomi Rosenblum, *A World History of Photography*, 1997, Abbeville Press, ISBN 0896594386.

[71] Hugh Salzberg, *From Caveman to Chemist*, 1991, American Chemical Society.

[72] John Stillman, *The Story of Early Chemistry*, 1924, D. Appleton & Co..

[73] Graham Taylor, Patricia Sudnik, *Du Pont and the International Chemical Industry*, 1984, Twayne Publishers, ISBN 0805798056.

[74] Anthony Travis, *The Rainbow Makers*, 1993, Lehigh University Press, ISBN 0934223181.

[75] Kenneth Warren, *Chemical Foundations: The Alkali Industry in Britain to 1926*, 1980, Clarendon Press, ISBN 0198232314.

[76] Trevor Williams, *A Biographical Dictionary of Scientists*, 1969, Wiley-Interscience, ISBN 0198232314.

Memetics

[77] Richard Dawkins, *The Selfish Gene*, 1976, Oxford University Press, ISBN 0192860925.

[78] Susan Blackmore, *The Meme Machine*, 1999, Oxford University Press, ISBN 019286212X.

Paper

[79] Vance Studley, *The Art and Craft of Handmade Paper*, 1990, Dover Publications, ISBN 0486264211.

[80] Sophie Dawson, *The Art and Craft of Papermaking*, 1992, Running Press, ISBN 01561381586, ISBN 1887374248.

[81] Jules Heller, *Paper-Making*, 1997, Watson-Guptill, ISBN 0823038424.

Photography

[82] *Coming into Focus*, Edited by John Barnier, 2000, Chronicle Books, ISBN 0811818942.

[83] John Burk, Walter Henry, Paul Messier, Timothy Vitale, *Albumen Photographs: History, Science, and Preservation (http://albumen.stanford.edu)*, 2000, Stanford University.

[84] Randall Webb, Martin Reed, *Alternative Photographic Processes*, 2000, Silver Pixel Press, ISBN 1883403707.

Rocks and Minerals

[85] David Barthelmy, *Mineralogy Database*, 2002, Webmineral.com (http://www.webmineral.com).

[86] Alan Holden, Phylis Morrison, *Crystals and Crystal Growing*, 1982, MIT Press, ISBN 0262580500.

[87] Tim Rast, Mike Melbourne, *Knappers Anonymous Website (http://www.geocities.com/knappersanonymous/)*, 2003, Mike Melbourne.

Soap

[88] Ann Bramson, *Soap: Making It, Enjoying It*, 1975, Workman Publishing, ISBN 0911104577.

[89] Susan Cavitch, *The Soapmaker's Companion*, 1997, Storey Publishing, ISBN 0882669656.

[90] Catherine Failor, *Making Transparent Soap*, 2000, Storey Publishing, ISBN 158017244X.

Textiles

[91] Rachel Brown, *The Weaving, Spinning, and Dyeing Book*, 1978, Alfred A. Knopf, ISBN 0394715950.

[92] J. N. Liles, *The Art and Craft of Natural Dyeing*, 1990, University of Tennessee Press, ISBN 0870496700.

Glossary

Acid

A substance which donates H^+ (proton) to other substances.

Aerobic Process

A process which occurs in the presence of oxygen.

Air

🜁 That which is hot and moist. Alchemical air symbolizes daring and skill. In this book Air speaks for skepticism and honesty.

Alkali

A substance which accepts H^+ (proton) from other substances. An alkali reacts with water to produce OH^- in solution.

Anaerobic Process

A process which occurs in the absence of oxygen.

Base

An alkali.

Calcination

A procedure by which the ***bejeezus*** is driven from a substance, often by intense heat.

Catalyst

An agent which speeds up a reaction. A catalyst is not consumed in the reaction it catalyzes and so does not appear as a reactant or product in the balanced chemical equation.

Combustion

A process by which a material combines with oxygen, usually releasing heat.

Compound

A pure substance which can be decomposed into two or more other pure substances.

Condensation

A reaction in which two molecules join together, spitting out a small molecule, usually water, in the process.

Conservation of Mass

A generalization of the observation that the mass of a sealed container does not change.

Dissolution

A procedure by which one substance is dissolved in another to form a solution.

Distillation

A procedure by which the bejeezus is collected and isolated.

Earth

🜃 That which is cold and dry. Alchemical earth symbolizes wealth and bounty. In this book Earth speaks for business and industry.

Electrolyte

A substance which dissociates into ions when dissolved. Because these ions are free to move about, an electrolyte solution conducts electricity.

Element

A pure substance which has never been decomposed into two or more other pure substances.

Equation

A relationship between two things which are equal. In a chemical equation, the two things must have the same number and kinds of elements. In a mathematical equation, the two things must have the same units.

Experiment

A procedure for making an observation which will be consistent with only one of two or more theories.

Fermentation

A procedure by which one substance is converted to another by decay, often by the action of microorganisms.

Fire

🜂 That which is hot and dry. Alchemical fire symbolizes mastery and will. In this book Fire speaks for basic research and experimentation.

Formula

A symbol, e.g. H_2O, used to denote the relative amounts of elements in a pure substance.

Glossary

The section of the book you are reading right now.

Heterogeneous Matter

Matter whose composition varies from one place to another within a sample.

Homogeneous Matter

Matter whose composition does not vary from one place to another within a sample.

Inorganic Compound

A compound which does not contain carbon. There are exceptions to this definition; carbonates, bicarbonates, and carbon dioxide are generally considered as inorganic compounds.

Lixiviation

A process by which water is passed over a solid, taking with it any soluble materials.

Meme

A unit of culture which is replicated through imitation. Successful memes exhibit fidelity, fecundity, and longevity. Synonyms include *I-dea, inspiration,* and *spirit.*

Mercury

☿ That which flees the fire, descending from heaven to earth. Alchemical mercury symbolizes the spirit and takes its name from the element mercury, which may be distilled without leaving any ash. In this book Mercury provides the motivation for the Sulfur and Salt.

Metathesis

A procedure by which two substances exchange anions or cations.

Organic Compound

A compound which contains carbon. There are exceptions to this definition; carbonates, bicarbonates, and carbon dioxide are generally considered as inorganic compounds.

Oxidation

A half-reaction in which electrons appear as products.

pH

A measure of the hydrogen ion concentration in a solution. Solutions with pH less than 7 are acidic, which those with pH higher than 7 are alkaline.

Precipitation

A procedure by which a solid falls out of a liquid solution.

Recrystallization

A procedure for purifying a substance by precipitating a solid from a liquid.

Stoichiometric Coefficient

The number in front of each substance in a balanced chemical equation.

Stoichiometric Question

A question involving the relative weights of reactants and products in a chemical reaction.

Substance

Homogeneous matter whose composition is fixed.

Reduction

A half-reaction in which electrons appear as reactants.

Salt

Θ That which remains from the fire, neither rising to heaven nor descending to earth. Alchemical salt symbolizes the body and takes its name from those salts which are extracted from ashes. In this book Salt provides instructions for making or doing something.

Separation

A procedure by which one substance is separated from another. Examples of separations include recrystallization, distillation, and chromatography.

Solute

The less abundant component of a solution.

Solution

Homogeneous matter whose composition is variable.

Solvent

The more abundant component of a solution.

Sublimation

A procedure for purifying a substance by precipitating a solid from a gas.

Sulfur

🜍 That which is consumed in the fire, rising from earth to heaven. Alchemical sulfur symbolizes the mind and takes its name from the element sulfur, which burns without leaving any ash. In this book Sulfur provides the chemical concepts needed to understand the Salt.

Theory

A logically consistent set of principles which account for, explain, or render intelligible a set of observations.

Water

🜄 That which is cold and moist. Alchemical water symbolizes belief and foresight. In this book Water speaks for theory and prediction.

Index

Printed in the United States
95630LV00002B/313/A

9 781581 125665